Operational Modal Analysis

This book is a collection of selected contributions from the 10th International Operational Modal Analysis (OMA) Conference held in Naples, Italy on May 21–24, 2024. Each chapter of *Operational Modal Analysis: Developments and Applications* covers a relevant contribution to extend the classical domain of OMA toward the development of new sensing as well as structural health monitoring (SHM) technologies, digital twins, and innovative assessment procedures for materials and structures. The focus of these chapters is on the primary role of OMA in civil engineering as well as related industrial applications and fields. Thus, the book provides an up-to-date overview of OMA applications and related technologies and methodologies and can support researchers, practicing engineers, and technicians, as well as students in order to enhance their knowledge on this important topic.

Features

- Presents numerous application examples and case studies.
- Shows how advanced AI technologies contribute to the progress of methodologies and techniques of OMA and SHM in the field.
- Introduces the applicability of OMA for the development of structural digital twins.

Resilience and Sustainability in Civil, Mechanical, Aerospace and Manufacturing Engineering Systems

Series Editor: Mohammad Noori, Cal Poly San Luis Obispo

Published Titles

Resilience of Critical Infrastructure Systems
Emerging Developments and Future Challenges

Operational Modal Analysis
Developments and Applications
Edited by Mohammad Noori, Carlo Rainieri, and Carmelo Gentile

Series Editor Bio: Mohammad Noori is Emeritus Professor of Mechanical Engineering at Cal Poly, San Luis Obispo, and Visiting Professor at the School of Civil Engineering, University of Leeds, UK, where he also serves as a co-founding director of Intelligent and Resilient Infrastructure Center. Noori is Fellow and Life Member of the American Society of Mechanical Engineers. His work in modeling the complex hysteretic behavior of structural systems, including pinching phenomenon, cited in the literature as the Bouc-Wen-Baber-Noori (BWBN) model, has been widely utilized in nonlinear random vibrations for seismic response analysis of concrete structures and has been incorporated in OpenSees seismic analysis program. His works in non-zero mean and non-Gaussian response analysis of hysteretic systems are also original contributions. Noori has also carried out pioneering work in seismic isolation of secondary systems and holds two US patents in that area. He has also been a pioneer in the application of artificial intelligence methods for structural health monitoring. He has authored over 300 refereed papers, 7 scientific books, and 32 book chapters and has edited 25 scientific books and special journals and archival volumes. Noori has supervised over 100 graduate students and post-doc scholars and presented over 100 keynotes, plenaries, and invited talks. He is the founding Executive Editor of an international journal and has served on the editorial boards as the associate editor of over 12 other journals and as a member of numerous scientific and advisory boards. Noori directed the Sensors Program at the National Science Foundation in 2014 and has served as a member of several NSF-sponsored US delegation teams for the development of US-Japan and US-China cooperative research programs. He served as the dean of engineering at Cal Poly, as the Higgins professor and head of mechanical engineering at WPI and as the Reynolds Professor and head of mechanical and aerosphere engineering at NC State University. Noori also served as the Chair of the national committee of mechanical engineering department heads and was one of seven co-founders of the National Institute of Aerospace, in partnership with NASA Langley Research Center.

For more information about this series, please visit: https://www.routledge.com/Resilience-and-Sustainability-in-Civil-Mechanical-Aerospace-and-Manufacturing/book-series/ENG

Operational Modal Analysis

Developments and Applications

Edited by
Mohammad Noori, Carlo Rainieri,
and Carmelo Gentile

CRC Press
Taylor & Francis Group
Boca Raton London New York

CRC Press is an imprint of the
Taylor & Francis Group, an **informa** business

Designed cover image: Shutterstock.

First edition published 2026
by CRC Press
2385 NW Executive Center Drive, Suite 320, Boca Raton FL 33431

and by CRC Press
4 Park Square, Milton Park, Abingdon, Oxon, OX14 4RN

CRC Press is an imprint of Taylor & Francis Group, LLC

ISBN: 978-1-032-85175-4 (hbk)
ISBN: 978-1-032-85180-8 (pbk)
ISBN: 978-1-003-51694-1 (ebk)

DOI: 10.1201/9781003516941

Typeset in Times
by Apex CoVantage, LLC

Contents

About the editors

Mohammad Noori is emeritus professor of mechanical engineering at Cal Poly, San Luis Obispo, and visiting professor at the University of Leeds, UK where he is founding director of the Intelligent and Resilient Infrastructure Center. Prof. Noori is a fellow and life member of the American Society of Mechanical Engineers, has been a fellow of the Japanese Society for Promotion of Science, and has held distinguished visiting professorships at several universities abroad. His work in modeling the complex hysteretic behavior of structural systems, including pinching phenomenon, widely cited in the literature as the Bouc-Wen-Baber-Noori (BWBN) model, has been used in nonlinear random vibrations; for seismic response analysis of concrete structures and other applications, it has been incorporated in OpenSees seismic analysis program and ABAQUS finite element software. His work in non-zero mean, non-Gaussian response analysis of hysteretic systems in random vibrations is also an original contribution. He has also carried out pioneering work in seismic isolation using shape memory alloys and holds 2 US patents in that area. Prof. Noori has been a pioneer in the application of AI-based methods and wavelet transform for structural health monitoring. He has authored over 300 refereed papers, 7 scientific books, and 32 book chapters and has edited 25 scientific books and special journal volumes. Prof. Noori has supervised over 100 graduate students and post-doc scholars and has presented over 100 keynotes, plenaries, and invited talks. He is the founding executive editor of an international journal, has served on the editorial boards or as the associate editor of over 12 other journals, and is the founding editor of 3 scientific collections by CRC/Taylor & Francis, De Gruyter, and Momentum Press publishers, and has served on numerous scientific and advisory boards. Prof. Noori directed the Sensors Program at the NSF in 2014. He served as the Dean of Engineering at Cal Poly, as the Higgins Professor and Head of the ME Department at WPI, and as the Reynolds Professor and Head of the MAE Department at NC State University. Prof. Noori also served as the chair of the national committee of mechanical engineering department heads and was a co-founder of the National Institute of Aerospace in partnership with NASA Langley Research Center.

Carlo Rainieri is currently a senior researcher at the National Research Council of Italy. His research interests are in the fields of civil structural health monitoring, operational modal analysis, and smart materials. He has joined several research projects focused on civil structural health monitoring. He has been a member of the scientific committee of the IOMAC Association since 2015. In 2019, he received the IOMAC Scientific Award for his contribution to the development of operational modal analysis. He was a chair of the 8th Civil Structural Health Monitoring Workshop (2021) and the 10th International Operational Modal Analysis Conference (2024). In 2021, he was also appointed as council member of the International Society for Structural Health Monitoring of Intelligent Infrastructure. He is the author of the first book on operational modal analysis, *Operational Modal Analysis of Civil Engineering Structure: An Introduction and Guide for Applications* and approximately 200 scientific papers published in international peer-reviewed journals and national and international conference proceedings. His main achievements in the field of operational modal analysis and civil structural health monitoring concern the development of data processing methods for vibration-based structural health monitoring applications, including original automated operational modal analysis procedures and novel methods for compensation of environmental/operational influence on modal properties. He is also the founder and former CEO of S2X s.r.l. (www.s2x.it), a spin-off company of the University of Molise, aimed at providing highly qualified solutions and services in the fields of civil structural health monitoring and operational modal analysis.

Carmelo Gentile is full professor of structural engineering at the Department of Architecture, Built Environment and Construction Engineering (DABC), Technical University of Milan (Politecnico di Milano), Italy. He has been the author or co-author of more than 300 technical and scientific papers. His main research expertise and interests are bridge engineering, cultural heritage structures, dynamic tests and continuous monitoring of bridges and historic structures, earthquake engineering, modal and structural identification, structural dynamics, vibration-based damage assessment, and microwave remote sensing. Dr. Gentile serves as a member of the scientific committee of the main international conferences on structural and experimental dynamics (e.g., EURODYN, EVACES, IOMAC, SMART). He has also been a member of the scientific committee of the Laboratory of Testing Materials and Structures (LPM), Politecnico di Milano since 2019. He has been the scientific consultant of the Laboratory of Vibrations and Dynamic Monitoring of Structures (section of LPM) since 2006. He is also responsible for the Laboratory of Vibrations and Dynamic Monitoring of Structures for the full-scale testing and/or continuous dynamic monitoring of more than 200 bridges, including several arch bridges and cable-stayed bridges and the viaducts of the new A58 motorway. Dr. Gentile is the coordinator responsible for several projects of vibration-based SHM, such as the dynamic monitoring of the historic San Michele Bridge (2011–2015), the structural monitoring of the Milan Cathedral since 2018, the continuous dynamic monitoring of the Brivio Arch Bridge (1917) since 2020, and the dynamic monitoring of a steel-concrete composite bridge passing over the A4 motorway (since 2021). Dr. Gentile also serves as a member of the scientific committee and the board of the FABRE Consortium involved in research in the field of structural assessment and monitoring of existing bridges. He has delivered invited keynote lectures at various international conferences (e.g., SF&R 2022, JISDM 2019, IOMAC 2019, ICEM 2018, SF&R 2018, EVACES 2013, IABSE Conference Rotterdam 2013, and EURODYN 2011).

Contributors

Maher Abdelghani
Associate Professor, Department of Mechanical Engineering, City Taffala
University of Sousse, Higher Institute of Applied Sciences and Technology of Sousse

Wael A. Altabey
Department of Mechanical Engineering, Alexandria University, Alexandria, Egypt and Department of Civil and Architectural Engineering
Henan University of Technology, Zhengzhou, China

L. Auersch
Fed Inst Material Res and Testing Germany

A. Avramova
Politecnico di Milano, Italy

Elisa Bassoli
Department of Engineering "Enzo Ferrari"
University of Modena and Reggio Emilia, Italy

P. Cassese
Construction Technologies Institute
National Reseach Council, Italy

Cristina Castagnetti
DIEF, University of Modena and Reggio Emilia, Italy

L. Cieri
Construction Technologies Institute
National Reseach Council, Italy

Giuliano Coppotelli
University of Rome "La Sapienza"
Dipartimento di Ingegneria Meccanica e Aerospaziale, Via Eudossiana
Rome, Italy

C. Fortis
Politecnico di Milano, Italy

Carmelo Gentile
Politecnico di Milano, Italy

Giorgia Ghirelli
Department of Engineering "Enzo Ferrari"
University of Modena and Reggio Emilia, Italy

M. Girardi
Institute of Information Science and Technologies
"A. Faedo", ISTI–CNR, Pisa, Italy

D. Gorjup
University of Ljubljana
Faculty of Mechanical Engineering, Slovenia

G. Gurioli
Dipartimento di Matematica e Informatica "U. Dini"
Università degli Studi di Firenze
Florence, Italy

Soroosh Kamali
DICAM, University of Bologna, Italy

F. Mancini
DIEF, University of Modena and Reggio Emilia, Italy

Stefano Mariani
DICAM, University of Bologna, Italy

Alessandro Marzani
Department of Civil, Chemical, Environmental and Materials Engineering
University of Bologna, Bologna, Italy

N. Messina
Institute of Information Science and Technologies
"A. Faedo", ISTI–CNR, Pisa, Italy

M.G. Mulas
Politecnico di Milano, Italy

Mohammad Noori
Professor California Polytechnic State University, San Luis Obispo
California, USA
Department of Civil and Architectural Engineering
Henan University of Technology
Zhengzhou, China

C. Padovani
Institute of Information Science and Technologies
"A. Faedo", ISTI–CNR, Pisa, Italy

D. Pellegrini
Institute of Information Science and Technologies
"A. Faedo", ISTI–CNR, Pisa, Italy

M. Pirrò
Politecnico di Milano, Italy

Federico Ponsi
Department of Engineering "Enzo Ferrari", University of Modena and Reggio Emilia

Carlo Rainieri
Construction Technologies Institute
National Reseach Council, Italy

I. Rosati
Construction Technologies Institute
National Reseach Council, Italy

S. Said
Fed Inst Material Res and Testing, Germany

R. G. Sbarra
University of Rome La Sapienza, Italy

Ahmed Silik
Department of Civil and Architectural Engineering
Henan University of Technology
Zhengzhou, China

J. Slavič
University of Ljubljana
Faculty of Mechanical Engineering, Slovenia

Ghita Eslami Varzaneh
Department of Engineering "Enzo Ferrari"
University of Modena and Reggio Emilia, Italy

Loris Vincenzi
Department of Engineering "Enzo Ferrari"
University of Modena and Reggio Emilia, Italy

K. Zaletelj
University of Ljubljana, Faculty of Mechanical Engineering, Slovenia

1 High-speed camera-based methods for experimental and operational modal analysis

K. Zaletelj, D. Gorjup, and J. Slavič

1.1 INTRODUCTION

The high spatial density of information achievable using full-field image-based vibration measurement methods presents significant advantages in numerous fields such as structural health monitoring, model updating, and identification of structural parameters. However, accurately measuring full-field deflection shapes in high-frequency vibrations using conventional methods presents considerable challenges. Accelerometer usage yields limited spatial data [1], while laser vibrometers [2] and electronic speckle pattern interferometry [3] provide nonsimultaneous or partial displacement measurements, respectively. The use of high-speed cameras, increasingly popular in structural dynamics applications, enables comprehensive, concurrent 3D displacement mapping through methods like digital image correlation [4] and simplified optical flow [5]. These optical techniques have shown efficacy in civil engineering for lower-frequency displacements (below 100 Hz) and have been extended to capture dynamic responses up to 2.4 kHz. However, displacements at frequencies above 1 kHz—typically less than a micrometer and smaller than a camera's pixel—are often obfuscated in digital high-speed camera videos by considerable levels of image noise, rendering them indistinguishable [6]. Furthermore, a single high-speed camera captures only in-plane movements, while spatial measurements require additional equipment and complex measurement setups.

Another variant of experimental modal analysis (EMA) is strain experimental modal analysis [7]. Strain EMA is used to identify fatigue hotspots that are correlated with the critical locations on the structure [8]. The low spatial density of information when using traditional approaches (strain gauges) is often a limiting factor in the fast and extensive identification of critical locations on a structure. As an alternative, the thermoelasticity principle has been used to identify the full-field strain shapes [9, 10].

DOI: 10.1201/9781003516941-1

In this chapter, the state-of-the-art methods for enhancing full-field modal shape identification, extending the mode shapes into three dimensions, and identifying the strain mode shapes using the thermoelasticity principle are presented.

1.2 MODAL IDENTIFICATION ON HIGH-SPEED CAMERA DATA

The modal identification based on the high-speed camera data consists of two parts. First, the displacements are identified from sequences of digital images. From the resulting displacement data, the modal parameters are identified using appropriate identification algorithms.

1.2.1 The Simplified Gradient-Based Optical Flow Method

The simplified gradient-based optical flow (SGBOF) method [6] is applied for motion detection in image sequences, leveraging the linear relationship between pixel intensity changes and displacements for small, sub-pixel movements. This approach translates changes in pixel brightness into spatial shifts using numerically estimated spatial pattern gradients. SGBOF demonstrates high sensitivity in areas with distinct intensity gradients, facilitating precise movement tracking when the observed object's surface can be appropriately prepared before measurement. Figure 1.1 shows examples of a speckle pattern and a line pattern image, which are often used in surface preparation for digital image correlation and SGBOF displacement measurements, respectively.

The identified displacements in the direction of the spatial gradient ∇I are given by:

$$\Delta s = \frac{I\left(x,y,t\right)-I(x,y,t+\Delta t}{\left|\nabla I\right|},$$

where x and y are the discrete pixel locations, and t is time. The intensity gradient $\left|\nabla I\right|$ is determined by the numerical derivation of a selected reference image, often the first image in the acquired video sequence.

FIGURE 1.1 Examples of speckle pattern (a) and line pattern (b) images.

SGBOF's use of individual pixels as sensors reduces spatial blurring, a common issue in larger subset sizes found in methods such as digital image correlation (DIC). Assuming small displacements, the method simplifies displacement detection and accelerates processing. Sub-pixel resolution of less than one-thousandth of a pixel is possible [6], though image noise can adversely affect the achievable accuracy of the measurement. Techniques such as local averaging, a high-quality reference image, and longer image sequences help mitigate noise effects, enhancing SGBOF's performance.

1.2.2 The hybrid method

Camera-based displacement measurements can provide high-spatial-resolution response data but are typically less precise than accelerometers or laser vibrometers due to the lower dynamic range of the measured displacements, particularly in the higher frequency range. Although vibrations may be masked by noise, making them seem undetectable, modern identification techniques such as LSCF [11]/LSFD [12] can still identify modal data below the noise floor by incorporating responses across a broader frequency range, thereby reducing the noise impact on parameters.

Javh et al. [13] presented a hybrid method for identifying modal parameters from camera measurements. Instead of using only high-speed camera data to construct a stability chart, the eigenvalues, i.e. the transfer function's poles, were constructed using a single high-dynamic-range sensor, e.g., an accelerometer. From the accelerometer-based stability chart, the complex poles can be chosen in the frequency range up to 10 kHz, giving the operator information on the damping ratios and natural frequencies:

$$\lambda_r = -\xi_r \omega_r \pm \omega_r \sqrt{1-\xi_r}$$

Based on the identified poles, the LSFD method can be used to identify the modal constants ${}_rA_{jk}$ through least-squares optimization:

$$\alpha_{jk}(\omega) = \sum_{r=1}^{N} \left(\frac{{}_rA_{jk}}{i\omega - \lambda_r} + \frac{{}_rA_{jk}}{i\omega - \lambda_r} \right)$$

where $\alpha_{jk}(\omega)$ are the FRF identified based on the camera measurements. The identified modal constants can be used to reliably determine full-field deflection shapes from camera measurements, as well as reconstruct hybrid FRFs. A comparison of the reconstructed and directly measured FRFs is shown in Figure 1.2. Note that the reconstructed FRF is full-field and clearly represents the natural frequencies well below the noise floor of the camera measurement. While the averaged amplitudes of the reconstructed FRFs and the accelerometer measurements differ in Figure 1.2, the identified full-field modal shapes are consistent, as shown in the original publication [13].

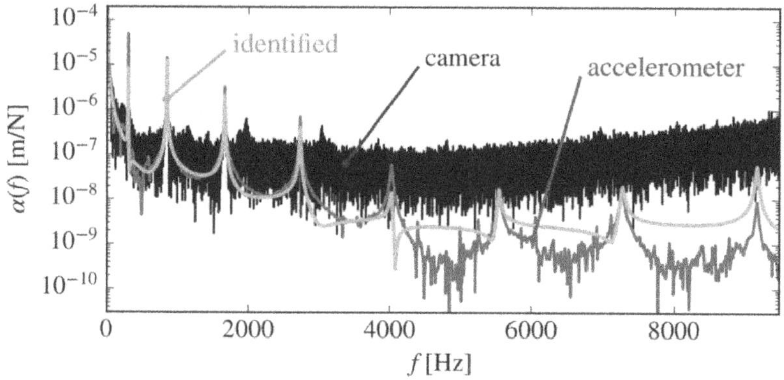

FIGURE 1.2 High-speed camera (black), accelerometer (gray), and reconstructed (light grey) FRFs.

1.3 FREQUENCY DOMAIN TRIANGULATION

A method of multiview vibration measurement using a single monochrome camera and spatial deflection shape triangulation in the frequency domain was proposed in [14]. Using a single moving camera, frequency-domain images of the deflection shapes of a time-invariant mechanical structure under stationary excitation are acquired from various viewpoints. Spatial operating deflection shapes $X(\omega)$ can be reconstructed by frequency-domain triangulation of image-based displacement data. The measurement of 3D operating deflection shapes using only a single monochrome still camera is also possible using the spectral optical flow Imaging method [15].

A camera view in a multiview measurement setup is defined by a transformation matrix P that projects the coordinates of a point in space U from a global coordinate frame onto an image of the point u [16]:

$$wu = PU$$

All physical points on the camera's optical axis are projected onto the same point in the image, making the perspective camera projection nonlinear in the Euclidean coordinate frame [16].

By identifying the position of a single point in space U in multiple images, we arrive at an overdetermined system of linear equations, the solution of which is the linear triangulation method [16]. Additional views add more equations to the already overdetermined system, which can be solved in a least-squares sense for the unknown coordinates U in the triangulation process.

The frequency-domain multiview deflection shape triangulation method relates the frequency-domain image of the deflection shape $\Delta u(\omega)$ and the 3D operating deflection shape $X(\omega)$ to be calculated by a perspective camera transform:

$$\Delta u(\omega) + u_{\text{REF}} = \frac{1}{w_{\text{REF}}} P\left(X(\omega) + U_{\text{REF}}\right)$$

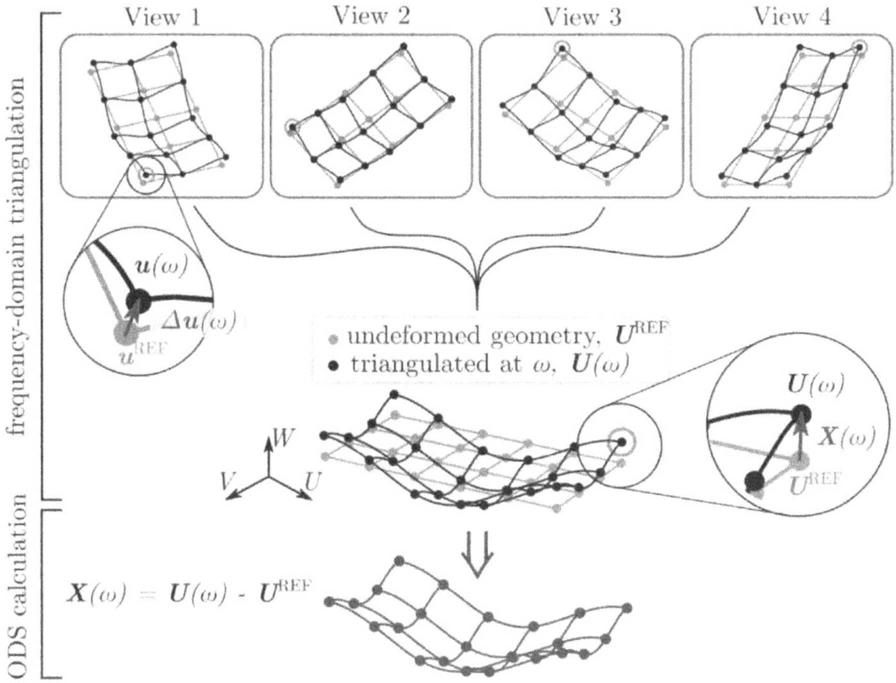

FIGURE 1.3 A schematic representation of the frequency-domain triangulation procedure.

where u_{REF} and U_{REF} represent the initial (reference) positions of the observed image points and the corresponding points in space, respectively, and $\Delta u(\omega)$ is the frequency-domain displacement measured in the image plane. Although the perspective camera transformation is nonlinear in Euclidean space, Gorjup et al. [14] show this relation to be valid, assuming small harmonic motion.

Multiple views of the displaced points at ω are now used to triangulate the deflected 3D shape $U(\omega)$. From that, the spatial deflection shape $X(\omega)$ can be obtained by subtracting the initial 3D positions of the undeformed point grid [14], as illustrated in Figure 1.3:

$$X(\omega) = U(\omega) - U_{REF}$$

Using the frequency domain triangulation method, arbitrarily many views, acquired using a single moving camera can be added to the measurement system, provided that each measurement point is observed in at least two views. This has multiple benefits, such as extending the measurement field of view while maintaining the measurement resolution, as well as reducing the measurement noise by overdetermining the multiview imaging system, as shown in Figure 1.4.

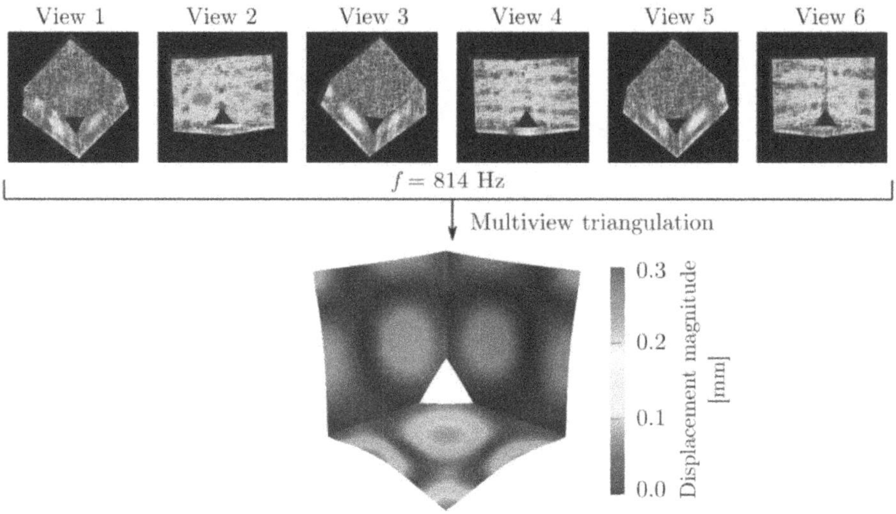

FIGURE 1.4 Multiview measurement noise reduction.

1.4 THERMOELASTICITY-BASED STRAIN EMA

Strain experimental/operational modal analysis is a variant of the classical modal analysis, where the strain mode shapes are identified [17]. The strain mode shapes can be used to identify the fatigue hotspots, which are correlated with the critical locations on the structure [8].

The traditional method of using strain sensors for measurement lacks the high spatial density necessary to accurately pinpoint critical locations or create a detailed map of damage intensity. Alternative, noncontact methods have been used to infer full-field strain shapes. However, these methods require an established relationship between kinematics and stress/strain. For instance, in the case of the Euler–Bernoulli beam, calculating the double spatial derivative introduces considerable uncertainty.

Alternatively, the thermoelastic principle [18] allows the direct measurement of the full-field stress/strain response of any structure using a high-speed infrared (IR) camera. The laws of thermoelasticity explain how changes in temperature are related to elastic stresses in materials. These laws come from combining the first law of thermodynamics, which deals with energy conservation, with the principles of continuum mechanics, which describe how materials deform and move. These laws assume that the materials are elastic and the processes are reversible, all happening within a closed system. Assuming an adiabatic process, an isotropic material, and plane stress isotropy, the simplified formulation of thermoelasticity is as follows [18]:

$$\Delta T = K_m \left(\Delta\sigma_x + \Delta\sigma_y \right),$$

where K_m is the thermoelastic coefficient:

$$K_m = \frac{-\alpha T_0}{\rho C_p}$$

ρ is the material's density; c_p is the specific heat at constant pressure; α is the thermal expansion coefficient; and T_0 is the ambient temperature.

For a beam under free-free boundary conditions (experiment details are presented in [10]), the first four strain modal constants are presented in Figure 1.5. Note that the (strain) modal constants are a scalar multiple of a (strain) modal shape.

According to Zaletelj et al. [10], the MAC value [19] between the thermoelasticity-based and kinematics-based strain modal constants exceeds 0.8 when the amplitudes of the surface stress exceed 100 kPa and 500 kPa for aluminum and steel, respectively. The relationship between the MAC value and the amplitude of the surface stress is presented in Figure 1.6. It can be seen that for aluminum, the required

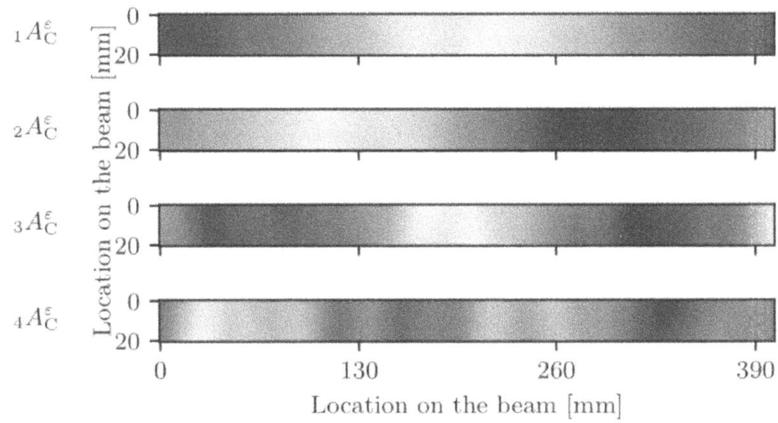

FIGURE 1.5 Strain modal constants based on the thermoelastic principle.

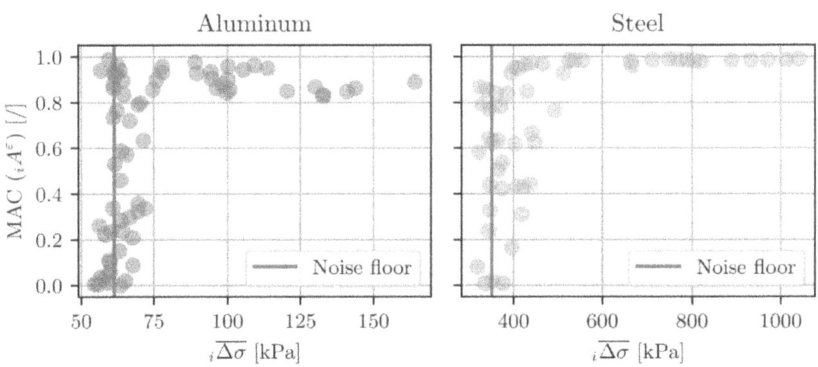

FIGURE 1.6 MAC value with respect to the amplitude of the surface stress.

surface stress amplitude to achieve a MAC value of 0.8 is much lower than for steel. For further explanation, refer to Zaletelj et al. [10].

1.5 CONCLUSION

In this chapter, advanced methods for experimental and operational modal analysis using high-speed cameras were presented. The simplified gradient-based optical flow (SGBOF) method, combined with the hybrid approach, was shown to enable precise deflection shape identification in the frequency range up to 10 kHz. By integrating high-speed camera measurements with accelerometer data, the hybrid method overcomes the limitations of individual systems, offering improved dynamic range and robustness against image noise, particularly for high-frequency, low-amplitude vibrations.

Furthermore, frequency-domain triangulation was introduced as a robust method for identifying full-field 3D spatial deflection shapes using high-speed camera data. By utilizing multiple views of the structure, acquired using a single moving monochrome camera, this method significantly enhances the image-based measurement capabilities of a single high-speed camera. The overdetermined nature of the triangulation process not only extends the field of view but also mitigates measurement noise, resulting in cleaner and more reliable data. These benefits make frequency-domain triangulation particularly useful for capturing complex dynamic responses in structures with intricate geometries or distributed excitation sources.

In addition to deflection shape identification, this study explored strain EMA through the thermoelasticity principle, which provides a noncontact method for obtaining full-field strain modal shapes using high-speed IR cameras. By directly measuring temperature changes induced by elastic stresses, the thermoelastic approach eliminates the need for indirect strain calculations, reducing post-processing complexity while maintaining high spatial resolution. Experimental results demonstrated that the MAC value between thermoelasticity-based strain shapes and kinematically derived strain shapes exceeds 0.8 for sufficiently high stress amplitudes, with aluminum requiring significantly lower stress levels compared to steel. This highlights the potential of thermoelasticity-based strain EMA for identifying fatigue hotspots and critical regions in structures with minimal measurement setup and processing time.

Overall, the methods presented in this work—SGBOF, hybrid modal analysis, frequency-domain triangulation, and thermoelasticity-based strain EMA—collectively push the boundaries of modal analysis capabilities using high-speed imaging technologies. These approaches provide significant improvements in spatial resolution, noise reduction, and operational simplicity, offering new possibilities for structural health monitoring, model updating, and parameter identification.

1.6 ACKNOWLEDGMENT

The authors acknowledge partial financial support from the Slovenian Research Agency (Research Core Funding No. P2–0263).

REFERENCES

[1] C. Schedlinski, J. Schell, E. Biegler, and J. Sauer, "Application of optical measurement techniques for experimental modal analyses of lightweight structures," *Proceedings of ISMA2016*, pp. 2301–2314, 2016.

[2] D. A. Ehrhardt, M. S. Allen, S. Yang, and T. J. Beberniss, "Full-field linear and nonlinear measurements using continuous-scan laser doppler vibrometry and high speed three-dimensional digital image correlation," *Mechanical Systems and Signal Processing*, vol. 86, pp. 82–97, 2017.

[3] O. J. Løkberg, "Electronic speckle pattern interferometry," in *Optical Metrology*. Springer, 1987, pp. 542–572.

[4] W. H. Peters and W. F. Ranson, "Digital imaging techniques in experimental stress analysis," *Optical Engineering*, vol. 21, p. 213427, Jun. 1982.

[5] B. D. Lucas and T. Kanade, "An iterative image registration technique with an application to stereo vision," in *Proceedings of the 7th International Joint Conference on Artificial Intelligence—Volume 2*, pp. 674–679, 1981.

[6] J. Javh, J. Slavič, and M. Boltežar, "The subpixel resolution of optical-flow-based modal analysis," *Mechanical Systems and Signal Processing*, vol. 88, pp. 89–99, May 2017.

[7] L. H. Yam, T. P. Leung, D. B. Li, and K. Z. Xue, "Theoretical and experimental study of modal strain analysis," *Journal of Sound and Vibration*, vol. 191, pp. 251–260, Mar. 1996.

[8] Y. Zhou and J. Tao, "Theoretical and numerical investigation of stress mode shapes in multi-axial random fatigue," *Mechanical Systems and Signal Processing*, vol. 127, pp. 499–512, Jul. 2019.

[9] Á. J. Molina-Viedma, L. Felipe-Sesé, E. López-Alba, and F. A. Díaz, "Thermoelastic effect in modal shapes at high frequencies using infrared thermography," *Measurement: Journal of the International Measurement Confederation*, vol. 176, May 2021.

[10] K. Zaletelj, J. Slavič, J. Šonc, and M. Boltežar, "Strain experimental modal analysis of an Euler–Bernoulli beam based on the thermoelastic principle," *Mechanical Systems and Signal Processing*, vol. 201, p. 110655, Oct. 2023.

[11] P. Guillaume, L. Hermans, and H. Van der Auweraer, "Maximum likelihood identification of modal parameters from operational data," in *Proceedings of the 17th International Modal Analysis Conference (IMAC17)*, pp. 1887–1893, 1999.

[12] B. Cauberghe, "Applied frequency-domain system identification in the field of experimental and operational modal analysis," PhD thesis, Vrije Universiteit Brussel, 2004.

[13] J. Javh, J. Slavič, and M. Boltežar, "High frequency modal identification on noisy high-speed camera data," *Mechanical Systems and Signal Processing*, vol. 98, pp. 344–351, 2018.

[14] D. Gorjup, J. Slavič, and M. Boltežar, "Frequency domain triangulation for full-field 3D operating-deflection-shape identification," *Mechanical Systems and Signal Processing*, vol. 133, p. 106287, 2019.

[15] D. Gorjup, J. Slavič, A. Babnik, and M. Boltežar, "Still-camera multiview spectral optical flow imaging for 3D operating-deflection-shape identification," *Mechanical Systems and Signal Processing*, vol. 152, p. 107456, 2021.

[16] R. Hartley and A. Zisserman, *Multiple View Geometry in Computer Vision*, 2nd ed. New York, NY, USA: Cambridge University Press, 2003.

[17] D. J. Ewins, *Modal Testing: Theory, Practice and Application*. John Wiley & Sons, 2019.

[18] W. Thomson, "XV. On the dynamical theory of heat, with numerical results deduced from Mr Joule's equivalent of a thermal unit, and M. Regnault's observations on steam," *Transactions of the Royal Society of Edinburgh*, vol. 20, no. 2, pp. 261–288, 1853.
[19] R. J. Allemang, "The modal assurance criterion—twenty years of use and abuse," *Sound and Vibration*, vol. 37, no. 8, pp. 14–21, 2003.

2 Vision-based dynamic monitoring of a steel footbridge adopting artificial targets of different geometries

Federico Ponsi, Ghita Eslami Varzaneh, Elisa Bassoli, Giorgia Ghirelli, and Loris Vincenzi

2.1 INTRODUCTION

In the past two decades, several methodologies and technological advancements have been explored and applied within the field of structural health monitoring (SHM). Among these, contactless technologies have gained increasing attention from civil engineers due to their ability to drastically reduce installation costs and efforts, as well as avoid the interruption of functionality in the monitored structure or infrastructure. This is especially important when monitoring structures and infrastructure that serve as key transportation routes are difficult or unsafe to access or are part of cultural heritage. Several studies on contactless technologies are proposed in the literature, including laboratory experimental measurements and real-world applications, conducted with the aim of monitoring physical structural parameters such as displacements, rotations, vibrations, and cracking. With the main purpose of monitoring the structural health—to avoid crises and collapses, to detect red flags of potential damage, or more simply for system identification and model updating—the reliance on an accurate system that reduces costs and traffic disruptions is definitely a breeding ground for today's civil engineering.

Contactless technologies available for civil monitoring purposes include, but are not limited to, global navigation satellite systems (GNSS) (Poluzzi et al. 2019), satellite remote sensing (Bassoli et al. 2023), terrestrial radar interferometry (Castagnetti et al. 2019; Guerzoni et al. 2023), and vision-based techniques (Fradelos et al. 2020).

Within these technologies, vision-based monitoring has the potential to overcome dependence on expensive industrial products (Xu et al. 2019), showing great capabilities in measuring both static and dynamic structural displacements even with consumer-grade equipment. Besides this, other advantages of vision-based methods

DOI: 10.1201/9781003516941-2

are the direct evaluation of displacements (without the need for double integration of accelerations) and the multipoint monitoring capability of a single device, with flexible placement in different locations depending on the specific application scenario (Buoli et al. 2023; Ponsi et al. 2024).

Although vision-based monitoring provides numerous benefits such as cost-effectiveness, time efficiency, ease of use, and reduced execution and management costs, its accuracy in measuring structural displacements is still under investigation and remains difficult to quantify, as it is not solely reliant on the inherent performance of the camera hardware. Indeed, special attention is also required for calibration and data processing, including the definition of camera parameters to correct lens distortion in each video frame, analyses for perspective adjustment, and specific measures to identify the scale factor, which is crucial for converting the displacements expressed in pixel units into physical units. Moreover, factors associated with environmental conditions are widely acknowledged as additional significant sources of errors and uncertainties. These encompass camera vibrations caused by wind or user interaction, fluctuating weather conditions, ambient light variations, and nonuniform air refraction due to temperature shifts between the camera and the monitored object.

The latest applications of vision-based monitoring are widely reviewed in Spencer et al. (2019), Dong and Catbas (2020), and Zona (2020), comprising tests on bridges (Feng and Feng 2017; Chen et al. 2018) and footbridges (Xu et al. 2018; Lydon et al. 2019). Moreover, studies assessing environmental uncertainties in vision-based monitoring (Ye et al. 2016) can be found in the literature, predominantly conducted through theoretical analysis and laboratory tests (Zona 2020). However, the impact of external factors on the accuracy of on-site testing remains unclear because of the absence of comprehensive outdoor full-scale tests. In this scenario, this chapter aims to present the outcomes of outdoor vision-based monitoring to evaluate its potential and criticality by analyzing video frames captured during environmental and forced vibrations. The research focuses on a deformable steel footbridge in Modena, Italy. A series of targets have been installed on the structure alongside a traditional network of accelerometers used for comparison purposes. To assess the impact of target shape on the monitoring results, two types of targets are installed on the footbridge: checkerboard targets and concentric circle targets. Additional targets are positioned on the ground—assumed to remain stationary throughout the tests—to help in measuring and correcting the effects of camera movement on the estimated structural displacements.

The chapter is organized as follows: in Section 2.2, the main aspects characterizing the vision-based monitoring procedure are described; Section 2.3 delineates the case of study and the monitoring campaign, during which videos were captured under vibrations induced by a jumping pedestrian. Next, Section 2.4 discusses the results, analyzed in the time and frequency domains. Finally, Section 2.5 draws conclusions and outlines future perspectives.

2.2 VISION-BASED PROCEDURE OVERVIEW

Vision-based monitoring relies on the fundamental concept that, assuming a stationary camera, changes in the position of artificial targets or distinctive features of the structure between consecutive frames correspond to the displacement of the related point on the structure over time. The proposed method for reconstructing the absolute structural displacement at each monitored position involves two main stages: the vision-based monitoring setup, which is tailored to the specific structure being observed, and the video post-processing phase, which serves as the essential connection between the raw data (represented by a sequence of video frames) and the displacement time series of the targets. In more detail, the procedure can be outlined in the following steps:

1. Monitoring campaign setup is based on the specific application scenario, considering the expected displacement to define the extent of the monitored structure portion. Accordingly, it is necessary to select optical lenses (with an appropriate range of focal lengths), frame rate, monitored points to reconstruct the structural motion, targets to be tracked, and camera-to-structure distance.
2. Calibration of key parameters to convert displacement from pixels to a metric scale, such as focal length, perspective correction, and scale factor.
3. Detection and extraction of key features from each frame of the recorded video, along with tracking their variation frame by frame. The key features to be tracked can essentially be selected between intrinsic distinctive elements of the structure (such as bolts or holes) and artificial targets installed on the structure. The former is extremely attractive when the structure is preserved as part of the cultural heritage, as well as when it is not easily or safely accessible. On the other hand, the specific geometry and high contrast of artificial targets lead to more accurate results, albeit requiring access to the structure.
4. Filtering the effects of camera shake, as indicated by the movement of the ground-based targets, which are supposed to be fixed.

In terms of video processing approaches, the most commonly used algorithms in structural engineering applications can be summarized as follows:

- Correlation-based template matching, where a reference rectangular region in the first frame is used to search for an area in subsequent frames that closely matches the reference. The primary objective is to identify the position within each frame where the template best matches, typically by maximizing a similarity measure such as intensity cross-correlation. One of the main advantages of this method is that it requires minimal user input, limited to defining the template region in the reference frame. However, template matching is limited by its sensitivity to variations in shading, lighting, and background conditions, as well as challenges in handling significant deformations, necessitating that the template undergo only rigid translation and rotation.
- Feature point matching does not rely on a complete area as a reference but instead uses sparse feature points, often referred to as key points.

A key point represents a small region (e.g., 5 × 5 pixels) characterized by distinctive and invariant features such as building corners, unique architectural details, or bolts in steel structures. Instead of using raw image intensities as in template matching, feature point matching employs a descriptor that provides a sophisticated representation derived from the shape and appearance of a small window surrounding the key point. This method can achieve higher accuracy than template matching, as descriptors make it less sensitive to changes in illumination, shape deformations, and scale variations. However, feature point matching requires target regions to have rich textures to achieve high accuracy.

- Shape-based tracking focuses on identifying specific geometries present in custom-made artificial targets or structural elements (e.g., like-line cables). The core concept is to detect the selected target pattern within each frame to reconstruct its motion over time. This process generally relies on image gradients or brightness thresholding to identify the edges of linear patterns, the outlines of circles, or the intersection points of squares in checkerboards. Although these algorithms are less general because they depend on the presence of a specific geometry to be monitored, they are unaffected by changes in lighting or background conditions, and they do not require high texture resolution.

In a real-world monitoring scenario, if it is possible to install artificial targets on the monitored structure, shape-based tracking proves to be the most suitable approach. Thus, it is the method employed for target tracking in the case study presented in Section 2.3. In the following, the identification of target features for both checkerboard (Section 2.1) and concentric circle (Section 2.2) targets is discussed in detail. Once such an identification procedure is applied to a target on the structure, the displacement time series of the corresponding structural point is obtained. However, it is crucial to note that this measurement is not absolute, as it represents the relative displacement between the camera and the monitored position. This implies that unwanted movements of the camera (caused by wind, for instance) might be inaccurately identified as motion of the structure itself, resulting in misleading outcomes. Hence, the outlined procedure (specifically described in Section 2.1 or 2.2, depending on the target shape) is also employed for ground-based targets to determine the average apparent motion of the stationary objects within the video. Finally, the displacements detected from the targets on the structure are corrected for unwanted camera shake by subtracting the apparent motion of the camera, as identified by the fixed targets.

Before the target tracking operation, the so-called region of interest (ROI) is defined, which is a rectangular portion of a frame that limits the area within which the target tracking is performed. The MATLAB Image Processing and Computer Vision Toolboxes (MathWorks) are employed for this purpose. Following these operations, the defined ROI is handled as a multidimensional matrix of pixels, where each pixel is defined by its 2D coordinates (with respect to the upper-left vertex of the frame) and by the RGB intensity level.

2.2.1 Checkerboard detection and displacement computation

The square corners of a checkerboard are key features for target detection, owing to their distinct RGB intensity contrast with adjacent pixels. Sub-pixel precision in corner detection is achievable, as checkerboard squares are generally not confined within an integer number of pixels. The image gradient, reflecting the change in RGB intensity between neighboring pixels, enables the identification of the (sub-pixel accurate) corner coordinates in each frame (Geiger et al. 2012).

For each frame, the average position of previously identified square corners is calculated. A projective transformation is then applied to mitigate perspective distortion effects. The transformation is based on the average inclination of vertically and horizontally aligned corners, specifically those detected within a time window under unforced conditions. After the projection, the mean corner coordinates in each frame are transformed into new projected mean coordinates, still expressed in pixel units.

These coordinates are then converted from pixels to real-world units. The conversion uses the ratio of the known dimension of the target square (in mm) to the average pixel distance between adjacent corners, differentiating between the vertical and horizontal directions. Finally, the displacement of the target is calculated as its frame-by-frame relative position with respect to a reference pair of coordinates, defined as the target's average position under unforced conditions.

2.2.2 Circles detection and displacement computation

The circular Hough transform (CHT) is used for circle detection, as outlined by Atherton and Kerbyson (1999). CHT is robust in handling noise, occlusions, and varying lighting conditions. The procedure identifies circles given a prescribed sensitivity, which is a scalar parameter whose value can range in [0, 1] and needs to be tuned to find reliable results. Circle identification results are highly dependent on the selected sensitivity: if the sensitivity is set too low, not all circles may be detected; if set too high, false detections may occur. Additionally, the optimal sensitivity may vary across different frames. To address this, an iterative procedure was developed for automatic sensitivity calibration. This process involves gradually increasing the sensitivity in steps of 0.01 until the prespecified number of circles is detected in each frame. The detection algorithm computes the circle center coordinates and the radii—in pixel units—for each frame. Concentric circles allow for a more precise estimation of the target center position compared to a single circle, since the final center is computed by averaging the coordinates of all the detected centers, one for each concentric circle.

Center coordinates are then converted from pixels to real-world units. The conversion is accomplished by comparing the mean radius R (mm) of the circles measured on the target and the mean identified radius across all the frames r (pixels). The resulting scale factor R/r (mm/pixel) is then multiplied by the variation of the detected coordinates in each frame to obtain the displacement in metric scale. To assess the target displacement, the frame-by-frame average position of the key features is calculated with respect to a reference position, namely the target's average position during the unforced ambient condition.

2.3 CASE OF STUDY

The case of study is a truss-girder steel footbridge crossing the Panaro River in Modena, Italy (Figure 2.1a). The footbridge is slender and lightweight, which makes it highly deformable and particularly sensitive to dynamic vibrations caused by pedestrians, cyclists, and wind. It presents a box-shaped cross section measuring 3.00 by 3.20 meters, composed of truss girders with hollow tubular sections. The structure has a total length of 160 meters, divided into three spans: 45 meters, 70 meters, and 45 meters, respectively. The monitoring campaign focuses on seven cross sections of the central 70-meter span, including the midspan and six additional sections symmetrically placed at 7.5-meter intervals on both sides. Each monitored position features checkerboard targets labeled with specific identification numbers, as shown in Figure 2.1b. To compare how the pattern geometry affects tracking

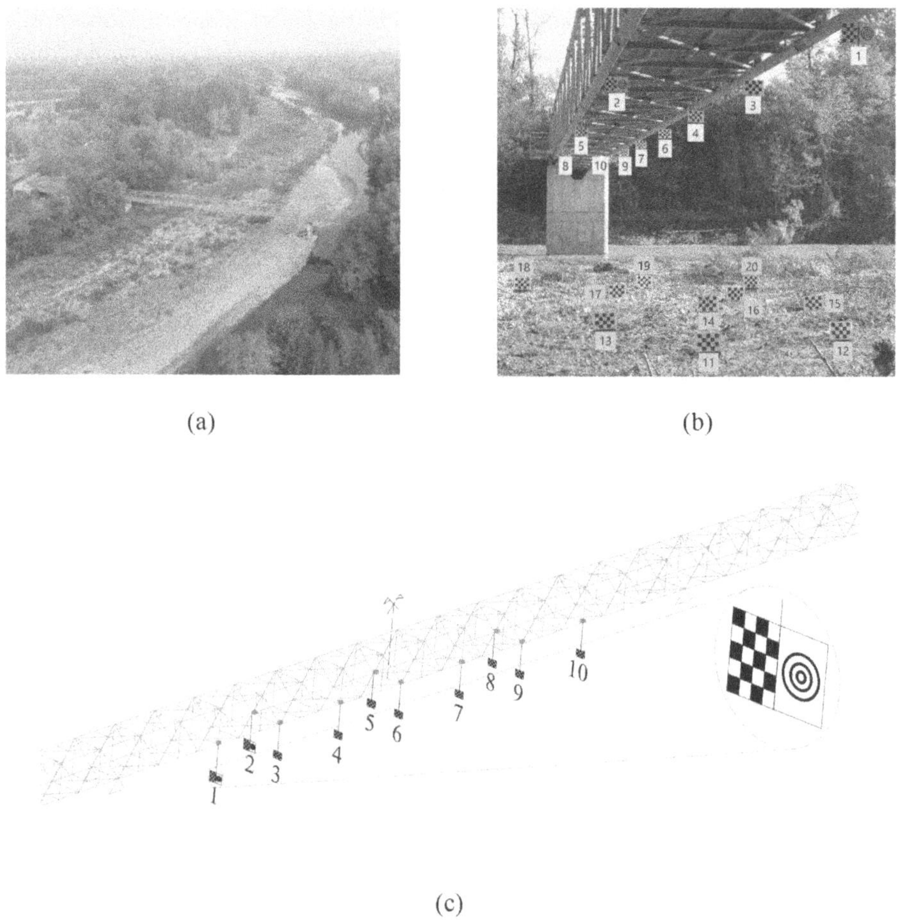

FIGURE 2.1 (a) Wide view of the footbridge, (b) target locations, and (c) schematic view of the measuring positions.

accuracy, concentric circles and checkerboard targets are placed close to each other in Sections 2.1 and 2.2, and the corresponding results are reported in Section 2.4.

The monitoring setup includes a Nikon D7500 camera mounted on a tripod and positioned outside the structure on the dry riverbed, capturing the view illustrated in Figure 2.1b. Videos are recorded in 4K resolution at 30 frames per second (FPS).

The initial and final seconds of each video—corresponding to the start and stop of recording—were trimmed to limit interference from manual operation of the camera.

Ten fixed targets, numbered 11 to 20 in Figure 2.1b, have been placed on the ground to quantify and mitigate unintended camera vibrations due to environmental factors. Assuming that the ground-based targets are motionless, any detected movement can be attributed solely to noise or camera vibrations. This allows filtering the displacement time series of the moving targets on the structure.

Additionally, a conventional monitoring system with ten Teleco biaxial MEMS accelerometers (Guidorzi et al. 2010) has been installed for comparison purposes. The accelerometers operate at a sampling frequency of 80 Hz, and they are positioned at approximately the same locations as the targets on the footbridge.

During the monitoring campaign, vibrations of the footbridge were induced in a controlled manner through a jumping excitation, with a metronome set at 138 and 132 beats per minute (BPM). These specific rates were chosen to excite the bridge's primary vertical and torsional modes, which have natural frequencies of 2.32 and 4.43 Hz, respectively (values obtained from preliminary ambient vibration tests and confirmed by the findings of Section 2.4). With reference to Figure 2.1c, the pedestrian performed jumps near location 5 for the 138 BPM test and near location 2 for the 132 BPM test. In each test, the acquired video included the phase before the excitation (to record ambient vibration data), the 10-second jumping period, and a subsequent two-minute interval containing the structural motion attenuation that followed the excitation.

2.4 RESULTS

This section presents the results of the tests described in Section 2.3 (with jumping cadences of 132 and 138 BPM, respectively), with specific focus on the two different artificial targets tracked: checkerboards and concentric circles.

The presented results refer to locations numbered 1, 2, 5, 6, and 8, as highlighted in Figure 2.1c. At each of these locations, a checkerboard target and an accelerometer are installed. Additionally, a concentric circle target is installed in sections 1 and 2, which are the closest to the camera.

Figure 2.2 illustrates the vertical displacement time series obtained by tracking the target placed on the structure and cleaned of unintended motion due to camera shaking. They are compared against the time series obtained by double integrating acceleration recordings and filtering low-frequency noise. The procedure is formulated as a regularization problem in which the acceleration is approximated by the second-order central finite difference of displacement. The final displacement is then

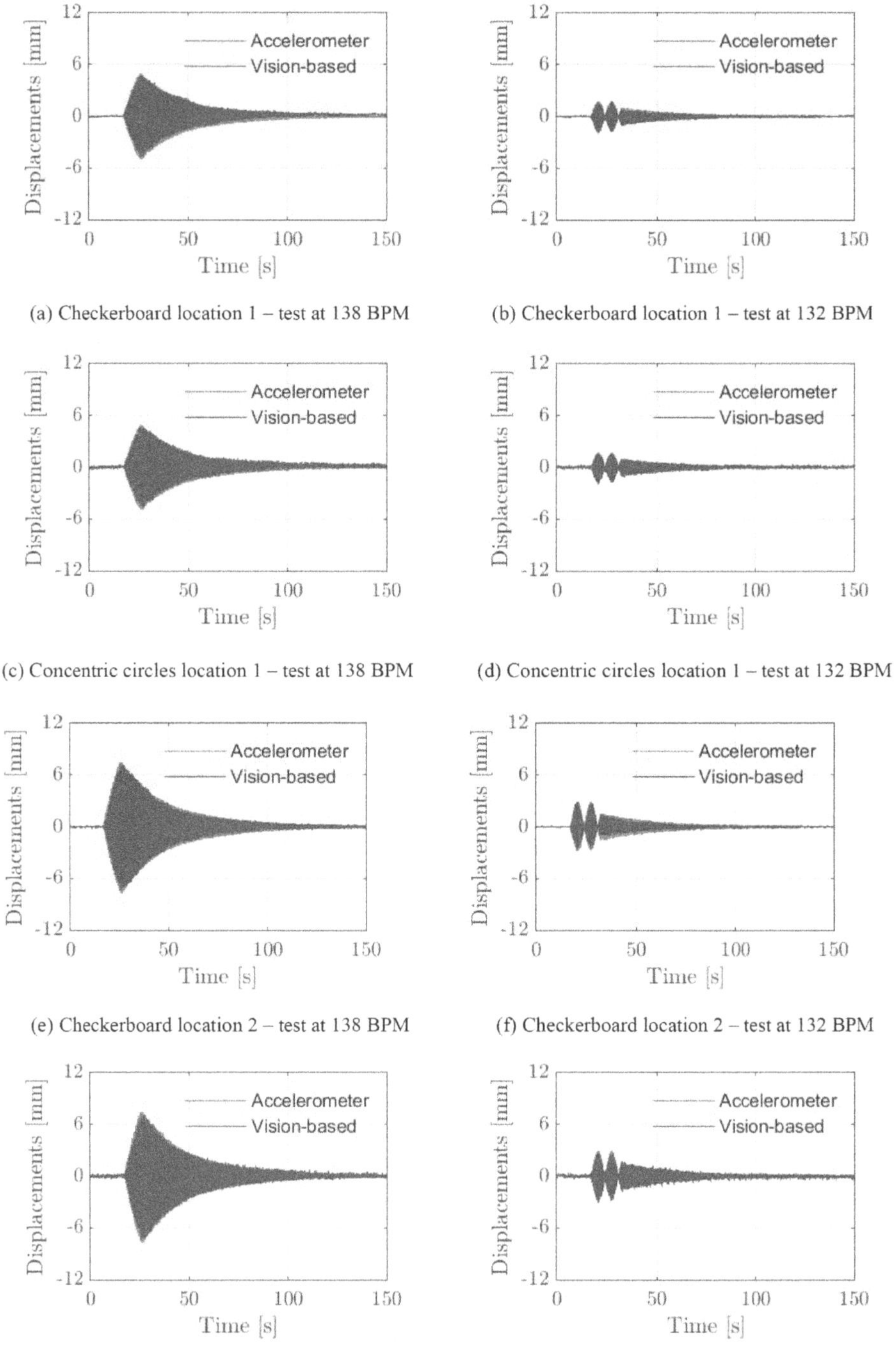

(a) Checkerboard location 1 – test at 138 BPM
(b) Checkerboard location 1 – test at 132 BPM
(c) Concentric circles location 1 – test at 138 BPM
(d) Concentric circles location 1 – test at 132 BPM
(e) Checkerboard location 2 – test at 138 BPM
(f) Checkerboard location 2 – test at 132 BPM
(g) Concentric circles location 2 – test at 138 BPM
(h) Concentric circles location 2 – test at 132 BPM

FIGURE 2.2 Vertical displacements detected for the two tests and for the targets located in sections 1, 2, 5, 6, and 8.

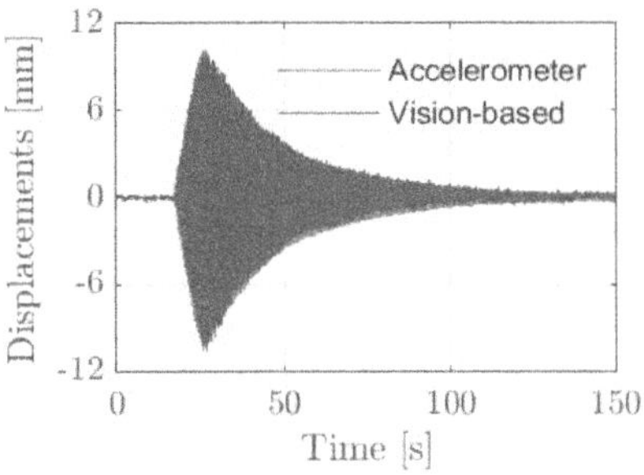

(i) Checkerboard location 5 – test at 138 BPM

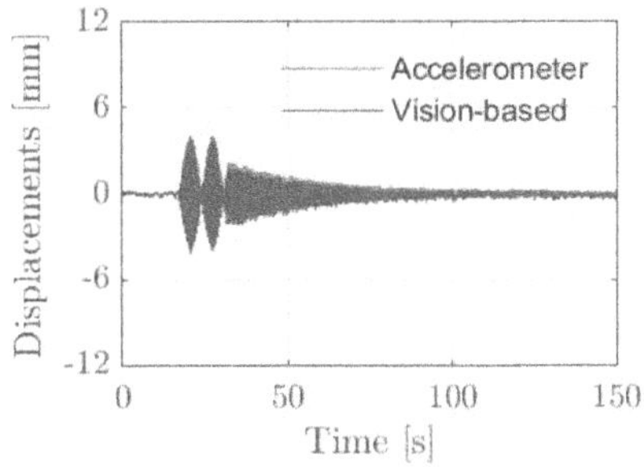

(j) Checkerboard location 5 – test at 132 BPM

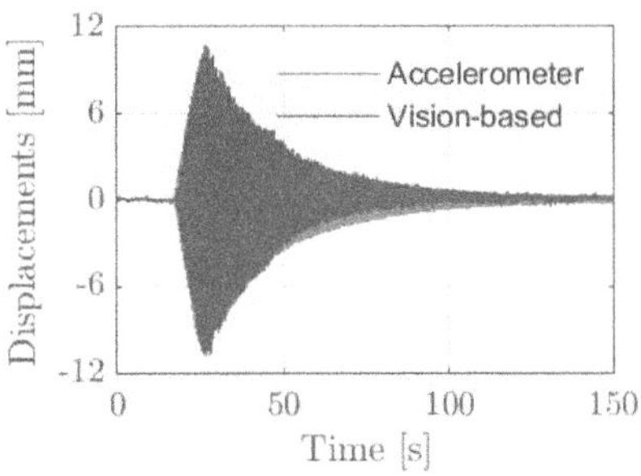

(k) Checkerboard location 6 – test at 138 BPM

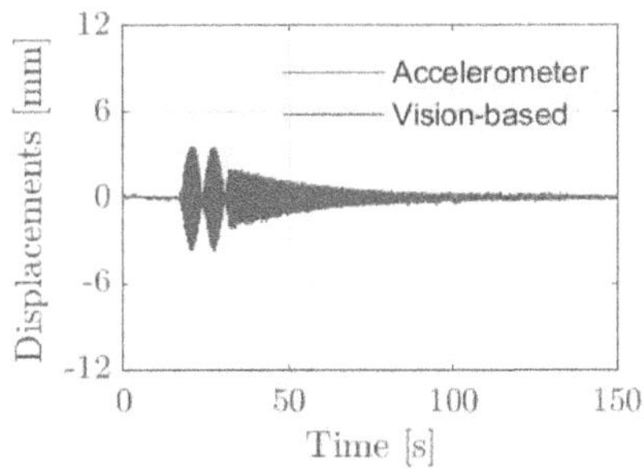

(l) Checkerboard location 6 – test at 132 BPM

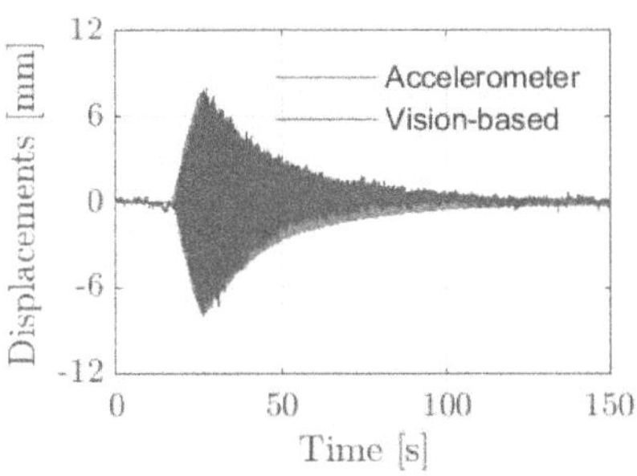

(m) Checkerboard location 8 – test at 138 BPM

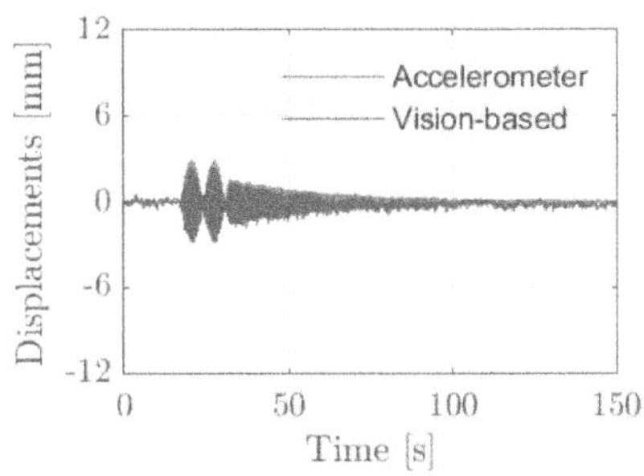

(n) Checkerboard location 8 – test at 132 BPM

FIGURE 2.2 (Continued)

reconstructed within an overlapping time window by minimizing the least-squared errors between the measured and approximated acceleration.

Each row in Figure 2.2 corresponds to one of the five distinct locations under examination. Note that for locations 1 and 2, results regarding concentric circles are also reported.

Moreover, the first column of Figure 2.2 deals with results of the test involving a pedestrian jumping at 138 BPM close to location 5; the human-induced forcing was designed to excite the first vertical mode of the footbridge, characterized by a natural frequency of 2.32 Hz. Similarly, the second column pertains to the test conducted with a jumping excitation of 132 BPM (corresponding to 2.20 Hz) applied close to location 2. Herein, the jumping frequency is intended to activate the first torsional mode of the structure, characterized by a natural frequency of 4.43 Hz. Displacements related to this test are characterized by amplitude modulation. This behavior is observed for both vision-based and accelerometer-based displacements, and it can be due to the fact that the exciting and modal frequencies are not exactly coincident. Further investigations in the frequency domain are carried out in the following.

Vertical displacements obtained by the two methods reveal a generally good correspondence in the time interval containing the largest vibrations. The same consideration applies to the comparison between the results of vision-based monitoring adopting different target shapes. In contrast, notable differences are observed in the noise level. Figure 2.3 shows the horizontal displacements and is organized similarly to Figure 2.2, focusing only on location 2. As expected, the horizontal motion resulting from bending vibrations is minimal, with deviations from zero primarily due to instrumental noise. In this regard, Tables 2.1 and 2.2, respectively, list the standard deviation of vision-based and accelerometer-based displacements (both

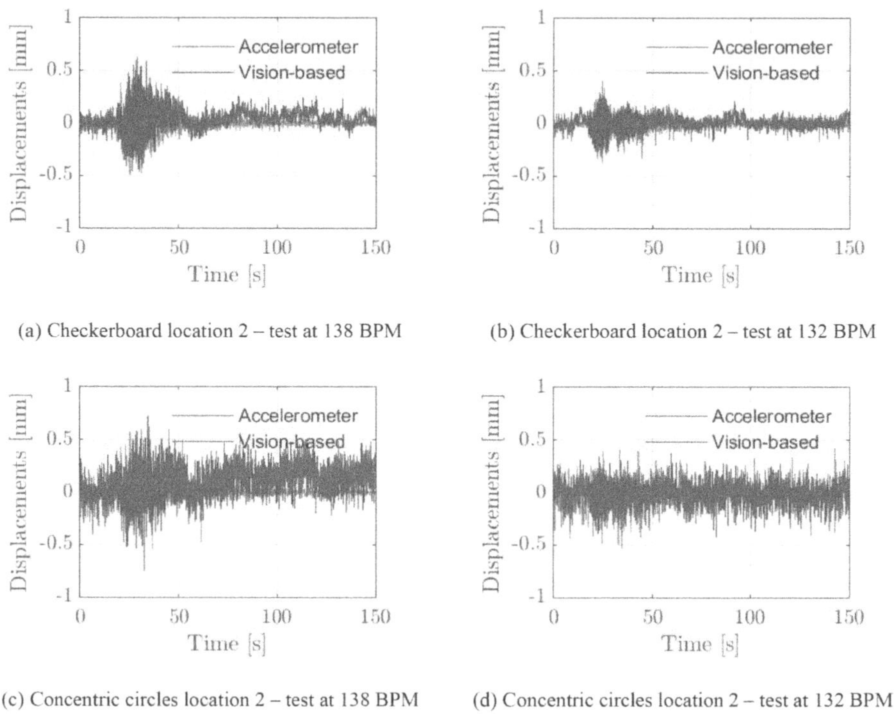

FIGURE 2.3 Horizontal displacements detected for the two tests and for the targets located in section 2.

TABLE 2.1
Ambient noise in vision-based displacements: target location ID number, camera-to-target distance D, target area A in square pixels, scale factor along vertical r_v and horizontal r_h directions, and standard deviation σ of vision-based displacements along vertical and horizontal directions (superscripts v and h, respectively) prior to the 138- or 132-BPM forcing (as indicated in the subscript).

ID	*target*	D (m)	r_v (mm/pxl)	r_h (mm/pxl)	σ^v_{138} (mm)	σ^h_{138} (mm)	σ^v_{132} (mm)	σ^h_{132} (mm)
1	Checkboard	15.0	2.52	2.38	0.029	0.063	0.030	0.042
1	Concentric circles	15.0	2.50	2.42	0.095	0.096	0.087	0.088
2	Checkboard	21.6	3.55	3.51	0.059	0.061	0.038	0.061
2	Concentric circles	21.6	3.53	3.55	0.119	0.138	0.119	0.122
5	Checkboard	36.5	6.01	5.96	0.089	0.097	0.105	0.100
8	Checkboard	52.2	8.54	8.49	0.223	0.139	0.169	0.167

TABLE 2.2
Ambient noise in accelerometer-based displacements: sensor location ID number, standard deviation σ of accelerometer-based displacements along vertical and horizontal directions (superscripts v and h, respectively) prior to the 138 or 132 BPM forcing (as indicated in the subscript).

ID	σ^v_{138} (mm)	σ^h_{138} (mm)	σ^v_{132} (mm)	σ^h_{132} (mm)
1	0.019	0.017	0.015	0.016
2	0.028	0.018	0.017	0.016
5	0.036	0.019	0.020	0.019
8	0.027	0.017	0.016	0.015

vertical and horizontal) under operational conditions during the period preceding the jumping excitation. These values are assumed to be representative indications of the noise level.

The noise level of displacements reconstructed from accelerations exhibits similarity in all directions, sensors, and tests, with slight differences depending on the magnitude of the environmental vibrations. When displacements are computed through video processing, operational noise remains quite independent of direction and test but varies from target to target depending on the target's geometric pattern. Indeed, the accuracy in vision-based monitoring is not simply related to the target-to-camera distance, but it depends on the number of pixels occupied by the object

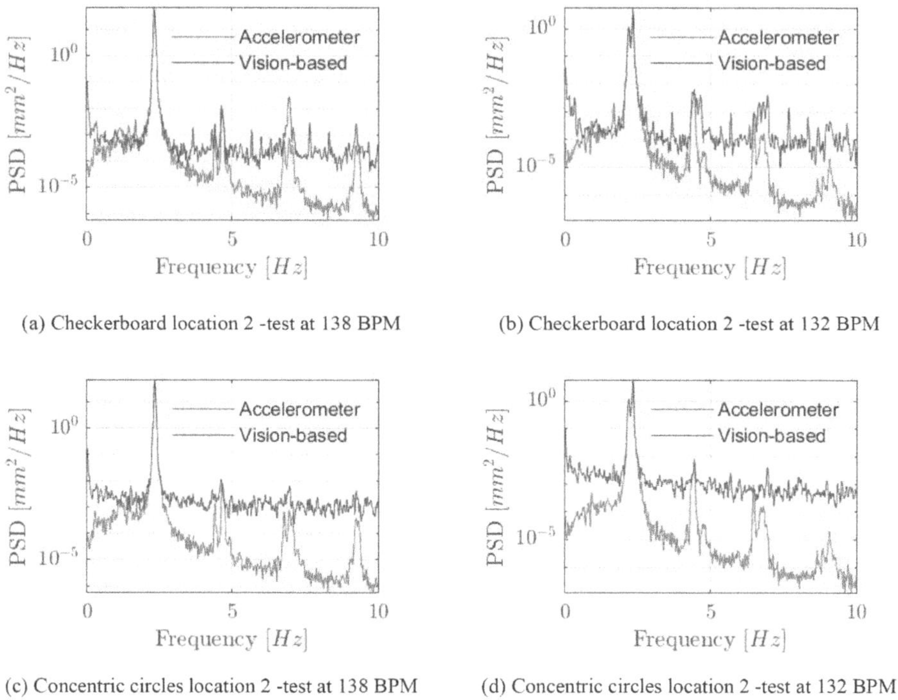

(a) Checkerboard location 2 -test at 138 BPM
(b) Checkerboard location 2 -test at 132 BPM
(c) Concentric circles location 2 -test at 138 BPM
(d) Concentric circles location 2 -test at 132 BPM

FIGURE 2.4 Comparison in the frequency domain of the displacements detected for different targets located in section 2.

being monitored and on the target shape. Focusing the comparison on locations 1 and 2, the scale factor of the checkboard and concentric circle target are approximately the same, but the noise level of displacements obtained by tracking the checkboard is remarkably lower.

Figure 2.4 represents the comparison between accelerometer and vision-based vertical displacements in the frequency domain, considering the example of location 2. The complete displacement time series are considered to characterize both modal and forcing frequencies. The amplitude of the power spectral density (PSD) peaks depends on the exciting frequency of the test, with an evident distinction between the peak related to the natural frequency and the peak of the forcing frequency in the case of the 132 BPM test (see the second column of Figure 2.4).

Although the pedestrian motion was not directly measured, it is reasonable to assume that the two distinct peaks observed in the interval 2.20–2.30 Hz can be attributed to two different sources: the first peak to the external forcing frequency and the second peak to the footbridge's natural frequency. This assumption is supported by the fact that the second peak was consistently observed across all tests, including those conducted under environmental excitation.

The noise level associated with a high-frequency content is clearly larger for the vision-based displacements. This is in accordance with the observations about the

TABLE 2.3
Footbridge modal properties identified from vision-based displacements: mode ID number, natural frequency *f*, and mode shape.

ID	*f* (Hz)	Mode shape
1	2.32	First bending
2	4.43	Torsional
3	4.65	Second bending

results in the time domain. Moreover, the low-frequency content of displacements reconstructed from acceleration is characterized by very low amplitudes, as a consequence of the filtering performed by the improved double-integration method.

However, the dominant peaks exhibit approximately the same PSD amplitude and frequency for the two methodologies, providing further evidence of the effectiveness of vision-based techniques in the field of structural monitoring.

Finally, vision-based displacements are processed through the enhanced frequency domain decomposition (EFDD) method, an identification algorithm widely adopted in operational modal analysis that operates in the frequency domain (Brincker et al. 2001).

Considering the time window after the jumping excitation, three vibration modes are identified, and they are summarized in Table 2.3. They are two bending modes and one torsional mode, with natural frequencies ranging from 2.32 Hz to 4.65 Hz.

2.5 CONCLUSIONS

A vision-based monitoring approach based on target tracking has been designed and tested on a real-scale case study. Jumping tests designed to trigger diverse modes of the structure have been performed. In this way, displacements due to both bending and torsional vibrations can be estimated (the latter being an aspect still not largely investigated in the vision-based monitoring literature). Checkerboard targets and accelerometers have been installed at each location of interest, while in some locations, concentric circle targets have been added to assess the influence of a different target shape.

Fixed targets have also been disposed on the dry riverbed. This has enhanced the robustness of outdoor monitoring setups, enabling the correct distinction between the actual displacement of the structure itself and any apparent motion detected in the target due to undesired external disturbances (such as camera shaking due to wind).

The level of noise in the estimated displacements is selected as a key parameter in defining the accuracy of the results. Different accuracies are obtained as a function of the target pixel area and the target shape, with that of checkerboard targets being remarkably lower for the same pixel area.

The comparison with the displacements reconstructed from accelerations is deemed satisfactory in both time and frequency domains, especially considering the substantial savings in terms of installation effort, time, and cost associated with vision-based monitoring. Finally, the first three natural modes of the footbridge have been identified from the vision-based displacements, proving the potential of vision-based techniques in the modal analysis of large-scale structures.

2.6 ACKNOWLEDGMENTS

This work was supported by the FAR 2023 Project—FOMO line (vision-based approaches for the structural health monitoring of existing bridges, VIS4SHM). The financial support of the University of Modena and Reggio Emilia and the "Fondazione di Modena" is gratefully acknowledged.

REFERENCES

Atherton, T. J., Kerbyson, D. J.: Size invariant circle detection. Image and Vision Computing 17, 795–803 (1999)

Bassoli, E., Vincenzi, L., Grassi, F., Mancini, F.: A multi-temporal DInSAR-based method for the assessment of the 3D rigid motion of buildings and corresponding uncertainties. Journal of Building Engineering 73, 106738 (2023)

Brincker, R., Ventura, C. E., Andersen, P.: Damping estimation by frequency do-main decomposition. In: Proceedings of IMAC 19: A Conference on Structural Dynamics, pp. 698–703, Kissimmee, Florida, USA (2001)

Buoli, E., Bassoli, E., Eslami Varzaneh, G., Ponsi, F., Vincenzi, L.: Vision-based dynamic monitoring of a steel footbridge. In: XII International Conference on Structural Dynamics, JPCS, Delft, The Netherlands (2023)

Castagnetti, C., Bassoli, E., Vincenzi, L., Mancini, F.: Dynamic assessment of masonry towers based on terrestrial radar interferometer and accelerometers. Sensors 19(6), 1319 (2019)

Chen, J. G., Adams, T. M., Sun, H., Bell, E. S., Buyukozturk, O.: Camera-based vibration measurement of the world war I memorial bridge in Portsmouth, New Hampshire. Journal of Structural Engineering 144(11), 04018207 (2018)

Dong, C. Z., Catbas, F. N.: A review of computer vision–based structural health monitoring at local and global levels. Structural Health Monitoring 20(2), 692–743 (2020)

Feng, D., Feng, M. Q.: Experimental validation of cost-effective vision-based structural health monitoring. Mechanical Systems and Signal Processing 88, 199–211 (2017)

Fradelos, Y., Thalla, O., Biliani, I., Stiros, S.: Study of lateral displacements and the natural frequency of a pedestrian bridge using low-cost cameras. Sensors 20(11), 3217 (2020)

Geiger, A., Moosmann, F., Car, O., Schuster, B.: Automatic camera and range sensor calibration using a single shot. In: 2012 IEEE International Conference on Robotics and Automation, pp. 3936–3943, IEEE, Saint Paul, MN, USA (2012)

Guerzoni, G., Faghand, E., Vitetta, G., Vincenzi, L., Mehrshahi, E.: Novel movement-based methods for the calibration of colocated multiple-input multiple-output radars. IEEE Access 11, 116090–116108 (2023)

Guidorzi, R., Diversi, R., Vincenzi, L., Mazzotti, C., Simioli V.: Structural monitoring of a tower by means of MEMS-based sensing and enhanced autoregressive models. European Journal of Control 20(1), 4–13 (2014)

Lydon, D., Lydon, M., Taylor, S., Del Rincon, J. M., Hester, D., Brownjohn, J.: Development and field testing of a vision-based displacement system using a low cost wireless action camera. Mechanical Systems and Signal Processing 121, 343–358 (2019)

Poluzzi, L., Barbarella, M., Tavasci, L., Gandolfi, S., Cenni, N.: Monitoring of the Garisenda tower through GNSS using advanced approaches toward the frame of reference stations. Journal of Cultural Heritage 38, 231–241 (2019)

Ponsi, F., Buoli, E., Eslami Varzaneh, G., Bassoli, E., Briseghella, B., Vincenzi, L.: Vision-based approach for the static and dynamic monitoring of bridges. In: FABRE Consortium Conference on Existing Bridges, Viaducts and Tunnels: Research, Innovation and Applications, Procedia Structural Integrity, Genoa, Italy (2024)

Spencer, B. F., Hoskere, V., Narazaki, Y.: Advances in computer vision-based civil infrastructure inspection and monitoring. Engineering 5(2), 199–222 (2019)

The MathWorks, Inc.: Computer Vision Toolbox. Natick, MA, United States https://it.mathworks.com/products/computer-vision.html

The MathWorks, Inc.: Image Processing Toolbox. Natick, MA, United States https://it.mathworks.com/products/image.html

Xu, Y., Brownjohn, J. M., Huseynov, F.: Accurate deformation monitoring on bridge structures using a cost-effective sensing system combined with a camera and accelerometers: Case study. Journal of Bridge Engineering 24(1), 05018014 (2019)

Xu, Y., Brownjohn, J. M., Kong, D.: A non-contact vision-based system for multipoint displacement monitoring in a cable-stayed footbridge. Structural Control and Health Monitoring 25(5), e2155 (2018)

Ye, X. W., Yi, T. H., Dong, C. Z., Liu, T.: Vision-based structural displacement measurement: System performance evaluation and influence factor analysis. Measurement 88, 372–384 (2016)

Zona, A.: Vision-based vibration monitoring of structures and infrastructures: An overview of recent applications. Infrastructures 6(1), 4 (2020)

3 Structural displacement estimation through data fusion of GNSS and acceleration data

Federico Ponsi, Cristina Castagnetti, Ghita Eslami Varzaneh, Elisa Bassoli, F. Mancini, and Loris Vincenzi

3.1 INTRODUCTION

Accelerometers are the main components of traditional vibration monitoring systems, which can record the response of a structure to various sources of excitation (Rainieri and Fabbrocino 2010). The extraction of modal features from the recorded accelerations (Ubertini et al. 2013) and the detection of changes during the monitoring period are key steps in well-established damage detection methods (Das et al. 2016; Huang et al. 2018; Ponsi et al. 2023). Displacement measurements have the potential to provide more valuable and direct information about structural health in various applications, as they can serve as an operational condition index and allow the rapid detection of permanent deformations.

Global navigation satellite systems (GNSS) (Poluzzi et al. 2019), linear variable displacement transducer (LVDT) (Joshi et al. 2019), satellite remote sensing (Talledo et al. 2022; Bassoli et al. 2023), terrestrial radar interferometry (Castagnetti et al. 2019, Guerzoni et al. 2023), and vision-based techniques (Feng and Feng 2018, Micozzi et al. 2023; Ponsi et al. 2024) are the main technologies that can be used for direct displacement measurement in structural monitoring. Each of these technologies has its own unique potential, but they also have some disadvantages that restrict their widespread use in structural monitoring. The sampling frequency of GNSS receivers is generally low, limiting their efficiency in dynamic applications. However, the availability of multi-constellation, high-frequency GNSS receivers holds the promise of providing even more valuable support for the monitoring of three-dimensional structural displacement. A fixed base is necessary for LVDT installation, but it is rarely available, especially when attempting to monitor common infrastructure movements. Direct displacement measurement can be achieved with low-cost vision-based techniques, but these are susceptible to light and weather conditions. In order to enhance their potential, further research is required.

Double integration of accelerations can be used to estimate displacements, but it has disadvantages such as the unknown initial conditions for integration and the

 DOI: 10.1201/9781003516941-3

severe amplitude of measured acceleration noise during integration. Stable and accurate filtering techniques have been proposed by Lee et al. (2010) and Hong et al. (2010) to suppress low-frequency noise. Linear dynamic displacement reconstruction is the only possible application of these techniques, as the filters also prevent low-frequency displacement components, including pseudo-static displacement caused by moving loads on a bridge or permanent displacement caused by damage to a structural element.

Data fusion is an emerging approach to overcome these limitations, which involves combining different measurements in order to produce more accurate estimates (Park et al. 2013). Accelerations are typically combined with other information captured by different sensors, including strain sensors (Ma et al. 2021), GNSS (Koo et al. 2017), video cameras (Park et al. 2018), inclinometers (Ozdagli et al. 2017), and millimeter-wave radar (Ma et al. 2023). The fusion of these data is typically accomplished using a multi-rate Kalman filter (Smyth and Wu 2007; Kim et al. 2014) or a finite impulse response (FIR) filter (Hong et al. 2013; Park et al. 2018).

The purpose of this chapter is to examine the potential of data fusion for structural health monitoring by combining GNSS displacement measurements with measurements obtained through a traditional accelerometer-based monitoring system. Although the GNSS receiver acquires displacements at lower sampling rates than traditional monitoring systems, they can still be used to adjust post-processed displacements and eliminate the uncertainties caused by the integration procedure through an approach based on the Kalman filter. The application of the data fusion approach is performed with reference to two experimental tests. The first is a laboratory structure for which a dynamic excitation is combined with an imposed permanent displacement. The accurate estimation of the latter is a key aspect of structural health monitoring. Since only a limited part of the research in the field of data fusion has been dedicated to real-scale structures, the second test is carried out on a steel footbridge crossing the Panaro River in Modena. For this test, the impact of noise in GNSS data and the GNSS sensor sampling rate are discussed.

The chapter is organized as follows: Section 3.2 introduces the Kalman filter formulation for data fusion, and Section 3.3 describes the GNSS- and accelerometer-based monitoring systems. Sections 3.4 and 3.5 present the experimental tests carried out on a laboratory-scale frame and on a full-scale footbridge, respectively, along with the results of the data fusion approach. Finally, conclusions are drawn in Section 3.6.

3.2 KALMAN FILTER FOR DATA FUSION

The derivation of the Kalman filter expression for data fusion relies on modeling the measurement process in a state-space equation form:

$$\dot{\mathbf{x}} = \mathbf{A}\mathbf{x} + \mathbf{B}\mathbf{u} + \mathbf{w} \tag{3.1}$$

$$z = \mathbf{H}\mathbf{x} + v \tag{3.2}$$

where the state vector $\mathbf{x} = \left[x\,\dot{x}\right]^T$ includes the displacement x and the velocity $\dot{x}$. The variables $z = x_m$ and $u = \ddot{x}_m$ are the measured displacement and acceleration,

$\mathbf{w} \sim (\mathbf{0}, \mathbf{Q})$ is a white noise process with covariance matrices **Q**, and $v \sim (0, r)$ denotes a random variable representing the uncertainty in displacement measurements with variance *r*. We assume $\mathbf{Q} = \begin{bmatrix} 0 & 0 \\ 0 & q \end{bmatrix}$ because the only source of noise in Eq. (3.1) is the measured acceleration where *q* is the acceleration, variance. The coefficient matrices are

$$\mathbf{A} = \begin{bmatrix} 0 & 1 \\ 0 & 0 \end{bmatrix}, \mathbf{B} = \begin{bmatrix} 0 \\ 1 \end{bmatrix}, \mathbf{H} = \begin{bmatrix} 1 & 0 \end{bmatrix} \tag{3.3}$$

Supposing that accelerations are acquired at a sampling rate T_a, Eqs. (3.1)–(3.2) are discretized as follows:

$$\mathbf{x}_{k+1} = \mathbf{A}_d \mathbf{x}_k + \mathbf{B}_d u_k + \mathbf{w}_k \tag{3.4}$$

$$z_k = \mathbf{H}\mathbf{x}_k + v_k \tag{3.5}$$

where the coefficient matrices altered by the discretization process are

$$\mathbf{A}_d = \begin{bmatrix} 1 & T_a \\ 0 & 1 \end{bmatrix}, \mathbf{B}_d = \begin{bmatrix} T_a^2 / 2 \\ T_a \end{bmatrix} \tag{3.6}$$

while **H** remains unchanged. The covariance matrices of the discrete noise sequences are

$$\mathbf{Q}_d = \begin{bmatrix} qT_a^3 / 3 & qT_a^2 / 2 \\ qT_a^2 / 2 & qT_a \end{bmatrix}, \mathbf{R}_d = \mathbf{R} / T_a \tag{3.7}$$

Introducing the state covariance matrix **P** and the state estimate $\hat{\mathbf{x}}$, the forward Kalman filter algorithm can be summarized as follows:

- Initialization: $\mathbf{x}_0 = \mathbf{0}; \mathbf{P}_0 = \mathbf{I}$
- Time update (a priori estimate)

$$\hat{\mathbf{x}}_{k+1|k} = \mathbf{A}_d \hat{\mathbf{x}}_{k|k} + \mathbf{B}_d u_k \tag{3.8}$$

$$\mathbf{P}_{k+1|k} = \mathbf{A}_d \mathbf{P}_{k|k} \mathbf{A}_d^T + \mathbf{Q}_d \tag{3.9}$$

- Measurement update (a posteriori estimate)

$$\hat{\mathbf{x}}_{k+1|k+1} = \hat{\mathbf{x}}_{k+1|k} + \mathbf{K}_{k+1} \left(z_{k+1} - \mathbf{H}\hat{\mathbf{x}}_{k+1|k} \right) \tag{3.10}$$

$$\mathbf{P}_{k+1|k+1} = \left(\mathbf{I} - \mathbf{K}_{k+1} \mathbf{H} \right) \mathbf{P}_{k+1|k} \tag{3.11}$$

where the Kalman gain $\mathbf{K}_{k+1}$ is given by

$$\mathbf{K}_{k+1} = \mathbf{P}_{k+1|k}\mathbf{H}^T\left(\mathbf{H}\mathbf{P}_{k+1|k}\mathbf{H}^T + \mathbf{R}_d\right)^{-1} \tag{3.12}$$

It is very common that acceleration and displacement are measured at different sampling rates, with a larger one for acceleration. Denoting with T_a and T_d the sampling intervals of acceleration and displacement, respectively, it occurs that $T_d > T_a$.

In this context, data fusion requires a multi-rate Kalman filter. The complete forward Kalman filter algorithm with time update (Eqs. (3.8)–(3.9)) and measurement update (Eqs. (3.10)–(3.11)) is achieved if both acceleration and displacement measurements are available.

The only time update is realized when acceleration measurements are available but displacement ones are not, namely between kT_d and $(k+1)T_d$ with k integer. In this case, the optimal estimate $\hat{\mathbf{x}}_{k+1|k+1}$ is equal to $\hat{\mathbf{x}}_{k+1|k}$ computed through Eq. (3.8).

The final displacement time series is then obtained by applying a smoothing procedure. Further details can be found in Smyth and Wu (2007) and Kim et al. (2014).

3.3 MONITORING SYSTEM

The present section describes the monitoring system employed for data acquisition in the two experimental tests. The monitoring system is composed of a GNSS system and an accelerometer-based monitoring system.

GNSS systems are currently employed to monitor slow displacements in long-span bridges, offering easy installation on decks and reliable performance, making them reliable for long-term applications. Although advancements in receiver and antenna technologies have enabled GNSS sensors to achieve sampling rates of up to 100 Hz, it is important to note that an increase in the operational frequency does not necessarily result in a proportional boost in accuracy. Casciati and Fuggini (2009) provided evidence that the accuracy remains within the sub-centimetric range, as verified through both static and dynamic tests. However, the level of accuracy provided may be insufficient for applications related to structural monitoring, especially when dealing with the low vibrations of bridges induced by environmental conditions.

The GNSS instrumentation used in this research includes a receiver placed on the structure and a master ground station. The employed measuring stations are of the STONEX S900A model, operating at an updated frequency of 20 Hz. Additionally, the instrument features control terminals linked via Wi-Fi Direct to a remote device, facilitating parameter control during measurements. Data from the GNSS receiver were acquired at a rate of 20 Hz, and both relative kinematic positioning and precise point positioning (PPP) post-processing methods were employed. The former is used to obtain the results presented in the following.

The accelerometer-based system is composed of a control unit and ten digitally connected biaxial micro-electro-mechanical systems (MEMS) accelerometers. Each sensing unit is capable of measuring temperature and acceleration with a sampling frequency ranging from 20 Hz to 80 Hz. Through the implementation of local digital filtering techniques and oversampling rates, these units can achieve a noise floor of approximately 0.3 mg, ensuring a high level of reliability. Notably, the system features digital data transmission and the capability to conduct certain system analyses directly on board the sensors, enabling the transfer of only processed synthetic data to the main control unit (Guidorzi et al. 2014).

3.4 LABORATORY-SCALE FRAME

The first test is performed on a laboratory-scale frame composed of steel elements assembled through bolt connections. A picture of the frame is shown in Figure 3.1a. The frame is 153 cm high and has 94 cm × 102 cm plane dimensions. The rectangular base of the frame is made with pairs of L-shaped elements joined together by means of bolds. The columns have a thin rectangular cross section and are connected at their top by paired L-shaped beams that are oriented in the direction of the column's maximum stiffness. Single L-shaped profiles ensure the connection of the beams in the orthogonal direction.

During the test, the response of the frame to an external excitation is measured using the monitoring system described in Section 3.3. Two MEMS accelerometers are installed on the top of the frame at the opposite extremities of one beam (Figure 3.1a). The GNSS receiver was placed close to an accelerometer, as shown in Figure 3.1b.

The excitation involves a dynamic and a permanent component. The dynamic excitation consists of the free vibration of the frame in the direction of minimum stiffness after an impulse excitation, while the permanent displacement is simulated by shifting the frame base during free vibrations.

(a) (b)

FIGURE 3.1 (a) Scale frame in the test configuration and (b) detail of the GNSS receiver and one accelerometer.

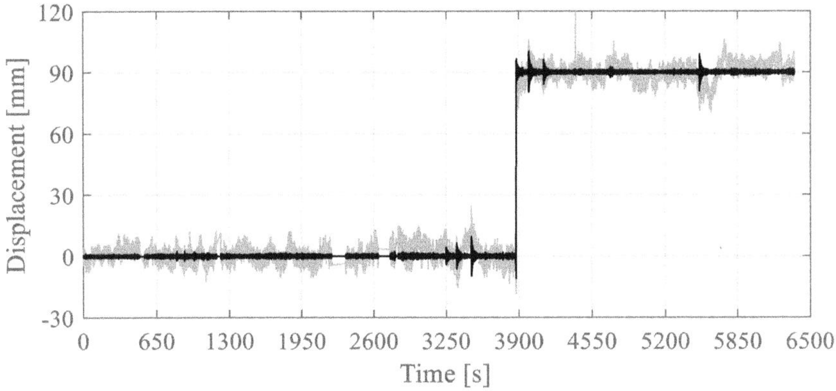

FIGURE 3.2 Comparison between raw GNSS data (grey) and the corresponding preprocessed data (black).

3.4.1 Results

In this section, the results of the monitoring and the data fusion are presented. Figure 3.2 shows the raw GNSS displacement acquired during the test on the steel frame. The shift of the frame base of about 10 cm is clearly detected, but the signal is characterized by a long-term trend and a remarkable noise level that makes a large part of the dynamic vibrations undetectable.

Data fusion is first performed with raw GNSS data and accelerations, and the result is depicted in Figure 3.3. The base shift is still visible, and the accuracy of the dynamic displacement is improved. However, the long-period trend derived from the raw GNSS data persists after the data fusion operation. This proves the necessity of preprocessing GNSS data before the application of the Kalman filter.

An elliptical band-pass filter is applied to the GNSS data in order to filter the frequency content outside the interval 2–4 Hz, which contains the main structural frequency characterizing the response. In this way, the long-term trend is filtered out, but even the base shift is removed, as shown in Figure 3.4. The filtered GNSS data are compared in the same figure with the displacement reconstructed from the acceleration alone through an improved double integration procedure (Lee et al. 2010). The applied method is formulated as a regularization problem in which the acceleration is approximated by the second-order central finite difference of displacement. The displacement is reconstructed within an overlapping time window by minimizing the least-squared errors between the measured and approximated acceleration.

The obtained displacement has a very low noise level, such that small vibrations can be detected. This is due to the filtering of the low-frequency noise involved in the method, which, however, also determines the elimination of the permanent component.

For this reason, a different solution involving the fusion of GNSS and acceleration data needs to be developed. The permanent component of the structural response

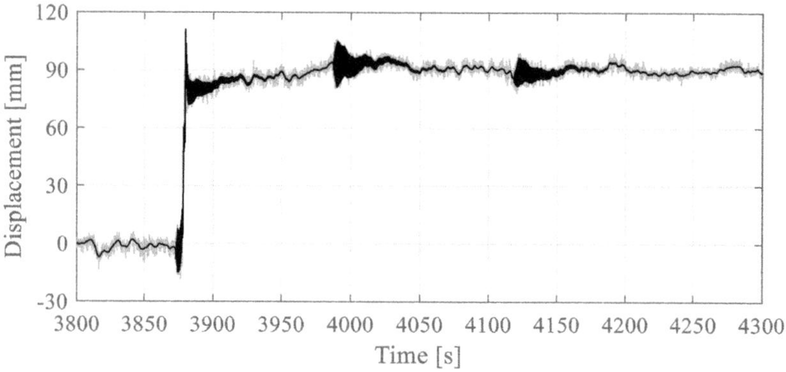

FIGURE 3.3 Data fusion results (black) employing raw GNSS data (grey).

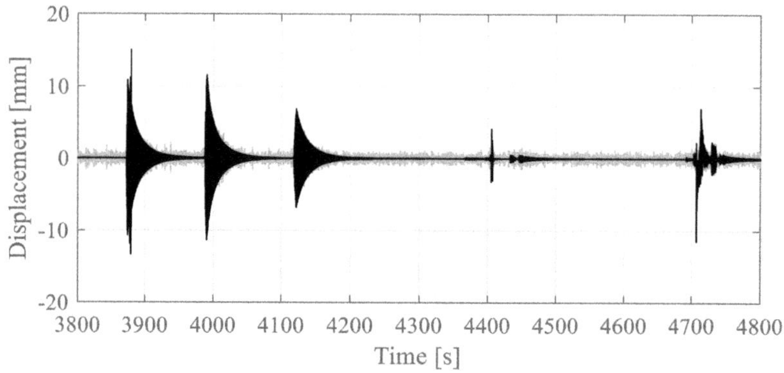

FIGURE 3.4 Comparison between filtered GNSS data (grey) and displacement reconstructed from acceleration alone (black)..

must be extracted from the GNSS data before the application of the bandpass filter. In this regard, a step function of the form

$$f(t) = \begin{cases} f_1 & \text{if } t < \bar{t} \\ f_2 & \text{if } t \geq \bar{t} \end{cases} \tag{3.13}$$

is fitted to the raw data of Figure 3.2. The step function quantifies the extent of the permanent displacement $(f_2 - f_1)$ and the time instant $\bar{t}$ at which it occurs. Their values are determined through a least-squares fitting of the raw GNSS data. At this point, the obtained step function can be summed with the filtered data (Figure 3.4), obtaining the displacement represented in green in Figure 3.2. The long-term trend has been removed, and at the same time, the permanent component is maintained.

Finally, data fusion employing the preprocessed GNSS data is carried out, and the comparison between the two time series is shown in Figure 3.5. The peculiarities of

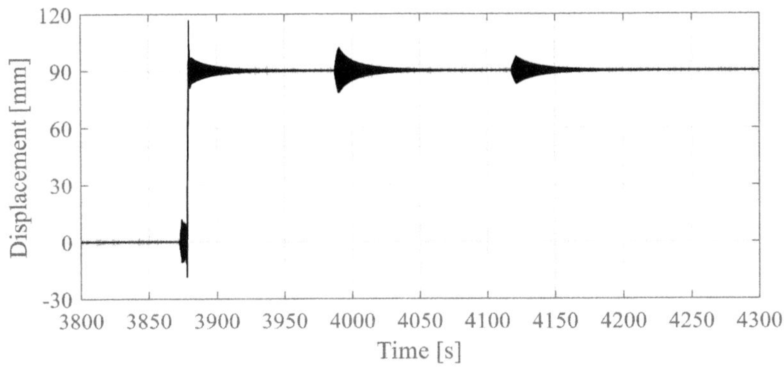

FIGURE 3.5 Data fusion results (black) employing the preprocessed GNSS data (grey).

the GNSS data are maintained, but noticeable differences are found in the dynamic component, which is more accurate in the result of the data fusion. The use of accelerations within the data fusion approach determines a significant reduction of the GNSS noise, and it allows the identification of low-amplitude vibrations.

3.5 FOOTBRIDGE ON THE PANARO RIVER

The second test is performed on a real-scale structure, a steel footbridge crossing the Panaro River in Modena, Italy. It is a truss girder with a box cross section of dimensions 3.00 × 3.20 meters (Figure 3.6a). It is composed of hollow tubular elements, and it has a total length of 160 meters. It is supported by two piles located 45 meters from its extremities, which divide the footbridge into three spans. The central span is the longest and it measures 70 meters. The footbridge is characterized by a slender and lightweight design. For this reason, it exhibits significant flexibility and high sensitivity to dynamic vibrations caused by pedestrians, cyclists, and wind.

The structure has been excited by a pedestrian jumping at an imposed rate of 138 and 132 beats per minute (BPM), controlled by a metronome. These values have been chosen with the aim of exciting the main vertical and torsional modes, whose natural frequencies are 2.32 Hz and 4.43 Hz, respectively. They were evaluated based on preliminary ambient vibration testing.

A schematic view of the footbridge is presented in Figure 3.7, where the locations of the accelerometers are denoted by the letters A'-G', B", D", and G". They correspond to seven cross sections—one at the midspan and six additional positions symmetrically arranged around the midspan at intervals of 7.5 meters each. The pedestrian jumped close to D" and B" for the tests at 138 BPM and 132 BPM, respectively.

An example of vertical acceleration recorded from the accelerometer placed at D' is illustrated in Figure 3.8. It has been caused by a pedestrian jumping at the midspan at 138 BPM. The two events presented in Figure 3.8 were caused by the same forcing conditions, but the first event was interrupted as soon as it started and was repeated after the footbridge vibration attenuated.

FIGURE 3.6 (a) View of the steel footbridge and (b) detail of the GNSS receiver (left) and the accelerometer (right) placed at midspan.

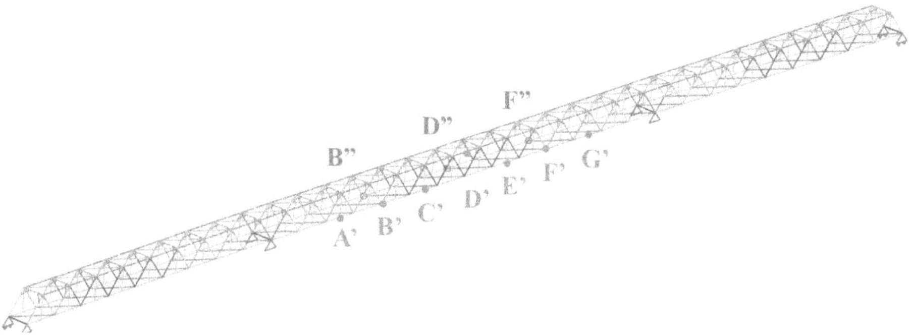

FIGURE 3.7 Schematic representation of the footbridge and indication of the monitored sections.

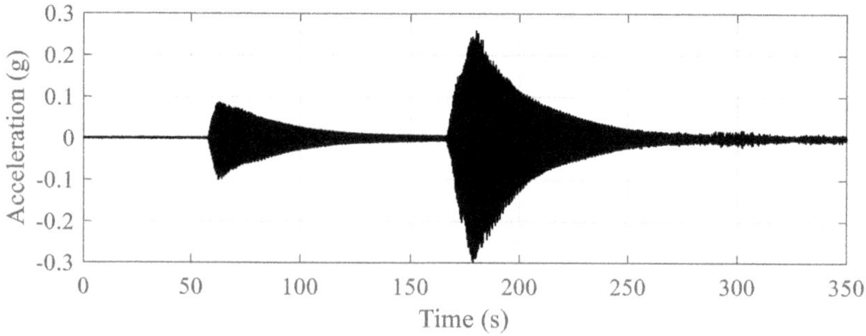

FIGURE 3.8 Example of measured acceleration.

The GNSS receiver was placed at D’, close to an accelerometer, as shown in Figure 3.6b. The fusion algorithm utilized ellipsoidal height data (representing the vertical oscillation component of the structure) to assess displacements induced by pedestrians.

3.5.1 Results

This section outlines the results and the main observations related to the fusion of data measured at the footbridge midspan, where both a GNSS receiver and a biaxial accelerometer were installed (Figure 3.6b). The data analyzed in the following correspond to the events shown in Figure 3.8. Two aspects of the procedure have been investigated: the filtering of GNSS data before the fusion process and the effect of their sampling rate on the results.

The application of a high-pass filter is necessary to eliminate low-frequency trends within the GNSS-acquired time window, as already highlighted in Section 3.4 for the test on the laboratory frame. Moreover, as the GNSS measurements are affected by considerable noise, different filters have been adopted and the results compared in terms of both noise reduction and vibration estimates. Noise is effectively reduced by applying a highly aggressive filter, but this implies a disadvantage. Indeed, the displacement peaks are decreased by the filter, which removes part of the structural response. On the other hand, this drawback does not occur with a mild filter, which at the same time retains a higher level of noise.

For this reason, the impact of a series of increasingly aggressive filters on the outcome of the fusion process was examined. As an illustrative example, the effects on the GNSS data of a Chebyshev filter and an adaptive least mean square (LMS) filter are illustrated in Figure 3.9.

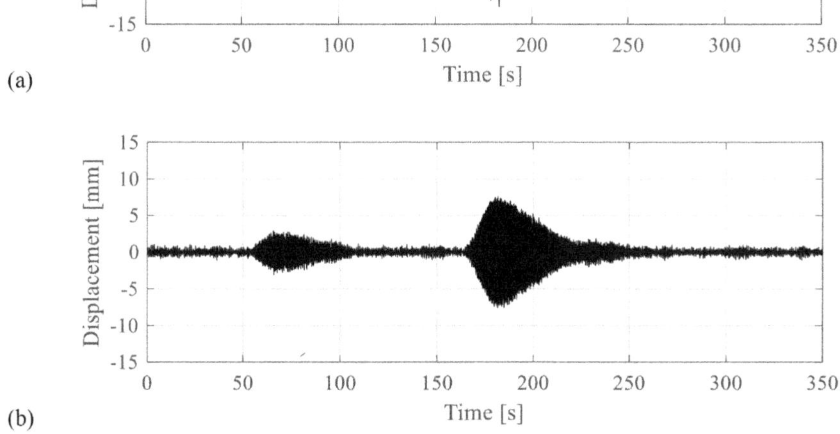

FIGURE 3.9 Displacements from GNSS data applying (a) a Chebyshev filter and (b) an adaptive LMS filter.

The standard deviations of the processed signals for a time window that precedes the jumping are considered as representative values of the residual noise. They are 1.50 mm and 0.34 mm for the Chebyshev and the adaptive LMS filter, respectively. Unfortunately, noise reduction is accompanied by a reduction in the displacement peak value. If we focus on the event characterized by the largest vibrations, the maximum displacement of the series processed by the adaptive LMS filter is about 7 mm—half of the maximum displacement (about 15 mm) for the series obtained with the Chebyshev filter.

The comparison between the data fusion results utilizing GNSS signals processed through the Chebyshev filter and the adaptive LMS filter is presented in Figure 3.10. No substantial differences are observed, with nearly identical patterns in the time series. The average difference value is approximately 0.03 mm, while the maximum differences are lower than 0.2 mm in the time interval containing the largest vibrations and increase up to 0.4 mm in the other windows.

The GNSS sampling rate is typically lower than that of accelerations. For this application, the GNSS sampling rate is 20 Hz, a fairly high value for GNSS instrumentation, which commonly adopts lower rates of 10 and 5 Hz. Therefore, the data

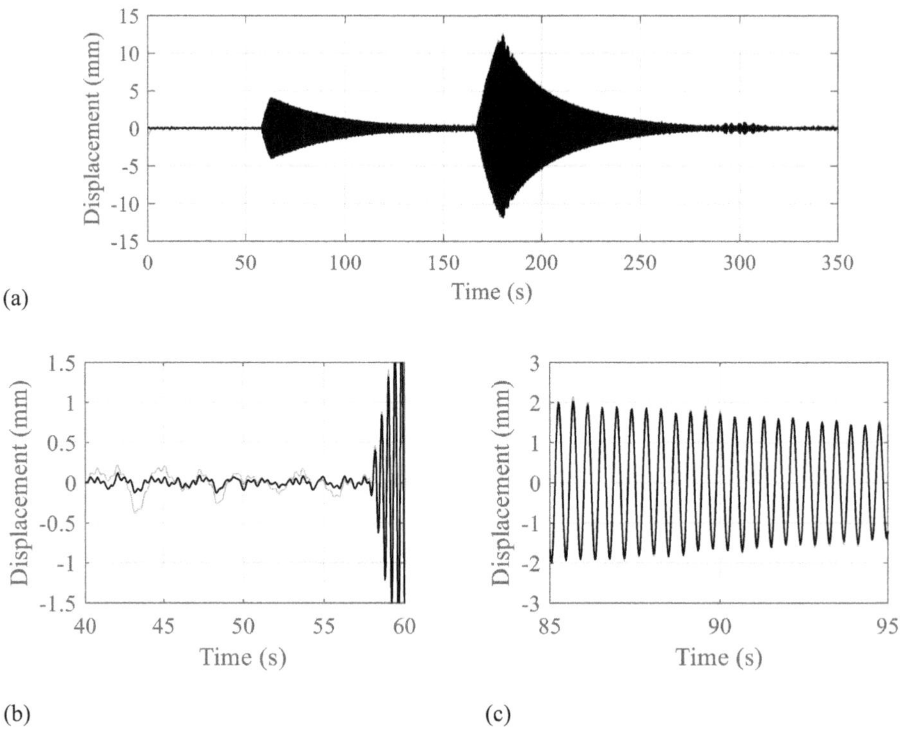

FIGURE 3.10 Comparison of post-data-fusion displacements obtained from the GNSS signal filtered with the Chebyshev (black) and the adaptive LMS filter (grey). (a) Overview of the complete time series and (b, c) zoomed sections of the time series.

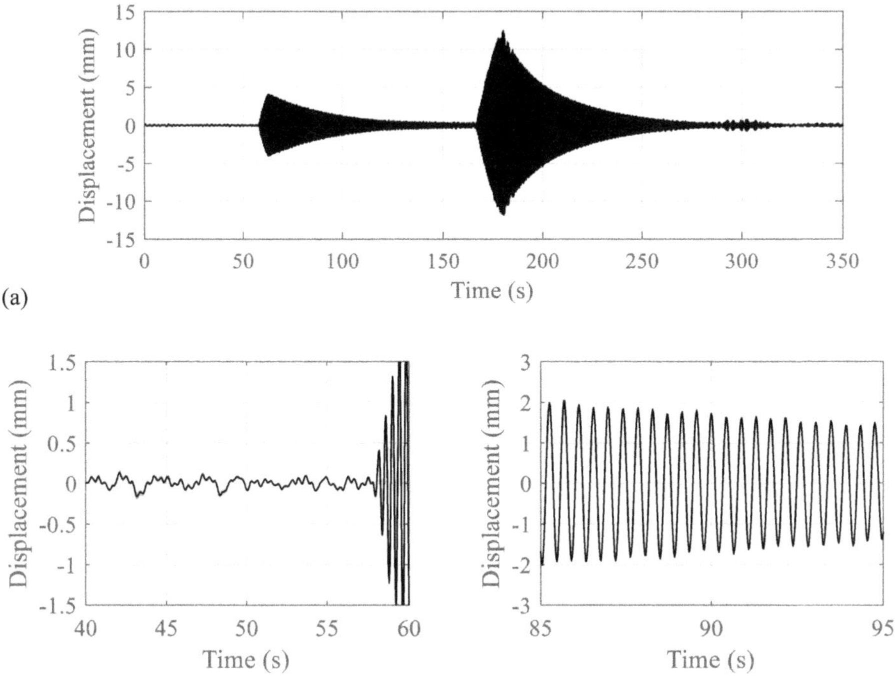

(a)

(b) (c)

FIGURE 3.11 Comparison of post-data-fusion displacements for different GNSS sampling rates: 20 Hz (light grey), 10 Hz (dark grey), and 5 Hz (black). (a) Overview of the complete time series; (b, c) zoomed segments of the time series.

are additionally processed at these rates to investigate if a sparser sampling affects the performance of the Kalman filter. Displacement series resulting from the fusion of GNSS data sampled at rates of 20, 10, and 5 Hz are shown in Figure 3.11. The very limited discrepancies among the series reveal that the data fusion process is marginally affected by the sampling frequency of the GNSS data. As the GNSS data are noisier than the accelerometer data, the shape of the processed displacement is primarily influenced by the acceleration time series. The main role of the GNSS data is to define the quasi-static form and to provide the amplitude of the displacement.

Finally, the data fusion results are compared with the displacement time series obtained by double integration of acceleration recordings using the method of Lee et al. (2010), already mentioned in Section 3.4. The selection of this algorithm was driven by the fact that it produces accurate results when the displacement time series has a globally zero mean value. The displacement time series represented in Figure 3.12 show very good agreement. The integration of high-rate acceleration data with the relatively low-sampled displacement by means of a multi-rate Kalman filter leads to a significant enhancement in signal estimation.

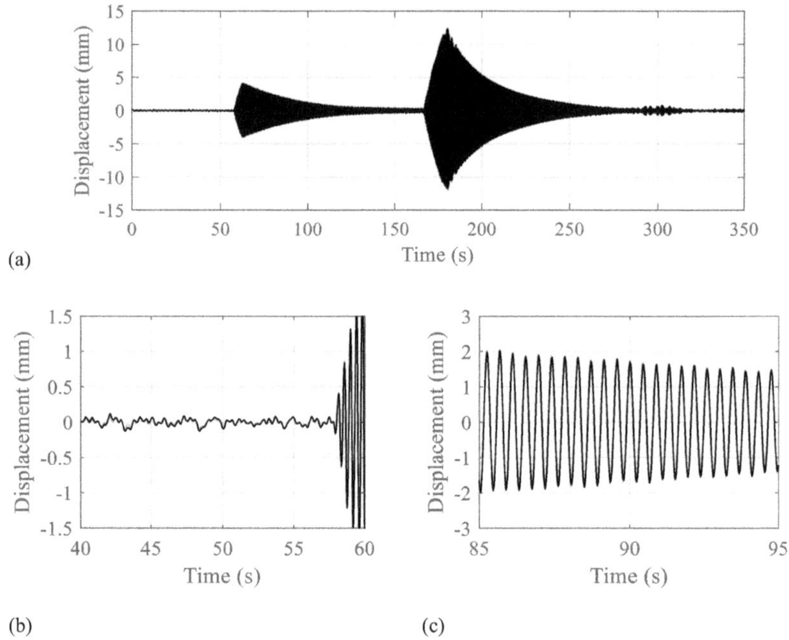

FIGURE 3.12 Comparison of post-data-fusion displacements (black) and displacements obtained from accelerations (grey). (a) Overview of the complete time series and (b, c) zoomed sections of the time series.

3.6 CONCLUSIONS

This chapter focused on the application of a data fusion procedure for structural displacement estimation. The procedure makes use of a multi-rate Kalman filter to combine acceleration and GNSS data. This kind of approach aims to overcome the limitation related to the use of a single sensor measurement in the estimation of structural displacements. GNSS data generally have a low sampling rate and a non-negligible noise level, which makes them inappropriate for dynamic applications. On the contrary, displacement reconstruction from acceleration recordings needs the removal of low-frequency noise components. This can be achieved through stable filtering procedures, which at the same time remove low-frequency components of the displacement such as quasi-static or permanent displacements.

The potential of data fusion is investigated in two experimental tests: a laboratory-scale frame and a real-scale steel footbridge. The results confirm the improved accuracy of displacement time series reconstructed by means of the data fusion approach. Preprocessing of GNSS data is recommended in order to filter low-frequency trends and reduce noise, thereby ensuring higher-quality data fusion results.

The permanent displacement simulated for the laboratory frame has been accurately estimated by combining GNSS data and accelerations. In this case, the preprocessing phase also involves fitting a step function to the GNSS data to extract the permanent component before applying the Kalman filter.

Displacement measures can be very useful in the context of structural health monitoring, as they ensure a rapid detection of the permanent deformations and represent an index of structural health condition. Future investigations will focus on developing an automated procedure to detect and extract the permanent component of displacement. The aim is to distinguish between short-term and long-term deformations caused by progressive damage and to correctly identify them in the estimated displacement time series.

3.7 ACKNOWLEDGMENTS

This research was carried out within the REVISION research project—Cascade Call within the RETURN research project (multi-Risk sciEnce for resilienT comUnities undeR a changiNg climate—Spoke 6 TS2—multi-risk resilience of critical infrastructures). The financial support of the Italian Ministry of University and Research (MUR) and the European Union—NextGenerationEU—project are gratefully acknowledged.

REFERENCES

Bassoli, E., Vincenzi, L., Grassi, F., Mancini, F.: A multi-temporal DInSAR-based method for the assessment of the 3D rigid motion of buildings and corresponding uncertainties. Journal of Building Engineering, 106738 (2023) https://doi.org/10.1016/j.jobe.2023.106738

Casciati, F., Fuggini, C.: Engineering vibration monitoring by GPS: Long duration records. Earthquake Engineering and Engineering Vibration 8, 459–467 (2009) https://doi.org/10.1007/s11803-009-9058

Castagnetti, C., Bassoli, E., Vincenzi, L., Mancini, F.: Dynamic assessment of Masonry towers based on terrestrial radar interferometer and accelerometers. Sensors 19(6), (2019) https://doi.org/10.3390/s19061319

Das, S., Saha, P., Patro, S. K.: Vibration-based damage detection techniques used for health monitoring of structures: A review. Journal of Civil Structural Health Monitoring 6, 477–507 (2016)

Feng, D., Feng, Q. M.: Computer vision for SHM of civil infrastructure: From dynamic response measurement to damage detection—a review. Engineering Structures 156, 105–117 (2018) https://doi.org/10.1016/j.engstruct.2017.11.018

Guerzoni, G., Faghand, E., Vitetta, G., Vincenzi, L., Mehrshahi, E.: Novel movement-based methods for the calibration of colocated multiple-input multiple-output radars. IEEE Access 11, 116090–116108 (2023) https://doi.org/10.1109/ACCESS.2023.3325995

Guidorzi, R., Diversi, R., Vincenzi, L., Mazzotti, C., Simioli, V.: Structural monitoring of a tower by means of MEMS-based sensing and enhanced autoregressive models. European Journal of Control 20(1), 4–13 (2014)

Huang, Y., Shao, C., Wu, B., Beck, J., Li, H.: State-of-the-art review on Bayesian inference in structural system identification and damage assessment. Advances in Structural Engineering 22, 136943321881154 (2018) https://doi.org/10.1177/1369433218811540

Hwa Hong, Y., Gun Lee, S., Sung Lee, H.: Design of the FEM-FIR filter for displacement reconstruction using accelerations and displacements measured at different sampling rates. Mechanical Systems and Signal Processing 38(2), 460–481 (2013) https://doi.org/10.1016/j.ymssp.2013.02.007

Hwa Hong, Y., Kim, H., Sung Lee, H.: Reconstruction of dynamic displacement and velocity from measured accelerations using the variational statement of an inverse problem. Journal of Sound and Vibration 329(23), 4980–5003 (2010) https://doi.org/10.1016/j.jsv.2010.05.016

Joshi, S., Madhav Harle, S.: Linear Variable Differential Transducer (LVDT) & its applications in civil engineering. International Journal of Transportation Engineering and Technology 3(4), 62–66 (2017) https://doi.org/10.11648/j.ijtet.20170304.13

Kim, J., Kim, K., Sohn, H.: Autonomous dynamic displacement estimation from data fusion of acceleration and intermittent displacement measurements. Mechanical Systems and Signal Processing 42(1), 194–205 (2014) https://doi.org/10.1016/j.ymssp.2013.09.014

Koo, G., Kim, K., Chung, J. Y., Choi, J., Kwon, N., Kang, D., Sohn, H.: Development of a high precision displacement measurement system by fusing a low cost RTK-GPS sensor and a force feedback accelerometer for infrastructure monitoring. Sensors 17(12), (2017) https://doi.org/10.3390/s17122745

Lee, H. S., Hong, Y. H., Park, H. W.: Design of an FIR filter for the displacement reconstruction using measured acceleration in low-frequency dominant structures. International Journal for Numerical Methods in Engineering 82(4), 403–434 (2010) https://doi.org/10.1002/nme.2769

Ma, Z., Choi, J., Yang, L., Sohn, H.: Structural displacement estimation using accelerometer and FMCW millimeter wave radar. Mechanical Systems and Signal Processing 182, 109582 (2023) https://doi.org/10.1016/j.ymssp.2022.109582

Ma, Z., Chung, J., Liu, P., Sohn, H.: Bridge displacement estimation by fusing accelerometer and strain gauge measurements. Structural Control and Health Monitoring 28(6), e2733 (2021) https://doi.org/10.1002/stc.2733

Micozzi, F., Morici, M., Zona, A., Dall'Asta, A.: Vision-based structural monitoring: application to a medium-span post-tensioned concrete bridge under vehicular traffic. Infrastructures 8(10), art. no. 152 (2023)

Ozdagli, A., Gomez, J., Moreu, F.: Real-time reference-free displacement of railroad bridges during train-crossing events. Journal of Bridge Engineering 22, (2017) https://doi.org/10.1061/(ASCE)BE.1943-5592.0001113

Park, J., Moon, D., Yoon, H., Gomez, F., Spencer Jr., B. F., Kim, J. R.: Visual–inertial displacement sensing using data fusion of vision-based displacement with acceleration. Structural Control and Health Monitoring 25(3), e2122 (2018) https://doi.org/10.1002/stc.2122

Park, J., Sim, S., Jung, H.: Displacement estimation using multimetric data fusion. Mechatronics, IEEE/ASME Transactions 18, 1675–1682 (2013) https://doi.org/10.1109/TMECH.2013.2275187

Poluzzi, L., Barbarella, M., Tavasci, L., Gandolfi, S., Cenni, N.: Monitoring of the Garisenda tower through GNSS using advanced approaches toward the frame of reference stations. Journal of Cultural Heritage 38, 231–241 (2019) https://doi.org/10.1016/j.culher.2019.02.008

Ponsi, F., Bassoli, E., Vincenzi, L.: Mitigation of model error effects in neural network-based structural damage detection. Frontiers in Built Environment 8, (2023) https://doi.org/10.3389/fbuil.2022.1109995

Ponsi, F., Buoli, E., Varzaneh, G. E., Bassoli, E., Ghirelli, G., Briseghella, B., Vincenzi, L.: Dynamic monitoring of a steel footbridge based on computer vision techniques. Lecture Notes in Civil Engineering, 515, pp. 292–303 (2024)

Rainieri, C., Fabbrocino, G.: Automated output-only dynamic identification of civil engineering structures. Mechanical Systems and Signal Processing 24(3), 678–695 (2010)

Smyth, A., Wu, M.: Multi-rate Kalman filtering for the data fusion of displacement and acceleration response measurements in dynamic system monitoring. Mechanical Systems and Signal Processing 21(2), 706–723 (2007) https://doi.org/10.1016/j.ymssp.2006.03.005

Talledo, D. A., Miano, A., Bonano, M., Di Carlo, F., Lanari, R., Manunta, M., Meda, A., Mele, A., Prota, A., Saetta, A., Stella, A.: Satellite radar interferometry: Potential and limitations for structural assessment and monitoring. Journal of Building Engineering 46, 103756 (2022)

Ubertini, F., Gentile, C., Materazzi, A. L.: Automated modal identification in operational conditions and its application to bridges. Engineering Structures 46, 264–278 (2013)

4 Stochastic modal appropriation method
Time-domain extension for MDOF systems

Giuliano Coppotelli and Maher Abdelghani

4.1 INTRODUCTION

The estimate of the modal parameters of a dynamic system in operating conditions can provide engineers and researchers with additional insights into its dynamic properties, as the effects of artificial boundary conditions, typical of laboratory tests, no longer affect the experimental data. Moreover, the modal parameters estimated in operational conditions reflect the structure in the proper operating configuration, thus considering the possible effects of the loading conditions. Indeed, such loading conditions can change the system's equilibrium configuration—where small vibrations are evaluated—from the one where no loading is applied. This experimental characterization has been made possible by the development of operational modal analysis (OMA) methods [1, 2]. Generally speaking, OMA methods are distinguished into two main groups based on how response data are analyzed: time or frequency domain methods [3–5]. The stochastic modal appropriation (SMA) method assumes that random and uncorrelated inputs excite the structure at all available degrees of freedom. The corresponding auto- and cross-correlation functions evaluated among the responses are therefore seen as a representation of the impulse response functions that could be used to evaluate the modal appropriation condition when the system is excited by a dummy harmonic excitation [6, 7]. In Coppotelli et al [8], this approach has been enriched with the possibility of estimating the mode shapes characterizing a multi-degree-of-freedom (MDOF) system, demonstrating that mode shapes can be estimated from the real part of the cross-power spectral density (PSD) functions evaluated from the output responses under quadrature conditions. In this chapter, mode shapes are directly derived from the time-domain data without the formal need to perform a time-to-frequency domain signal transformation, thus completing the theoretical formulation of the SMA method in the time domain. Several numerical analyses and experimental investigations are presented to evaluate the accuracy of this new SMA formulation concerning the acquisition parameters. Among all the tested structures, results related to a cantilever beam and a cantilever plate are presented and discussed for brevity.

DOI: 10.1201/9781003516941-4

4.2 THEORETICAL BACKGROUND

The SMA method is an OMA method originally developed in the time domain for single degree of freedom (SDOF) systems, allowing for the identification of the natural frequencies and damping ratios only [7, 9]. In this chapter, the extension of the SMA method to deal with MDOF systems is analytically derived, thus allowing the estimation of the mode shapes of general MDOF systems. It is worth noting that both the original formulation and this extension of the SMA method are based on the main hypothesis that the system responses at frequencies close to one of its natural frequencies are primarily due to the contribution of only the excited mode. In other words, the SMA method is based on the SDOF dynamic behavior hypothesis in the modal domain. This hypothesis is generally satisfied when the system is lightly damped and/or characterized by well-separated modes.

4.2.1 SMA algorithm extension to MDOF system: Estimate of the natural frequencies and damping ratios

Consider a generic multiple input-multiple output (MIMO) system, characterized by M mode shapes in the frequency range of interest. The system is randomly excited at N_i input locations, whereas the output time responses are recorded at N_0 outputs. The $N_{0\times} N_0$ cross-correlation matrix evaluated among the output responses is:

$$R(t)=\begin{bmatrix} R_{y1y1}(t) & \cdots & R_{y1yN_0}(t) \\ \vdots & & \vdots \\ R_{yN_0y1}(t) & \cdots & R_{yN_0yN_0}(t) \end{bmatrix} \tag{4.1}$$

Under the hypothesis of an SDOF behavior in the modal domain, or equivalently assuming the dynamic system is lightly damped and characterized by well-separated modes, the SMA method in its original formulation is applied to the response time histories, which are, in turn, filtered with a band-pass filter centered at each expected r-th natural frequency. This information can be easily obtained from the analysis of the resonance peaks identified in the corresponding power spectral densities. Therefore, the band-pass filtered correlation function corresponding to the ij-th entry of Eq. (4.1), i.e., the correlation function corresponding to the i and j response signals, can be written as

$$R_{ij}(t)={}_r c_{ij} e^{-\varsigma_r \omega_r t}\left(\cos\left(\omega_{d_r} t\right)+\frac{\varsigma_r}{\sqrt{1-\varsigma_r^2}}\sin\left(\omega_{d_r} t\right)\right) \tag{4.2}$$

being ω_{dr}, ω_r, and ζ_r the r-th damped natural frequency, the natural frequency, and the damping ratio respectively, whereas ${}_r c_{ij}$ is the contribution of the r-th mode to the displacement at the i-th DOF when the structure is excited at the j-th DOF. The values of such natural frequency and damping ratio can be estimated by applying the SMA method in its original form. By introducing the stretching and rotating parameter α_r corresponding to each mode, the phase shift $\vartheta(\alpha_r, \omega_{hr})$ between the generic output response and the harmonic input excitation of frequency ω_{hr} is

$$\vartheta(\omega_r, \alpha) = \arctan\left[\frac{\omega_r^3 - \omega_n^2\left(1 - 4\zeta_n^2\right)\omega_r + 2\zeta_n \alpha \omega_n^2}{\left(\omega_n^2 - \omega_r^2\right)\left(2\zeta_n \omega_n - \alpha\omega_r\right) + 2\omega_r^2 \zeta_n \omega_n}\right] \tag{4.3}$$

The modal appropriation condition defined by the SMA method [7], that is, $\vartheta(\omega_r, \alpha) = 0$, allows the estimation of the natural frequency and damping ratio by finding the frequency ω_{hr} for which the modal appropriation condition is met.

Specifically:

$$\vartheta(\omega_r, \alpha) = 0 \quad \Leftrightarrow \quad \begin{cases} \omega_{h_r} = \omega_r \\ \alpha_r = 2\zeta_r \end{cases} \tag{4.4}$$

The full r-th unknown pair of natural frequencies and damping ratios are thus estimated by applying the SMA modal appropriation condition to each r-th set of the output response signals.

4.2.2 SMA algorithm extension to MDOF system: Estimate of the mode shapes

The previously identified modal appropriation condition can also be used to estimate the mode shapes from the same correlation function matrix of Eq. (4.1). Considering the response corresponding to the k-th degree of freedom as the reference signal, it is possible to evaluate the response of the system when excited by a harmonic excitation with a frequency equal to the r-th natural frequency. By definition, the response of the system to this kind of excitation is the mode shape. Because the r-th natural frequency, the damping ratio, and the parameter α_r have already been estimated from the previous step, the N_o output responses $\bar{y}_k(\omega_r, 2\zeta_r, t)$, $k = 1, \ldots, N_o$, collected in the vector $\bar{y}(\omega_r, 2\zeta_r, t)$, evaluated from the convolution between the stretched and rotated (by the factor α_r) correlation function $R_{yiyk}(2\zeta_r; t)$ and the unitary harmonic excitation $\sin(\omega_r t)$, are equivalent to the motion of the i-th component following the r-th mode shape $\phi^{(r)}$. Therefore, the mode shape can be estimated from the amplitudes with their corresponding signs of the harmonic responses evaluated at each N_o output DOF at a prescribed time $\bar{t}$ (usually set to zero), that is,

$$\phi^{(r)} = \begin{pmatrix} \phi_1^{(r)} \\ \vdots \\ \phi_{N_0}^{(r)} \end{pmatrix} = \begin{pmatrix} sgn\left[\angle \bar{y}_1\left(\omega_r, 2\zeta_r, \bar{t}\right)\right]\left\|\bar{y}_1\left(\omega_r, 2\zeta_r, \bar{t}\right)\right\| \\ \vdots \\ sgn\left[\angle \bar{y}_{N_0}\left(\omega_r, 2\zeta_r, \bar{t}\right)\right]\left\|\bar{y}_{N_0}\left(\omega_r, 2\zeta_r, \bar{t}\right)\right\| \end{pmatrix} \tag{4.5}$$

in which $sgn(\cdot)$, $\angle(\cdot)$, and $\|(\cdot)\|$ are the sign, phase, and amplitude operators, respectively.

4.3 EXPERIMENTAL INVESTIGATION

4.3.1 Cantilever Beam

The considered cantilever beam has dimensions 50 cm × 2.5 cm × 1.3 cm and is made of a steel alloy with E = 210 GPa, $\rho = 7500$ Kg/m^3, and $\mu = 0.3$. Five accelerometers have been distributed along the shear center of the beam to measure the output responses. The analysis frequency range was [0 – 1024] Hz, whereas the sampling frequency was 0.5 Hz, corresponding to an observation time of 2 s. Tapping fingers provided the random dynamic excitation to the beam. The corresponding average PSD and autocorrelation functions, evaluated among the five output responses, are reported in Figure 4.1.

Following the SMA algorithm, the modal characteristics can be easily estimated. Specifically, from the analysis of the previous PSD shown in Figure 4.1a, three peaks of resonance are identified. As a consequence, the frequencies of the dummy excitation will be in the ranges [40–50] Hz, [270–290] Hz, and [775–795] Hz for the three (expected) modes, respectively. For the first mode, the stretching parameter α has been varied in the range [0.001–0.06] with a resolution of 5.9×10^{-4}. From Figure 4.2a, the quadrature condition for the first mode shape is identified

(a) Average PSD (b) Average Auto-correlation Function

FIGURE 4.1 Experimental cantilever beam—response properties.

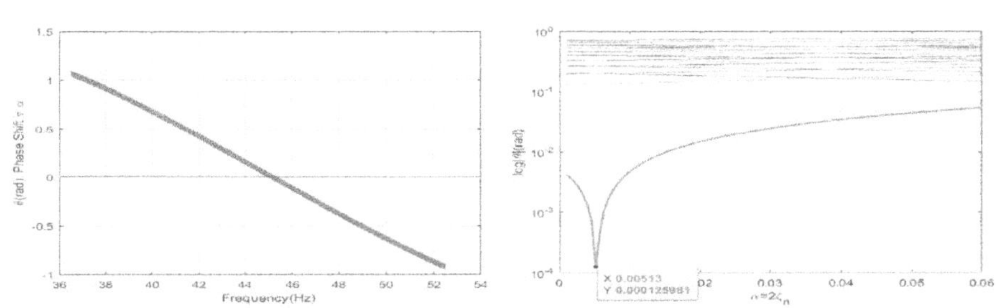

(a) Phase shift ϑ (rad) vs Frequency (Hz) (b) Natural logarithm of phase shift $\log|\vartheta|$ vs design parameter α

FIGURE 4.2 Experimental cantilever beam—natural frequency and damping ratio estimates for the first mode.

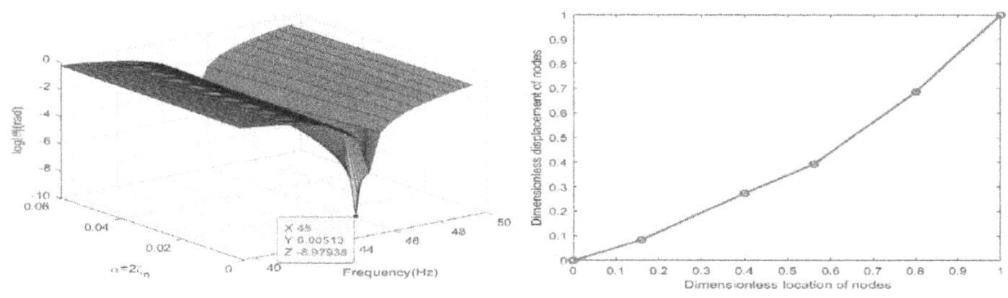

(a) Phase shift ϑ (rad) vs Frequency (b) Mode shape (Hz) and stretching parameter α

FIGURE 4.3 Experimental cantilever beam—estimation of the first mode characteristics.

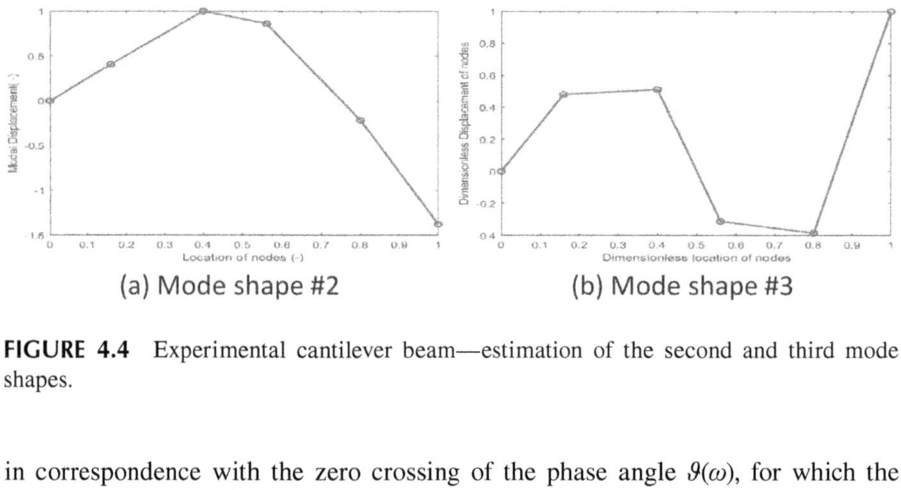

(a) Mode shape #2 (b) Mode shape #3

FIGURE 4.4 Experimental cantilever beam—estimation of the second and third mode shapes.

in correspondence with the zero crossing of the phase angle $\vartheta(\omega)$, for which the parameter α is identified at the practically minimum value in ϑ of Figure 4.2b. The full representation of the dependency of the phase shift ϑ on both the frequency, ω, and the stretching parameter, α, is shown in Figure 4.3a. The quadrature condition is achieved in correspondence with the minimum value of the phase shift ϑ, which gives the value of the natural frequency and the damping ratio. The SMA time-domain extension allows the estimation of the mode shape depicted in Figure 4.3b, in which the dimensionless modal displacement at each node is plotted at its corresponding dimensionless location along the beam.

The estimated second and third modes are reported in Figure 4.4. It is worth recalling that these modal components have been estimated by applying the MDOF time-domain extension of the SMA method. For the first column of the correlation matrix has been chosen as the reference input DOF, and the convolution between each component of the modified correlation vector and the unitary harmonic excitation, both evaluated at the appropriation condition, has been calculated. As discussed in Section 4.2, the mode shapes are easily estimated from the initial amplitudes and the relative phases at each degree of freedom of the modal response. The estimated

TABLE 4.1
Experimental cantilever beam—comparison between the natural frequency estimates.

Mode #	$f_{nexpI/O}$ (Hz)	$f_{nexp\,SMA}$ (Hz)	Δf_n (%)	$\zeta_{nexpI/O}$ (%)	$\zeta_{nexpSMA}$ (%)	$\Delta\zeta_n$ (%)
1	45.0	45.0	0.00	0.16	0.25	36.00
2	278.6	279.0	0.14	1.01	1.23	17.89
3	785.6	787.0	0.18	1.82	1.94	6.19

modal parameters have been finally compared with those from the traditional input/output experimental modal analysis. A modal survey of the same structure was carried out using an electrodynamic shaker placed at the beam tip, fed with a white noise excitation signal and by keeping the same acquisition parameters as the previous OMA analysis.

For this input/output analysis, the modal parameters were estimated using Siemens Test.LAB's PolyMAX approach. As one can see in Table 4.1, where the superscripts *I/O* and *SMA* refer to the input/output and OMA estimates, respectively, an excellent agreement has been found between the considered approaches due to the practically negligible differences in the estimates of the natural frequencies and acceptable differences in the damping ratios. Also, the mode shapes are in excellent correlation, resulting in a practically unitary modal assurance criterion (MAC) value between corresponding modes.

4.3.2 Cantilever plate

The second experimental investigation has been carried out on a cantilever plate made of aluminum alloy with a Young's modulus E = 70 GPa, mass density ρ = 2700 Kg/m^3, Poisson's ratio μ = 0.3, and dimensions 0.263 m × 0.2 m × 0.0028 m. Twelve accelerometers, uniformly distributed on the plate surface, were used to measure the transverse acceleration responses, whereas tapping fingers provided the random dynamic excitation. The considered analysis frequency band was [0–512] Hz, with an observation time of T = 2 s and a frequency resolution of 0.5 Hz. The corresponding average PSD and autocorrelation functions, evaluated among the 12 output responses, are reported in Figure 4.5, from which the presence of six possible modes can be foreseen. Only the estimates of the first four are reported for brevity.

Following the same approach as in the previous subsection, the natural frequencies and the damping ratios are obtained from the quadrature conditions around the four natural frequencies at which the resonance peaks have been identified. Different frequency ranges have been taken into account to define the dummy harmonic force of excitation, namely [27–32] Hz for the first natural frequency, [94–98] Hz for the second, [181–184] Hz for the third, and [326–330] Hz for the fourth natural frequency. The average auto-correlation function has been complexified by introducing the stretching parameter α in the range [0.002–0.06], with a resolution of 5.8×10^{-4} Hz

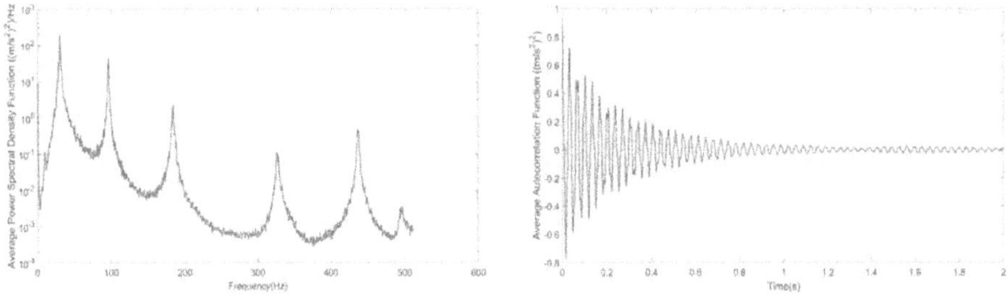

(a) Average PSD (b) Average Auto-correlation Function

FIGURE 4.5 Experimental cantilever plate—response properties.

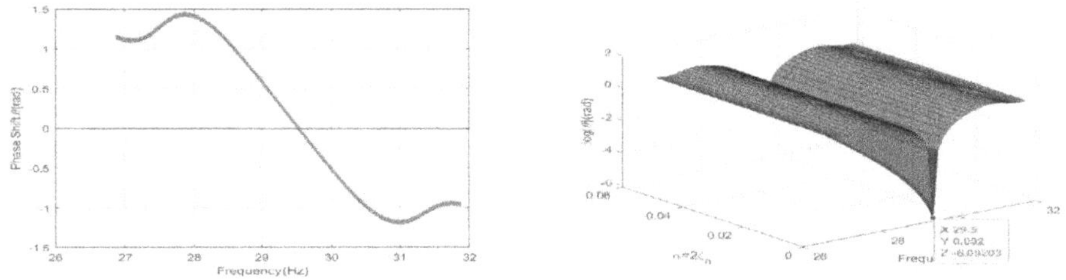

(a) Phase shift ϑ (rad) vs Frequency (Hz) (b) Phase shift ϑ (rad) vs Frequency (Hz) and stretching parameter α

FIGURE 4.6 Experimental cantilever plate—natural frequency and damping ratio estimates for the first mode.

for all the considered natural frequencies. Regarding the first mode, the quadrature condition is depicted in Figure 4.6, in which the presence of a local minimum of the phase shift function, $\vartheta(\omega,\alpha)$, is highlighted, making possible an efficient estimate of the natural frequency and the damping ratio. The same behavior is also observed when estimating the other modal parameters.

The SMA method MDOF extension in the time domain allowed the estimation of two bending modes and two torsional modes in the considered band [0–512] Hz, as shown in Figures 4.7 and 4.8. For this set of estimates, the response of the left tip corner of the plate has been considered as the reference channel, and the mode components are given by the amplitudes, with their relative signs, of the 12 initial time responses of the system under the appropriation conditions corresponding to each of the identified natural frequencies.

Also, for this experimental investigation, the modal parameter estimates have been compared with those from the input/output modal survey. For this investigation, an impact excitation in correspondence with the left tip corner of the plate was provided by a modal hammer. The acquisition parameters were the same as those used in the OMA analysis, whereas the PolyMax approach was used to estimate the modal parameters. The comparison of the natural frequencies is reported

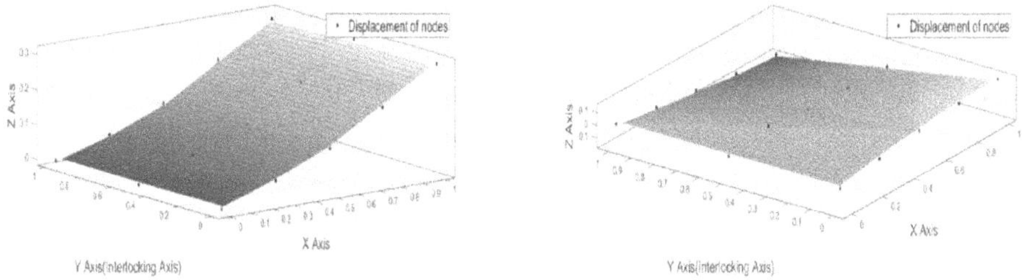

(a) Mode shape #1: First bending (b) Mode shape #2: First torsion

FIGURE 4.7 Experimental cantilever plate—estimations of the first and second mode shapes.

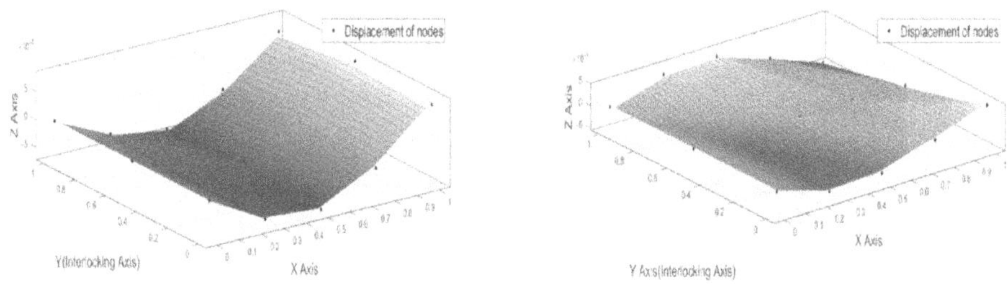

(a) Mode shape #3: Second bending (b) Mode shape #4: Second torsion

FIGURE 4.8 Experimental cantilever plate—estimation of the second and third mode shapes.

TABLE 4.2
Experimental cantilever plate—comparison between the natural frequency estimates.

Mode # & Type	$f_{nexpI/O}$ Hz	$f_{nexpSMA}$ Hz	Δf_n%
1st Bending	29.3	29.5	0.68
1st Torsional	95.5	96.3	0.83
2st Bending	182.7	182.5	0.11
2st Torsional	326.6	328.0	0.42

in Table 4.2, whereas the damping ratios are compared in Table 4.3. Once again, the results demonstrate the excellent capability of the time-domain MDOF extension of the SMA method to isolate the mode exactly at its natural frequency, since there is no practical difference from those estimated by the PolyMax method. The difference

Table 4.3
Experimental cantilever plate—comparison between the damping ratio estimates.

Mode #	$\zeta_{nexpI/O}$ %	$\zeta_{nexpSMA}$ %	$\Delta\zeta_n$%
1st Bending	0.22	0.10	54.54
1st Torsional	0.27	1.49	81.87
2st Bending	0.46	0.56	21.74
2st Torsional	0.59	0.39	33.89

between the estimates ranges from 0.83% for the first torsional mode down to the lowest value of 0.11% for the second bending mode. Also, the correlation between the damping ratios is very good. For this modal property, the difference reaches its maximum of 81.87% for the first torsional mode. The degree of consistency between the mode shapes is very high. The MAC between corresponding modes is 99% for all the modes except the first torsional mode, which has a MAC value as low as 96%.

4.4 CONCLUDING REMARKS

4.5 ACKNOWLEDGMENTS

This chapter is part of the activity carried out, thanks to the University of Rome "La Sapienza" grant "Progetto di Cooperazione Internazionale con Istituzioni Universitarie Partner in Paesi in via di Sviluppo", year 2019, and the grant "Identification of damping mechanism in a sloshing wing-like system", year 2020.

REFERENCES

1. W. Heylen, S. Lammens, P. Sas, Modal Analysis Theory and Testing. Leuven (1994).
2. L. Hermans, H. Van der Auweraer, *Modal testing and analysis of structures under operational conditions: Industrial applications*, Mechanical Systems and Signal Processing, 13(2) (1999) 193–216.
3. R. Brincker, L. Zhang, P. Andersen, *Modal identification from ambient responses using frequency domain decomposition*, XVIII IMAC, S. Antonio, TX, USA, Society for Experimental Mechanics, Inc., Bethel, CT, USA, pp. 625–630 (2000).
4. P. Van Overschee, B. De Moor, Subspace Identification for Linear Systems. Kluver Academic Publisher (1996).
5. A. Agneni, L. Balis Crema, G. Coppotelli, *Output-only analysis of structures with closely spaced modes*, Mechanical System and Signal Processing: Special Issue: Operational Modal Analysis, 24(5), (July 2010), 1240–1249.
6. M. Abdelghani, S. Chauhan, D. J. Inman, *A critical evaluation of the INOPMA algorithm for in-operation modal analysis*, In: Chouchane, M., Fakhfakh, T., Daly, H., Aifaoui, N., Chaari, F. (eds.), Design and Modeling of Mechanical Systems—II. Lecture Notes in Mechanical Engineering. Springer, Cham (2015). https://doi.org/10.1007/978-3-319-17527-0 56.

7. M. Abdelghani, M. I. Friswell, *Stochastic Modal Appropriation (SMA)*, In: Mains, M., Dilworth, B. (eds.), Topics in Modal Analysis & Testing, Volume 9. Conference Proceedings of the Society for Experimental Mechanics Series. Springer, Cham (2019). https://doi.org/10.1007/978-3-319-74700-21.
8. G. Coppotelli, G. Muti, M. Abdeghani, *Identification of modal parameters in rotating structures using Stochastic Modal Appropriation*, AIAA SciTech Forum 23–27 January 2023, National Harbor, MD & Online, AIAA SCITECH 2023 Forum, Paper No. 10.2514/6.2023-2070.
9. M. Abdelghani, *Physical only modes identification using the stochastic modal appropriation algorithm*, In: Jaoudi, A. A. (ed.), The Monte Carlo Methods—Recent Advances, New Perspectives, and Applications. Intechopen, London (2022).

5 Automated DBSCAN-based approach for the estimate of the modal parameters

R. G. Sbarra and Giuliano Coppotelli

5.1 INTRODUCTION

Investigating the dynamic behavior of aerospace systems under operational conditions is essential for advancing environmentally sustainable solutions in the aerospace industry. A critical part of this process is the successful execution of flight vibration tests, which assess system performance under operational stress and are essential for certifying new or modified systems [1, 2]. Operational modal analysis (OMA) methods are pivotal in this context, as they allow modal analysis solely based on the recording of the system responses. These methods operate under the assumption that the excitation force is a distributed stochastic load, constant over a broad frequency band—a condition inherently satisfied in aerospace applications due to the prevalence of white noise excitation. OMA methods typically identify the dynamic properties of a system by estimating poles at various values of an unknown model order, up to a predetermined maximum. These poles are estimated using the so-called stabilization diagram [3], which requires a user-dependent interpretation to identify consistent poles, thus introducing a degree of uncertainty. Automated approaches using unsupervised machine learning, such as clustering, have been proposed to reduce this subjectivity [4]. Various automated operational modal analysis (AOMA) methods have emerged, leveraging clustering algorithms such as K-means [5], K-means++ [6], and Gaussian mixture models (GMM) [7]. Despite their utility, these techniques face challenges such as, but not limited to, sensitivity to noisy poles in stabilization diagrams and the need to predefine the number of clusters. Addressing these limitations, recent AOMA methods employed the density-based spatial clustering of applications with noise (DBSCAN) [8, 9], which can automatically determine the number of clusters while effectively handling noise. However, even DBSCAN is not without issues, as the choice of its two input parameters—*eps* (maximum distance between points within the same cluster) and *MinPts* (minimum points in a cluster)—is crucial and often depends on user expertise. The sensitivity of DBSCAN to its parameters has spurred various studies proposing strategies to address this challenge. In this context, Zhang et al. [10] introduced a deep reinforcement learning–guided framework

 DOI: 10.1201/9781003516941-5

to automatically search for optimal DBSCAN parameters. Similarly, Star et al. [11] proposed a method to determine parameter values using sharp distance increases in the k-nearest neighbor graph. Bush et al. [12] leveraged the k-nearest neighbor density distributions to enhance parameter selection and clustering accuracy in overlapping density regions. Building on these advancements, this chapter focuses on refining the use of DBSCAN for OMA through a novel approach for defining the initialization parameters *eps* and *MinPts*. Specifically, the automated minimum order poles selection (MOPS) methodology is proposed to enhance the identification of physical poles. As previously demonstrated by the authors [13], prioritizing poles from lower model orders allows a significant reduction in the likelihood of selecting noisy poles. The effectiveness of the proposed method is demonstrated through numerical simulations and experimental validation, including the dynamic characterization of the Pazy wing, a benchmark relevant to modern flight configurations aimed at reducing fuel consumption and characterized by operational large displacements [14].

5.2 THEORETICAL BACKGROUND

5.2.1 STOCHASTIC SUBSPACE IDENTIFICATION

The stochastic subspace identification (SSI) method is a widely used approach for evaluating the dynamic behavior of systems. This technique employs the state-space formulation, which provides a robust framework for modeling and analyzing dynamic systems based on input-output data [15]. The core of the SSI method lies in estimating the state matrix A through the observability matrix, which can, in turn, be derived from either the time series or the spectral densities of the output recordings [16]. By computing the eigenvalues and eigenvectors of the state matrix A, it is possible to derive the modal parameters of the system, namely the natural frequencies, damping ratios, and mode shapes.

5.2.2 STABLE POLE DETERMINATION

Both time-domain and frequency-domain methods rely on an iterative procedure to estimate the unknown model order *N*, which corresponds to the number of modes needed for an accurate representation of the system's dynamic response. This procedure aims to identify the poles and determine the corresponding model order when the poles obtained from two consecutive iterations are found to be practically the same. The underlying hypothesis is that physical poles are independent of the model order, meaning that stable poles do not change with increasing model complexity.

A commonly used criterion for distinguishing stable poles from noisy poles is based on predefined thresholds for the relative differences in natural frequencies ε_{fn}, damping ratios $\varepsilon\zeta_n$, and mode shapes MAC_{thr} between two consecutive model orders. Specifically, a pole is considered stable if these differences remain below their respective thresholds.

The results of this iterative procedure are typically visualized in a stabilization diagram, where the identified poles are plotted for each model order up to a predefined maximum order, *MaxOrd*.

5.2.3 Density-based spatial clustering of applications with noise

The DBSCAN method is a density-based clustering algorithm that is particularly effective in identifying clusters with varying shapes and sizes while mitigating the impact of noise in data. Unlike other clustering algorithms, DBSCAN does not require specifying the target number of clusters. Instead, clusters are identified within dense, connected regions of data, defined by two parameters: the maximum distance between points within the same cluster (*eps*) and the minimum number of data points required to form a cluster (*MinPts*).

This algorithm categorizes data points into three types [17]: core points that have at least *MinPts* data points (including itself) within their *eps*-neighborhood, border points that are reachable from a core point but do not satisfy the *MinPts* condition within their *eps*-neighborhood, and noise points (outliers) that are neither core points nor border points.

Clusters are formed by grouping data points within the *eps*-neighborhood of core points, extending outward to include reachable border points while excluding noise.

DBSCAN is well suited for automating OMA methods due to its ability to operate without prior knowledge of the number of modes and its effectiveness in filtering noisy poles. These features make DBSCAN a powerful tool for identifying and clustering modal poles in OMA applications.

5.2.4 Convolutional neural network

Convolutional neural networks (CNNs) are a class of deep learning models widely used for tasks involving structured input data, such as images or spatially organized information, due to their ability to learn hierarchical feature representations. At their core, CNNs consist of convolutional layers that apply learnable filters to the input data, capturing local patterns and producing feature maps that preserve spatial relationships. These layers are typically followed by activation functions such as the rectified linear unit (ReLU), which adds nonlinearity to the network, enabling it to capture and represent complex patterns [18]. To further enhance performance, batch normalization is often employed to stabilize the learning process, thereby improving convergence and acting as a regularization technique [19]. Typically, pooling layers are employed to provide a compact representation of feature maps, minimizing the number of parameters and reducing the risk of overfitting [20]. Fully connected layers are used in the final stages to aggregate the learned features into a high-level representation suitable for regression or classification tasks. Training a CNN involves minimizing a loss function, such as mean squared error in regression tasks, using optimization algorithms like Adam [21], which combine the advantages of momentum-based and adaptive learning rate methods, enabling efficient convergence. Overall, CNNs are powerful tools for capturing spatial hierarchies and extracting meaningful features from data, making them versatile for various machine learning applications.

5.3 DEVELOPED ALGORITHM: AUTOMATED MINIMUM ORDER POLE SELECTION

Building on prior research utilizing DBSCAN for automating modal parameter estimation, the proposed automated minimum order pole selection (AMOPS) algorithm can be outlined in key stages.

- **OMA** involves analyzing the system's response, resulting in a set of poles derived from increasing model orders.
- **Modal damping filter** is applied to remove poles associated with non-physical modal damping.
- ***MinPts* initialization**: To effectively initialize DBSCAN for modal parameter estimation, the *MinPts* parameter was determined using a CNN. The proposed CNN architecture consists of layers starting with an input layer for image interpretation, followed by convolutional layers with increasing filter sizes to progressively extract hierarchical features. Each convolutional layer is followed by a batch normalization and a ReLU activation to stabilize training and introduce nonlinearity because of the complex pattern to be identified. Pooling layers are used to reduce spatial dimensions, preserving essential features while minimizing computational load. The global average pooling layer aggregates spatial features, feeding them into a fully connected layer and a ReLU activation for robust feature representation. This architecture was trained using the Adam optimizer with a learning rate schedule to ensure convergence. A synthetic dataset was generated to train the CNN. This dataset consists of 50,000 distributions of points simulating clusters with varying densities and noise levels. Each dataset was constructed by randomly choosing the number of clusters, with their centers spaced to avoid overlaps. The *MinPts* parameter was randomly assigned. Additionally, noise was added to mimic real-world disturbances. The resulting data points were normalized, and the distribution was transformed into a 2D image representation to capture local densities. Figure 5.1 presents a sample dataset demonstrating the spatial distribution of points, including added noise. The transformed image shown in Figure 5.2 illustrates the spatial distribution of points by highlighting local point densities, which helps the CNN identify cluster structures.

The dataset was split into training (70%), validation (15%), and testing (15%) subsets. To assess the influence of image resolution on model performance, a sensitivity analysis was conducted. The resolution of input images directly affects the level of detail captured by the convolutional layers, impacting the network's ability to discern fine-grained density patterns. Higher resolutions provide more detail but increase the computational cost and the risk of overfitting. Lower resolutions simplify the input, potentially reducing the network's capacity to learn meaningful features. The analysis explored four resolutions: 16×16, 32×32, 64×64, and 128×128. These values were chosen to balance computational efficiency and the

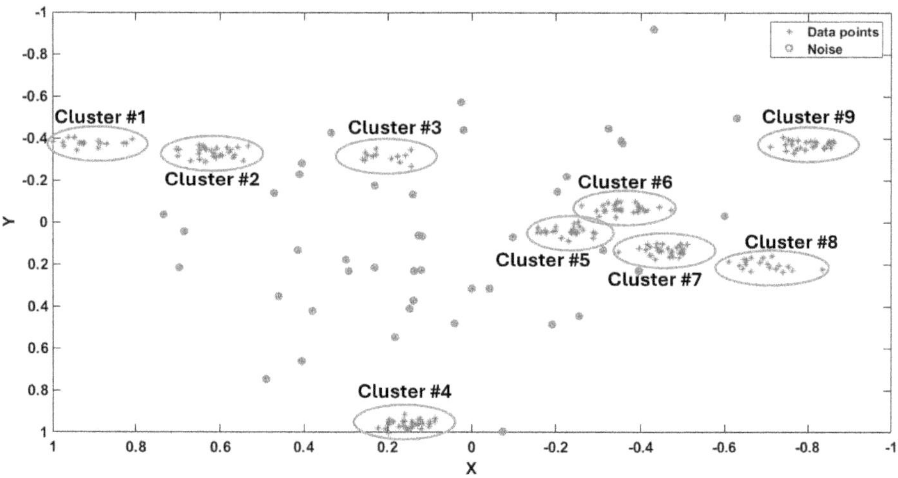

FIGURE 5.1 Example of a dataset with added noise.

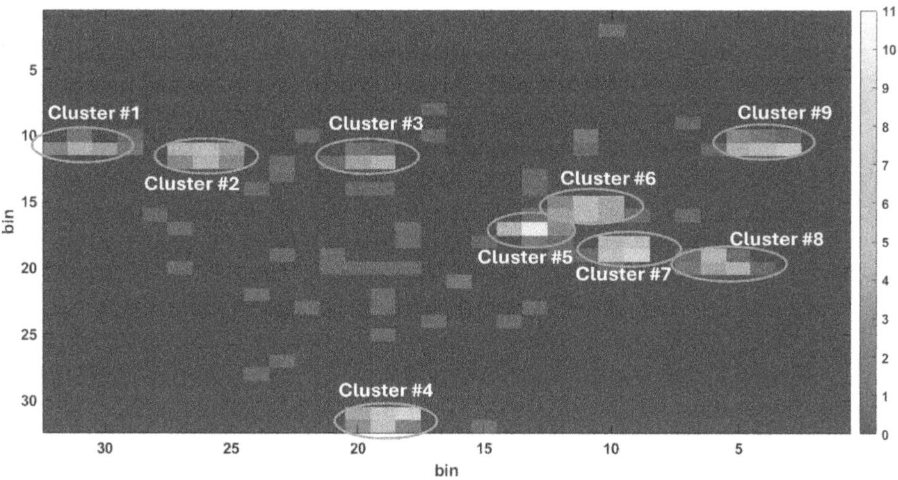

FIGURE 5.2 Image format of a dataset.

granularity of density representation. The performance was evaluated using the mean squared error (MSE) and R^2 metrics [22] for predicting *MinPts*, offering a comprehensive view of prediction accuracy and model fit. The results, reported in Table 5.1, demonstrate that the 32 × 32 resolution achieves the best balance between detail and computational efficiency, with the lowest MSE and highest R^2. The analysis confirms that increasing the resolution beyond 32 × 32 introduces diminishing returns, as higher resolutions result in a slight decrease in R^2 and an

TABLE 5.1
Evaluation of performance across different image resolutions for *MinPts* prediction.

Resolution	MSE	R^2
16 × 16	0.718	0.863
32 × 32	0.663	0.872
64 × 64	0.795	0.848
128 × 128	0.931	0.820

increase in MSE. Conversely, lower resolutions fail to capture sufficient details for optimal performance.

- ***eps* Initialization:** Regarding the *eps* parameter, no neural networks were used for its definition. This decision was made to keep the method simple and interpretable and because of the difficulty in establishing a direct and generalizable relationship between eps and the characteristics of the dataset. Instead, a criterion based on a modal similarity measure was adopted, which considers the relative differences in frequencies and modal shapes while excluding damping due to its high estimation uncertainty. The value of *eps* was set in relation to the stabilization diagram criteria, which state that the relative difference between frequencies, ε_{fn}, should be less than or equal to 1%, and the MAC should be greater than or equal to 0.95, ensuring high similarity between modal shapes. Consequently, the distance between two poles was defined as a combination of ε_{fn} and 1−MAC, balancing the two contributions with a weight to equalize their importance. Specifically, the distance was defined as $D = (1-MAC)/5 + \varepsilon_{fn}$, where the term (1−MAC)/5 and the relative frequency difference ε_{fn} are combined to reflect both modal aspects. This approach allowed *eps* to be set at 0.02, a value that complies with the stabilization criteria, ensuring that only poles with consistent dynamic and geometric characteristics are grouped together.
- **DBSCAN Clustering:** Recognizing that the stabilization criteria alone may not sufficiently eliminate all numerical poles, the DBSCAN clustering algorithm is employed to refine the process.
- **Pole Selection:** Once the clustering phase is completed, selecting a representative pole for each cluster becomes crucial for accurately determining the system's modal parameters. This selection process is guided by the observation that increasing the order of the system tends to generate more numerical poles. To minimize the risk of selecting a numerical pole, the method prioritizes poles with the lowest order within each cluster.

The structure of the proposed AMOPS algorithm, which includes all the previously described phases, is reported in Figure 5.3.

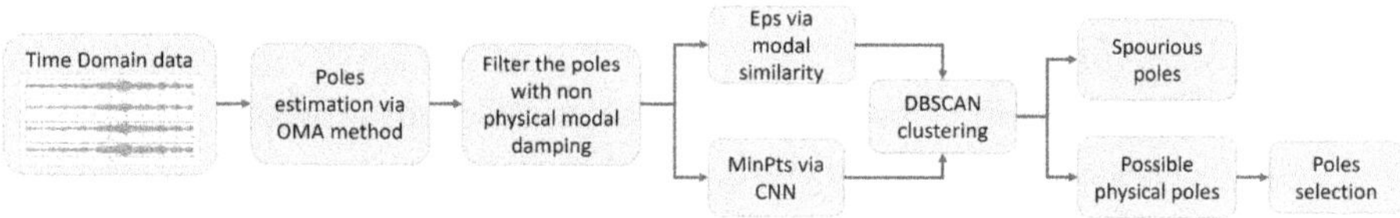

FIGURE 5.3 Flowchart of the proposed AMOPS algorithm.

5.4 NUMERICAL ASSESSMENT

To assess the effectiveness of the proposed methodology, a numerical simulation was conducted using a four degrees of freedom (DOF) mass-spring-damper system. The system's characteristics—mass, stiffness, and damping—were initially known, enabling the calculation of analytical modal parameters via the modal model. The considered 4-DOF numerical system was designed with two groups of closely spaced poles. This scenario presents a more challenging task for cluster analysis, as distinguishing between closely spaced groups of poles becomes increasingly difficult. The purpose of this setup is to rigorously test the robustness of the *eps* selection criterion, as the poles are intentionally located close to one another to evaluate whether the chosen distance metric effectively distinguishes between them. Furthermore, noise is introduced into the system to assess the reliability of the *MinPts* parameter estimation. Specifically, three distinct levels of additive white noise are introduced, corresponding to noise levels of 5%, 10%, and 20% of the root mean square response of the system.

For each case, the system responses to white noise excitation were generated, and the AMOPS algorithm was applied to the simulated responses to estimate the system's modal parameters. These estimates were subsequently compared to the analytical values.

The stabilization diagram was built considering a sufficiently high value for *MaxOrder* and $\varepsilon_{fn} \leq 1\%$, $\varepsilon\zeta_n \leq 2\%$, and $MAC_{thr} \geq 0.95$ for the stabilization criteria.

The assessment of the AMOPS algorithm was conducted within a probabilistic framework, where the total probability, *Ptot*, of obtaining imprecise estimates for the modal parameters was calculated. An estimate was considered imprecise if it met specific criteria related to damping, mode shapes, or cluster identification. These included cases where the relative error between analytically estimated modal damping and the damping estimated by AMOPS exceeded 30%, the MAC value between analytically estimated mode shapes and those obtained through AMOPS was less than 0.9, or the estimated number of clusters, *Nclust*, deviated from the known number of modes in the simulated system. The *Ptot* metric quantified the likelihood of at least one of these events occurring. Notably, errors in modal frequency estimates were not considered in the calculations due to their typically high precision. To implement this probabilistic approach, the simulated 4-DOF system generated five datasets, each differing only in the mass values while keeping stiffness and damping constants unchanged. Within each dataset, five distinct response signals were generated, yielding 25 subsets. These variations stemmed from the random nature

TABLE 5.2
Four-DOF close poles with noise analysis results.

Noise level %	Mean err %			Probability %			
	f_n	ζ_n	MAC	$err_{\zeta n}$>30%	MAC<0.9	err Nclust	Total
0	0.13	17.31	0.91	0	0	0	0
5	0.05	6.2	1.46	0	0	0	0
10	0.05	6.5	0.90	1	0	4	4.96
20	0.8	8.6	1.13	2	0	8	9.84

of the excitation. Each subset allowed for four modal estimates, resulting in a total of 100 estimates for the probability analysis. The response signals were generated with a sampling frequency f_s set at twice the maximum system frequency to prevent aliasing and considering $N_s = 2^{15}$ samples. The results reported in Table 5.2 underscore the robustness and accuracy of the modal parameter estimation process across various noise levels. The method consistently achieves low mean errors for modal frequencies and mode shapes, as indicated by the high MAC values and the absence of cases with MAC < 0.9. While the damping ratio remains a challenging parameter to estimate, the method performs well overall, maintaining errors within acceptable bounds. Notably, the *Ptot* probability of significant errors remains low, reflecting the robustness of the estimation process even in the presence of noise. However, higher noise levels introduce a risk of misidentifying modes as noise, leading to occasional inaccuracies in the cluster count and slightly elevated error probabilities. These findings emphasize the efficacy of the chosen initialization parameters and the overall reliability of the estimation methodology, with strong performance across natural frequencies and modal shapes, and acceptable accuracy for the damping ratio.

5.5 APPLICATION TO PAZY WING WIND TUNNEL TESTING

The AMOPS technique was applied to a real test case involving a self-assembled model of the Pazy Wing benchmark. The Pazy Wing was specifically designed to provide experimental data for a highly flexible aeroelastic test case, which can be used to validate numerical methods in the nonlinear regime with large deflections [23]. A replica of the wing was fabricated at the structural dynamics laboratory of the Department of Mechanical and Aerospace Engineering (DIMA) at the University of Rome La Sapienza. The wing's structure was made using an aluminum spar, while the ribs were 3D printed in ABS (acrylonitrile butadiene styrene). To prevent undesirable aerodynamic effects, eight accelerometers were placed inside the wing to measure accelerations in a direction orthogonal to the spar. The entire wing was then wrapped with heat-shrink film, commonly used in model aircraft construction. Several wind tunnel tests were conducted at DIMA to characterize the wing's dynamic behavior under various airflow velocities, angles of attack, and boundary conditions. Figure 5.4 shows the Pazy Wing during wind tunnel testing with rigid support constraints.

FIGURE 5.4 Wind tunnel test: configuration with rigid support.

The test measurements included nine different velocities ranging from 5 m/s to 34 m/s and five angles of attack ranging from 5° to 15°, resulting in a total of 45 datasets. The sampling frequency used during the tests was 1024 Hz, with an observation time of 64 seconds, resulting in 2^16 sampling points.

The recorded signals were analyzed using the natural input modal analysis (NIMA) routine, developed by DIMA's structural dynamics lab. NIMA is a Matlab-based tool for analyzing operational data using various OMA methods [24]. The SSI-fr OMA approach in NIMA was used to provide a comparison tool for the

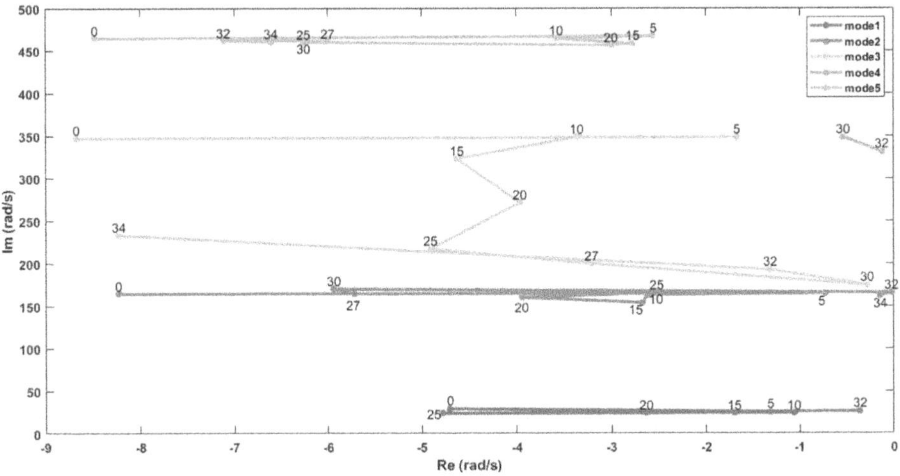

FIGURE 5.5 Root locus obtained with NIMA estimations.

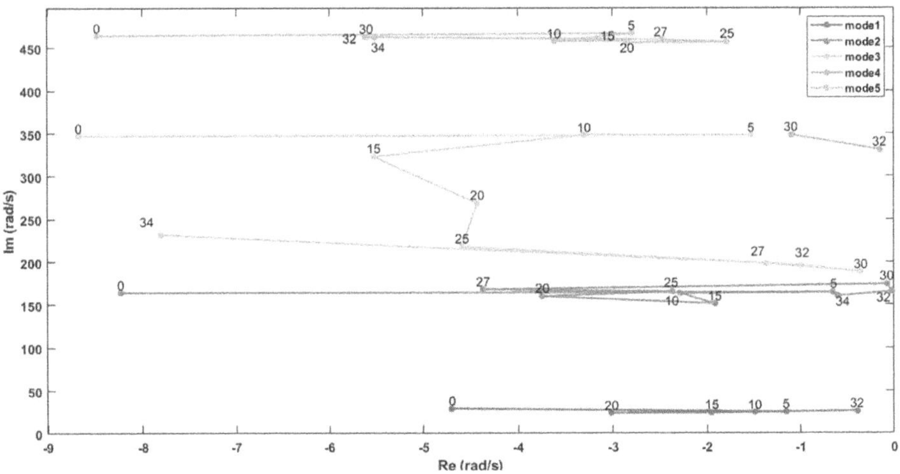

FIGURE 5.6 Root locus obtained with AMOPS estimations.

proposed AMOPS method. The root locus representation summarizes the poles (in terms of imaginary and real parts) estimated under different operational conditions, i.e., dynamic pressures. This method offers insights into the dynamic behavior of the Pazy Wing as the dynamic pressure varies while maintaining a fixed angle of attack. For brevity, only the case with an angle of attack of 15° is considered in this section. The root loci obtained using NIMA and AMOPS are shown in Figures 5.5 and 5.6, respectively. From these figures, it is evident that the root locus obtained from the AMOPS estimates aligns closely with that obtained through manual estimation.

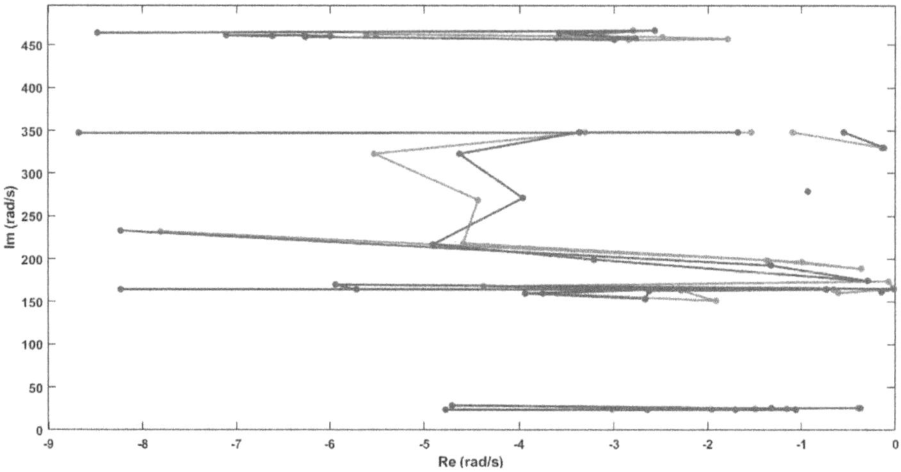

FIGURE 5.7 Overlapping root locus.

A clear similarity is observed in the imaginary part, corresponding to the natural frequency, while the real part, mainly controlled by the damping ratio, shows some discrepancy but remains comparable. Figure 5.7 overlays the root locus manually obtained using NIMA and the one from the AMOPS approach to highlight this comparison. Additionally, a comparison of the mode shapes has been performed, revealing strong consistency between those estimated using NIMA and those derived from the AMOPS approach. The automated method provides results comparable to the traditional method, with the key advantage being the significantly reduced analysis time while keeping the same accuracy of the estimates.

5.6 CONCLUSIONS

In this chapter, a novel methodology combining DBSCAN-based clustering with a machine learning framework has been proposed and validated for the automated interpretation of stabilization diagrams. The method leverages a density representation of stabilization data to predict clustering parameters using a CNN model. By addressing the challenge of parameter initialization, the approach enhances the reliability and robustness of pole selection while minimizing dependency on user-defined inputs. Extensive numerical analyses demonstrated that the method achieves high accuracy in estimating modal parameters. The resilience of the methodology to noise was further highlighted by reporting the ability of the developed approach to maintain accurate cluster identification, although higher noise levels occasionally led to the misclassification of modes as noise. This underscores the need for further advancements, such as employing deeper neural networks and training on larger datasets, to improve generalization and performance across diverse scenarios. In conclusion, the proposed method offers a reliable and efficient solution for automating stabilization diagram analysis, demonstrating strong potential for structural health monitoring and aerospace applications.

REFERENCES

[1] Heidersbach, Ross, Dhuree Seth, Matthew McCrink, and Moti Karpel. "Safe flutter flight testing of an unmanned aerial vehicle utilizing parametric flutter margin." In American Institute of Aeronautics and Astronautics Scitech 2023 Forum, p. 2071, 2023.
[2] Coppotelli, Giuliano, Elsabetta Conti, and Hans Van Tongeren. "Modal parameters estimate of AH-64D Apache helicopter from flight data and correlation with numerical model." In 2018 AIAA/ASCE/AHS/ASC Structures, Structural Dynamics, and Materials Conference, p. 1950, 2018.
[3] Wu, Chunli, Hanbing Liu, Xuxi Qin, and Jing Wang. "Stabilization diagrams to distinguish physical modes and spurious modes for structural parameter identification." Journal of Vibroengineering 19, no. 4 (2017): 2777–2794.
[4] Schwaiger, Manfred, and Otto Opitz, eds. Exploratory Data Analysis in Empirical Research: Proceedings of the 25th Annual Conference of the Gesellschaft für Klassfikation eV, University of Munich, March 14–16, 2001. Springer Science & Business Media, 2002.
[5] Hartigan, John A., and Manchek A. Wong. "Algorithm AS 136: A K-means clustering algorithm." Journal of the Royal Statistical Society: Series C (Applied Statistics) 28, no. 1 (1979): 100–108.
[6] Arthur, David, and Sergei Vassilvitskii. "K-means++ the advantages of careful seeding." In Proceedings of the Eighteenth Annual ACM-SIAM Symposium on Discrete Algorithms, pp. 1027–1035, 2007.
[7] Covioli, Jacopo Valentino, and Giuliano Coppotelli. "On the use of Gaussian Mixture Models for automated modal parameters estimation." In American Institute of Aeronautics and Astronautics Scitech 2021 Forum, p. 1035, 2021.
[8] Civera, Marco, Luigi Sibille, Luca Zanotti Fragonara, and Rosario Ceravolo. "A DBSCAN-based automated operational modal analysis algorithm for bridge monitoring." Measurement 208 (2023): 112451.
[9] Zhong, Qiang-Ming, Shi-Zhi Chen, Zhen Sun, and Lu-Chao Tian. "Fully automatic operational modal analysis method based on statistical rule enhanced adaptive clustering method." Engineering Structures 274 (2023): 115216.
[10] Starczewski, Artur, Piotr Goetzen, and Meng Joo Er. "A new method for automatic determining of the DBSCAN parameters." Journal of Artificial Intelligence and Soft Computing Research 10, no. 3 (2020): 209–221.
[11] Zhang, Ruitong, Hao Peng, Yingtong Dou, Jia Wu, Qingyun Sun, Yangyang Li, Jingyi Zhang, and Philip S. Yu. "Automating DBSCAN via deep reinforcement learning." In Proceedings of the 31st ACM International Conference on Information & Knowledge Management, pp. 2620–2630, 2022.
[12] Bushra, Adil Abdu, Dongyeon Kim, Yejin Kan, and Gangman Yi. "AutoSCAN: Automatic detection of DBSCAN parameters and efficient clustering of data in overlapping density regions." PeerJ Computer Science 10 (2024): e1921.
[13] Sbarra, R., and Giuliano Coppotelli. "DBSCAN-based approach for the automatic estimate of the modal parameters." In International Operational Modal Analysis Conference, pp. 618–630. Cham: Springer Nature Switzerland, 2024.
[14] Drachinsky, Ariel, and Daniella E. Raveh. "Nonlinear aeroelastic analysis of highly flexible wings using the modal rotation method." AIAA Journal 60, no. 5 (2022): 3122–3134.
[15] Van Overschee, Peter, and B. De Moor. Subspace Identification for Linear Systems: Theory—Implementation—Applications. Springer Science Business Media, 2012.

[16] Cauberghe, Bart. "Applied frequency-domain system identification in the field of experimental and operational modal analysis." (2004).
[17] Ester, Martin, Hans-Peter Kriegel, Jörg Sander, and Xiaowei Xu. "A density-based algorithm for discovering clusters in large spatial databases with noise." KDD 96, no. 34 (1996): 226–231.
[18] Rawat, Waseem, and Zenghui Wang. "Deep convolutional neural networks for image classification: A comprehensive review." Neural Computation 29, no. 9 (2017): 2352–2449.
[19] Ioffe, Sergey. "Batch normalization: Accelerating deep network training by reducing internal covariate shift." arXiv preprint arXiv:1502.03167 (2015).
[20] Scherer, Dominik, Andreas Müller, and Sven Behnke. "Evaluation of pooling operations in convolutional architectures for object recognition." In International Conference on Artificial Neural Networks, pp. 92–101. Berlin, Heidelberg: Springer, 2010.
[21] Ogundokun, Roseline Oluwaseun, Rytis Maskeliunas, Sanjay Misra, and Robertas Damaševičius. "Improved CNN based on batch normalization and Adam optimizer." In International Conference on Computational Science and Its Applications, pp. 593–604. Cham: Springer International Publishing, 2022.
[22] Rowe, W. Mean Square Error & R2 Score Clearly Explained. BMC, 2018.
[23] Ritter, Markus, Michael Fehrs, and Christoph Mertens. "Aerodynamic and static coupling simulations of the Pazy wing with transitional CFD for the Third Aeroelastic Prediction Workshop." In American Institute of Aeronautics and Astronautics SciTech 2023 Forum, p. 0762, 2023.
[24] Ameri, N., C. Grappasonni, G. Coppotelli, and D. J. Ewins. "Ground vibration tests of a helicopter structure using OMA techniques." Mechanical Systems and Signal Processing 35, no. 1–2 (2013): 35–51.

6 Modal, frequency-response, and moving-load analysis for system and damage identification of various structures

L. Auersch and S. Said

6.1 INTRODUCTION

The structural health monitoring of bridges and slender towers is usually done by modal analysis [1–5]. Recent research is focused on developing better tools for operational modal analysis [6–9]; removing environmental effects, especially the influence of temperature [10]; advancing automated monitoring [11, 12], statistical methods [13–15], and the use of artificial intelligence [16]. Many structures, however, are not as isolated as bridges or slender towers. They are coupled to other structures, such as a floor in a building, or coupled to the soil, such as buildings, foundations, or railway tracks. These structures need alternative methods for system and damage identification.

This chapter, which is an extended version of the contribution to the special session about "other structures" at the 10th International Conference on Operational Modal Analysis, presents some additional or complementary methods to modal analysis and their application to different structures. Section 6.2 derives a method of system identification based on the measured frequency response functions and the transformed and weighted system equations. This simple method is applied to the dynamic driving behavior of a car (Section 6.2) and to the modal analysis of beams and floors (Section 6.3). A special method to extract different mode shapes for closely spaced eigenfrequencies of neighboring floors, and its application to a large wooden multi-bay floor in a castle, is presented in Section 6.3. The problem of local hammer excitation, global ambient excitation, and global modes is addressed in Section 6.3 for the weakly coupled floor beams in the castle, and in Section 6.4, where the condition of a truss member in a roof construction had to be analyzed. In Section 6.5, a railway bridge is investigated using modal analysis and the moving-load method.

DOI: 10.1201/9781003516941-6

The quasi-static moving-load method with a slowly moving vehicle and measured inclinations is further demonstrated for a long six-span footbridge, a short single-span footbridge, and a laboratory beam in Section 6.6. In Section 6.7, damages to railway tracks are identified from the moving-load responses to normal train operations and by frequency response functions from hammer impacts. The railway tracks are coupled to the soil, and the radiation damping into the soil prevents or hides any resonance, so modal analysis is not an option. The soil-structure interaction is calculated using a combined finite-element boundary-element method, and the suitable soil properties as well as the extent of the damage are found through manual model updating. The manual model updating for the identification of suitable soil parameters is demonstrated in Section 6.8 for a small building that is excited by ground vibrations.

6.2 CAR MODELS FROM FREQUENCY RESPONSE FUNCTIONS

The identification of the driving model of a car (Figure 6.1a) is based on the normal form for linear systems

$$\frac{d}{dt}\mathbf{x} = \mathbf{A}\mathbf{x} + \mathbf{b}s \tag{6.1}$$

and in the frequency domain

$$i\omega\frac{\mathbf{x}}{s}(\omega) = \mathbf{A}\frac{\mathbf{x}}{s}(\omega) + \mathbf{b}. \tag{6.2}$$

These are $m \times n \times 2$ equations for m state components $\mathbf{x}$ and n frequencies $\omega = 2\pi f$, linear in the $n^2 + n$ unknown elements p of the system matrix $\mathbf{A}$ and the input vector $\mathbf{b}$, with the frequency response functions $\mathbf{x}/s$, which have been measured by a harmonic excitation s of the steering wheel. This overdetermined system of equations

$$\mathrm{L\,p} = \mathrm{r} \tag{6.3}$$

is solved by multiplying it with the transpose L^T

$$\mathrm{L}^\mathrm{T}\,\mathrm{L\,p} = \mathrm{L}^\mathrm{T}\,\mathrm{r} \tag{6.4}$$

and its inversion

$$\mathrm{p} = (\mathrm{L}^\mathrm{T}\,\mathrm{L})^{-1}\,\mathrm{L}^\mathrm{T}\,\mathrm{r}. \tag{6.5}$$

This is the least-squares approximation of the system Eq. (6.2) in the frequency domain. In Auersch [17], the weighting of the frequencies and of the state components $\mathbf{x}$ has been tested for a better representation of the frequency response

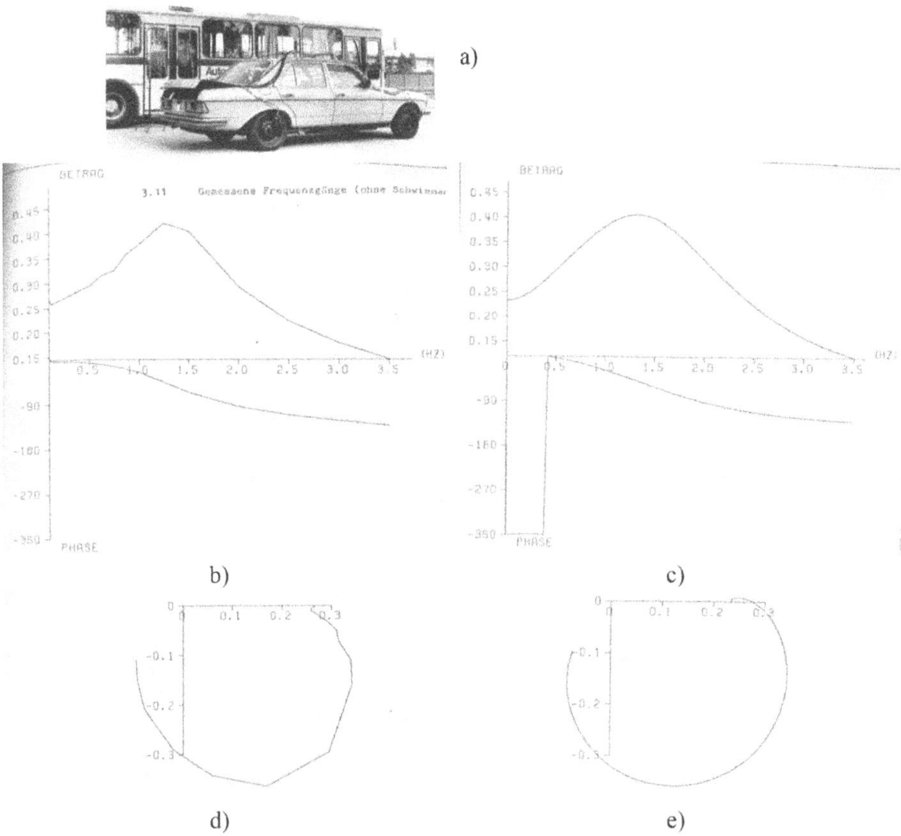

FIGURE 6.1 Frequency response function of a test car (a) as amplitude and phase (b, c) and in the complex plane (d, e), measurement (b, d) and approximation (c, e), and rotation speed around the vertical axis excited by the steering angle.

functions, and the model order m has been varied between 2 and 8, already showing good results for the two basic state components (see Figure 6.1).

The same idea can be applied to the representation of mechanical systems by the mass, stiffness, and damping matrices $\mathbf{M}$, $\mathbf{K}$, and $\mathbf{C}$:

$$\left(-\omega^2\mathbf{M}+\mathrm{i}\omega\mathbf{C}+\mathbf{K}\right)\frac{\mathbf{u}}{F}(\omega)=\mathbf{b} \tag{6.6}$$

and by the force F and displacement $\mathbf{u}$. The simplest application is for a single-degree-of-freedom system

$$\left(-\omega^2 m+\mathrm{i}\omega c+k\right)\frac{u}{F}(\omega)=1 \tag{6.7}$$

where only the three scalars m, k, and c have to be determined.

6.3 MODAL ANALYSIS FOR THE RESONANCE AND CONDITION OF FLOORS

The resonance of floors is the most important effect of train-induced building vibrations. Many floors in various buildings have been measured for their eigenfrequencies, damping, and maximum soil-to-floor amplification. A simple modal analysis has been established, which is based on the measured velocity v and force F, the corresponding admittance functions $H(\omega) = v/F(\omega)$, and the approximation of the system equation.

$$\frac{\left(-\omega^2 m + \mathrm{i}d + k\right)}{\mathrm{i}\omega} \frac{v}{F}(\omega) = 1 \tag{6.8}$$

The approximation of this linear equation in mass m, hysteretic damping d, and stiffness k results in a good fit for the impedance function $Z(\omega) = F/v(\omega)$. To get a good fit for the admittance function and the resonance behavior, the equation is weighted with the conjugate $H^*(\omega)$ of the measured admittance function

$$H^*(\omega) H(\omega) \frac{-m\omega^2 + \mathrm{i}d + k}{\mathrm{i}\omega} = H^*(\omega) \tag{6.9}$$

The solution (k, d, m) of this overdetermined system of $2n$ linear equations (the real and imaginary parts for each of the n frequencies) is easily obtained. The frequency weighting by $H^*(\omega)$ is essential to get reasonable results around the resonance maximum. This simple approximation method also provides the modal parameters—undamped eigenfrequency f_0

$$f_0 = \frac{1}{2\pi} \sqrt{\frac{k}{m}} \tag{6.10}$$

and damping ratio D

$$D = \frac{d}{2k}. \tag{6.11}$$

The method has been applied to an 8 m long steel beam in the laboratory (Figure 6.2a). The first and second eigenfrequencies, f_1 = 7.9 Hz and f_2 = 29 Hz, are well represented as amplitude curves (Figure 6.2b,c) and in the complex plane (Figure 6.2d,e). The low material damping of steel, D = 0.28 and 0.27%, is determined under laboratory conditions. Figure 6.3 shows the results for two short-span concrete floors in a new six-story residential building. The evaluated damping values of D = 2.8% at f_0 = 33 Hz and D = 4.1% at f_0 = 45 Hz are higher than the expected material damping of concrete.

The method has also been applied to wooden floor beams in a castle (Figure 6.4), but the results, with eigenfrequencies at about 9 Hz and a high modal damping of 8%, are not convincing (Figure 6.4). The simultaneous measurement of all six floor beams and the intermediate five wall beams revealed a cluster of natural frequencies

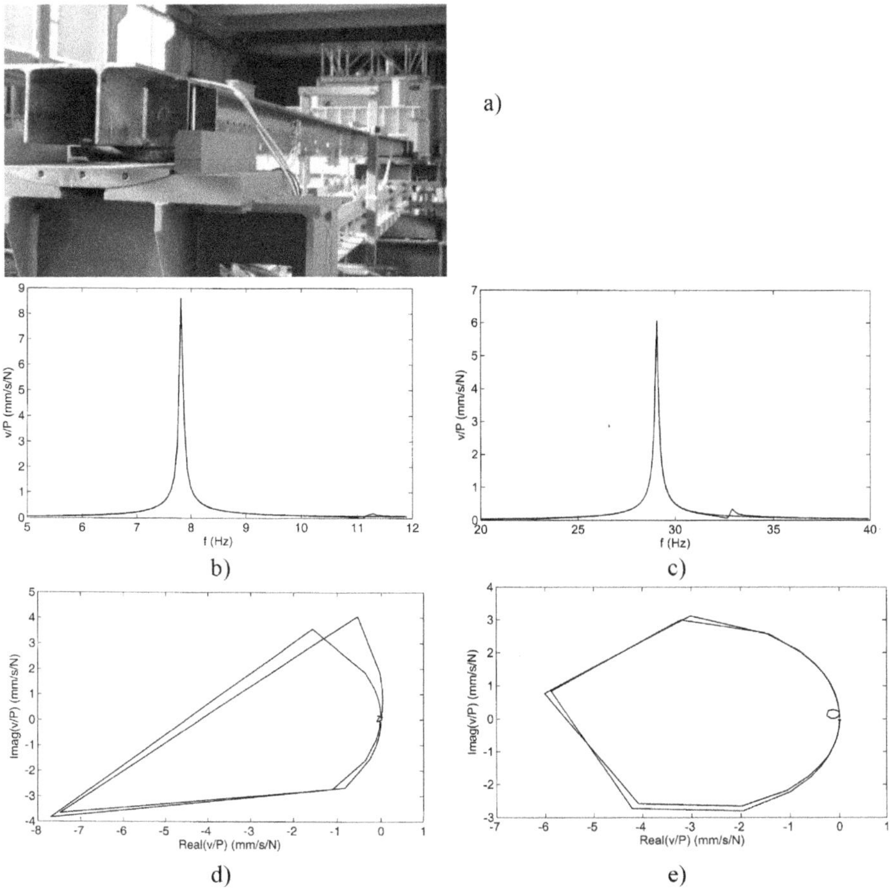

FIGURE 6.2 Measured and approximated frequency response functions (admittances) of a laboratory beam (a), (b, c) amplitude, (d, e) in the complex plane, (b, d) first eigenfrequency f_1 = 7.9 Hz with D = 0.28%, and (c, e) second eigenfrequency f_2 = 29 Hz with D = 0.27%.

between 8 and 9 Hz. Global modes of the six floor bays have been found from the ambient measurements, whereas the hammer tests resulted in more local vibration modes (Figure 6.5). The local mode from a hammer impact (a point load F_1 at x_1) is a superposition of several eigenmodes

$$\frac{v(x)}{F_1} = \sum_j \frac{w_j(x) w_j(x_1)}{\left(\omega_j^{\,2} - \omega^2\right) \int w_j^{\,2} \mu dx} \tag{6.12}$$

with μ the mass per unit length, ω_j the eigenfrequency, and w_j the corresponding mode shape. All excited modes are in phase and result in the maximum amplitude

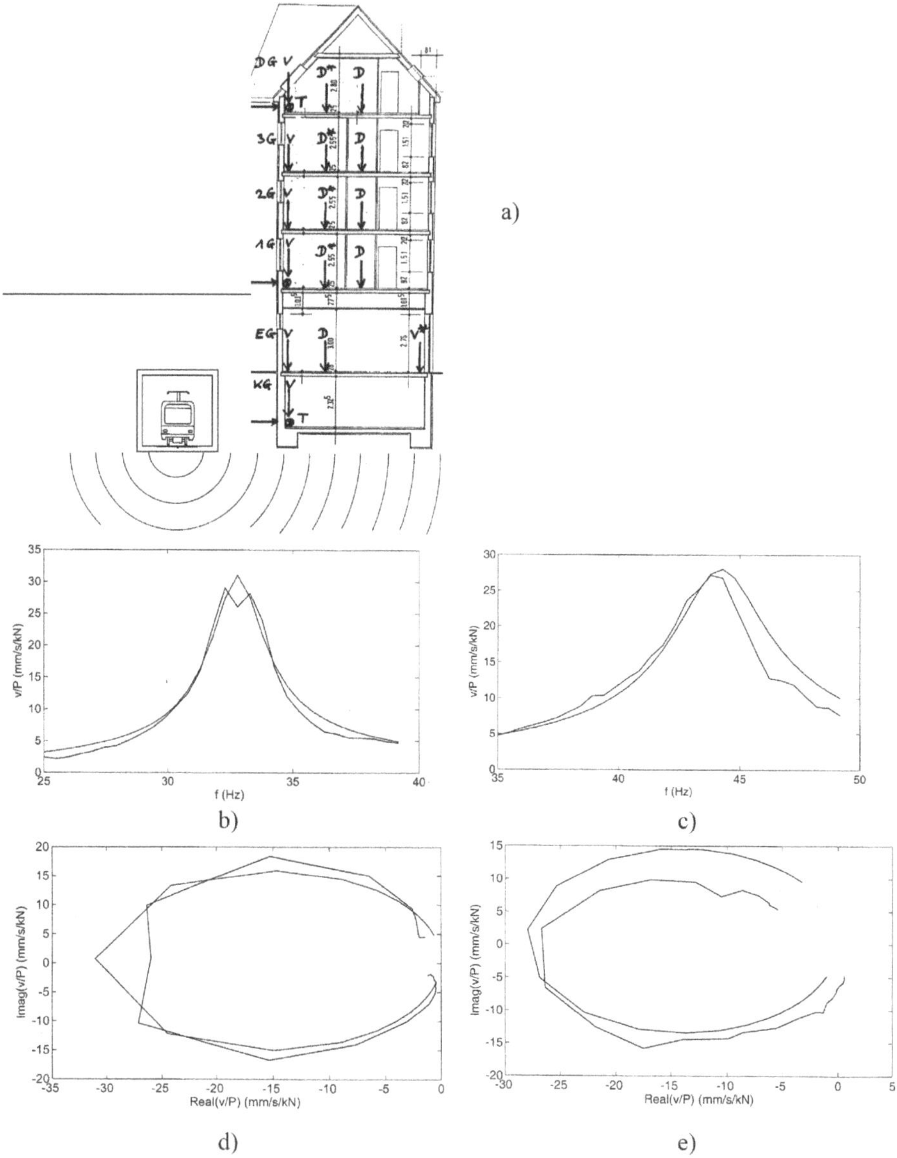

FIGURE 6.3 Measured and approximated frequency response functions (admittances) of concrete floors in a six-storey residential building (a), (b, c) amplitude, (d, e) in the complex plane, (b, d) longer floor span, $f_0 = 33$ Hz with $D = 2.8\%$, and (c, e) short floor span, $f_0 = 45$ Hz with $D = 4.1\%$.

at the excited floor, whereas some modes are in antiphase and reduce the amplitude at other floor bays.

If the difference between two vibration mode shapes at neighbouring frequencies is calculated,

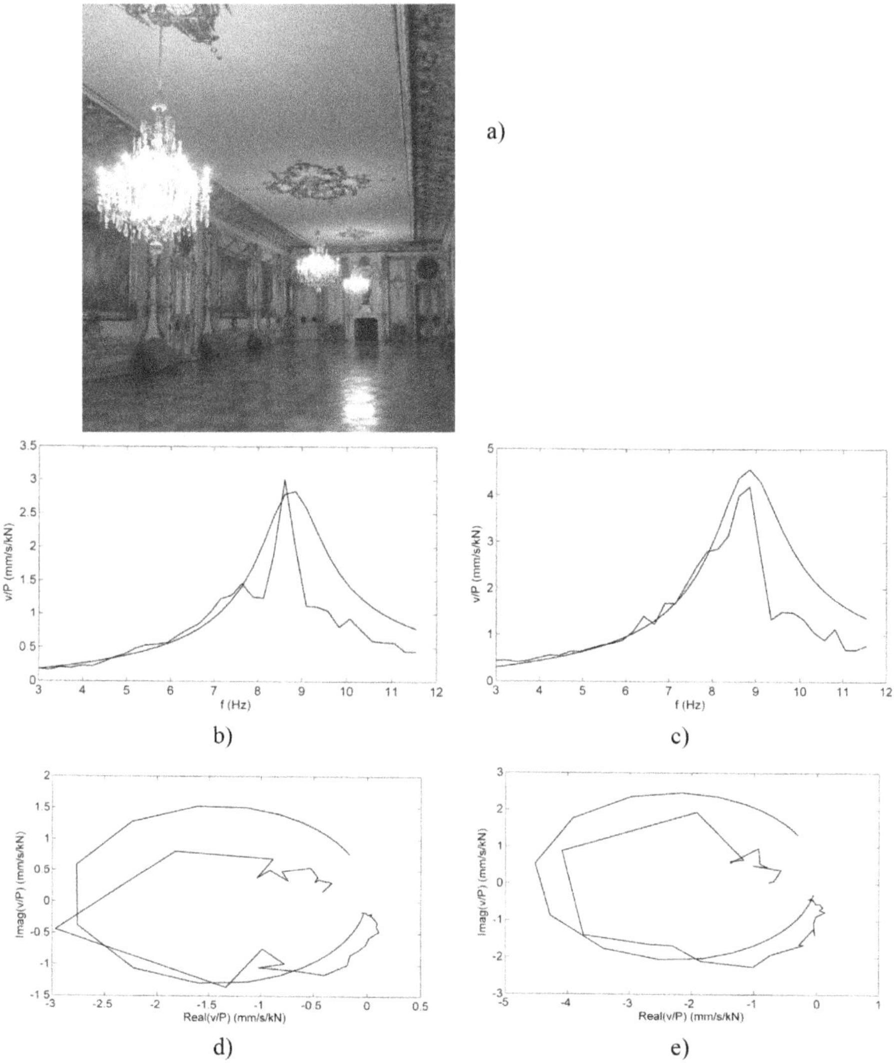

FIGURE 6.4 Measured and approximated frequency response functions (admittances) of two wooden floor beams in a castle (a), (b, c) amplitude, (d, e) in the complex plane, $f_0 = 9$ Hz with $D = 8\%$.

$$\frac{d}{d\omega}\frac{v(x)}{F_1} = \frac{d}{d\omega}\Sigma_j \frac{w_j(x)w_j(x_1)}{(\omega_1^2 - \omega^2)\int w_j^2 dm} = \Sigma_j \frac{w_j(x)w_j(x_1)}{\int w_j^2 dm}\frac{2\omega}{(\omega_j^2 - \omega^2)^2} \tag{6.13}$$

then the influence of the eigenmode at this frequency is increased. The function $1/(\omega_j^2 - \omega^2)$, which yields the resonance maximum, is squared and will become more dominant. This method has been applied to the vibration mode in Figure 6.5a, and

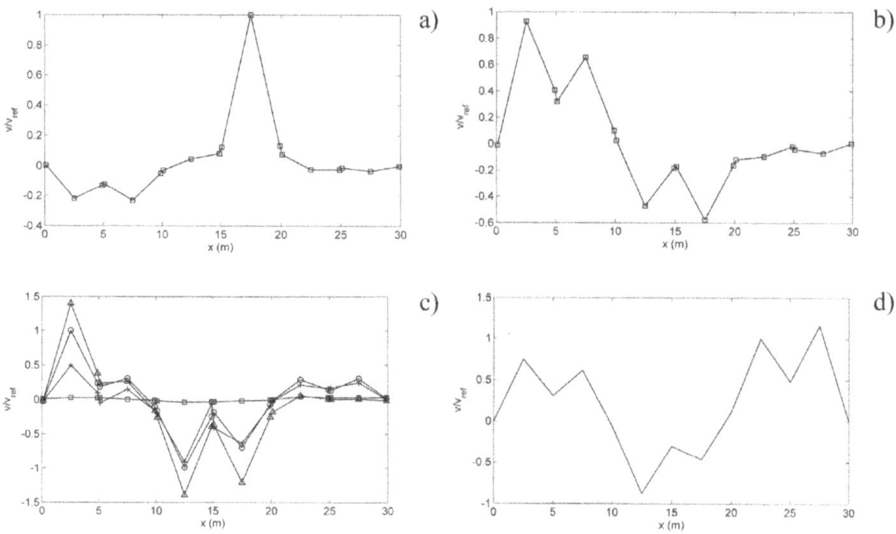

FIGURE 6.5 Mode shapes along the long axis of the wooden six-bay floor in the castle, (the measurement points are alternating on a floor beam and a wall beam), (a) local vibration mode from hammer excitation, (b) global differential mode from hammer excitation, (c) global vibration mode from ambient excitation (half- and quarter-span measurement lines) (d) calculated vibration mode from a coupled beam model.

whereas the original vibration mode is rather local at the excitation point, the differential mode in Figure 6.5b has contributions from all floor bays and is very similar to the mode from the ambient testing (Figure 6.5a) and to the calculated mode of a coupled beam model (Figure 6.5d).

6.4 MODAL ANALYSIS OF A TRUSS ROOF

A similar observation of local and global vibration modes has been made when the state of a truss member in a roof construction (Figure 6.6a) should be identified [18]. The local response from a hammer test consists of global modes due to the coupling by the rotational stiffness of the rusted hinges. This is shown for the eigenmodes at 17 and 20.2 Hz in Figure 6.6b,c, where the mode shapes from a hammer test differ only by the out-of-phase or in-phase vibration of a neighboring truss. Because of these problems with global modes, less complex truss structures have been investigated in the laboratory, and model and condition identification could be obtained by model updating [18]. In general, local modes for the analysis of a single truss or floor element can be difficult to obtain in the real environment of the building, and global modes must be considered.

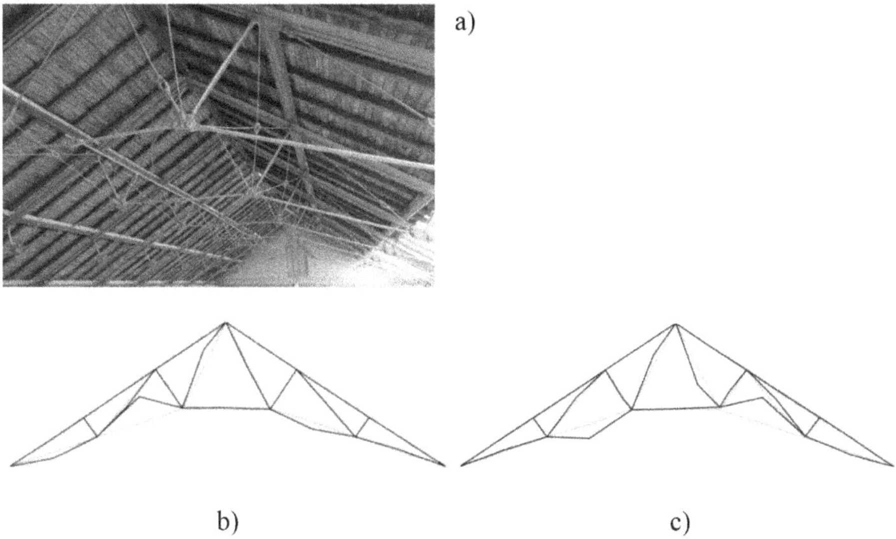

FIGURE 6.6 Eigenmodes of a steel trusswork in a roof construction (a), at (b) 17.0 Hz and (c) 20.2 Hz with almost the same mode shape except for the in-phase or out-of-phase vibration of the left central diagonal truss.

6.5 BRIDGE STUDIES BY MODAL ANALYSES AND MOVING-LOAD RESPONSES

At a three-span railway bridge on the Hanover-Würzburg high-speed line, a modal analysis has been performed using ambient and train excitation. The bridge consists of three continuous spans of 90, 120, and 90 m in length (Figure 6.7a). The mode shapes of the first eigenfrequencies are shown in Figure 6.7b for the theoretical modes and in Figure 6.7c-g for the measured modes. The fundamental mode is only the vibration of the longest mid-span. The second and third modes have maxima in the shorter side spans but also some contributions from the mid-span. The mode shapes and their spectra are of importance when the resonance amplification due to train passages must be predicted. For the present bridge, a resonance has been observed and calculated for a long freight train passing at 100 km/h [19]. The curvature of the mode shapes could be used for the identification of bridge damage if enough measurement points are available. A different method of damage detection has been tested with this three-span railway bridge and with a six-span footbridge—the measurement of inclinations during quasi-static moving-load tests.

6.6 DAMAGE IDENTIFICATION BY INFLUENCE LINES OF INCLINATIONS FROM MOVING LOAD TESTS

At the three-span railway bridge and the six-span continuous footbridge, the inclinations were measured at all bridge supports during the slow passage of a

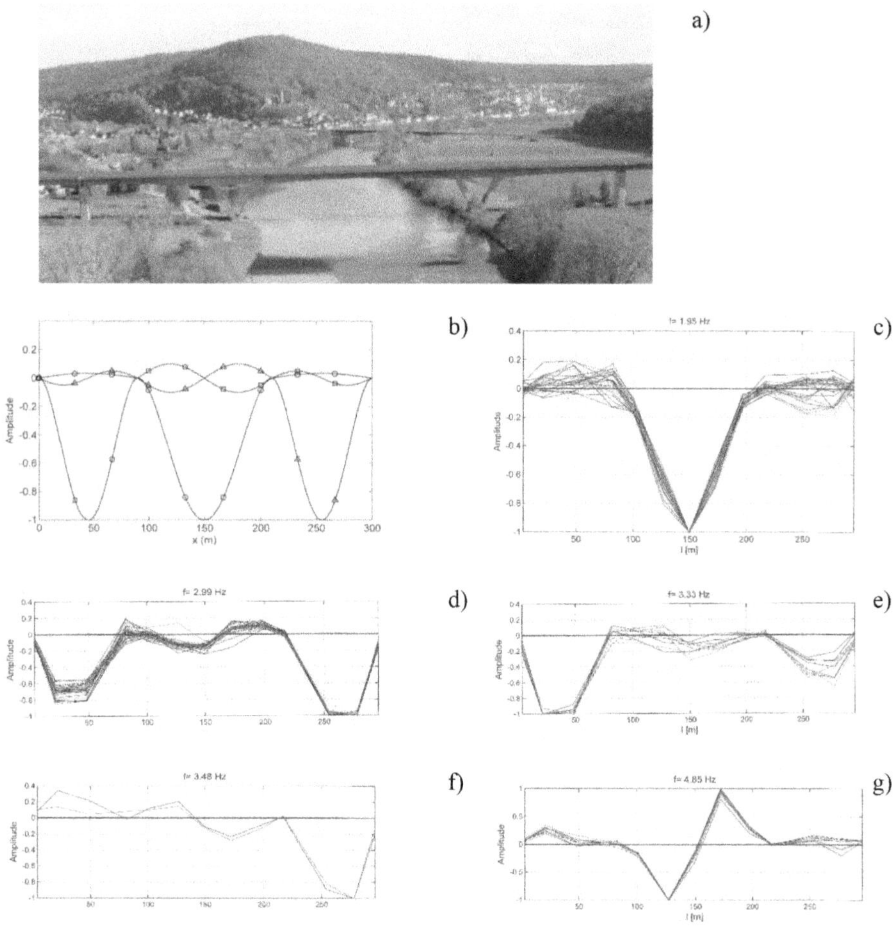

FIGURE 6.7 Mode shapes of the three-span railway bridge (a), (b) calculated O first, □ second and Δ third mode; (c) measured first mode, (d) mixed (second and third) mode, (e) second mode, (f) third mode, (g) second bending mode for the long mid span.

locomotive and a compaction roller. The corresponding influence lines are shown in Figure 6.8. The reproducibility of the results was checked through three measuring campaigns at the footbridge [20]. There are two advantages to the influence lines of inclinations. The influence lines from a moving load give continuous information along the bridge length, and the inclinations, as the first derivative of the displacements, are already closer to the curvature, which is considered a damage indicator. The differences in these influence lines can be used for the detection and localization of bridge damage. In Stöhr [21], stiffness modifications have been successfully identified with a special single-axle moving-load device on a laboratory beam (Figure 6.2a) and a slowly moving truck on a short single-span footbridge.

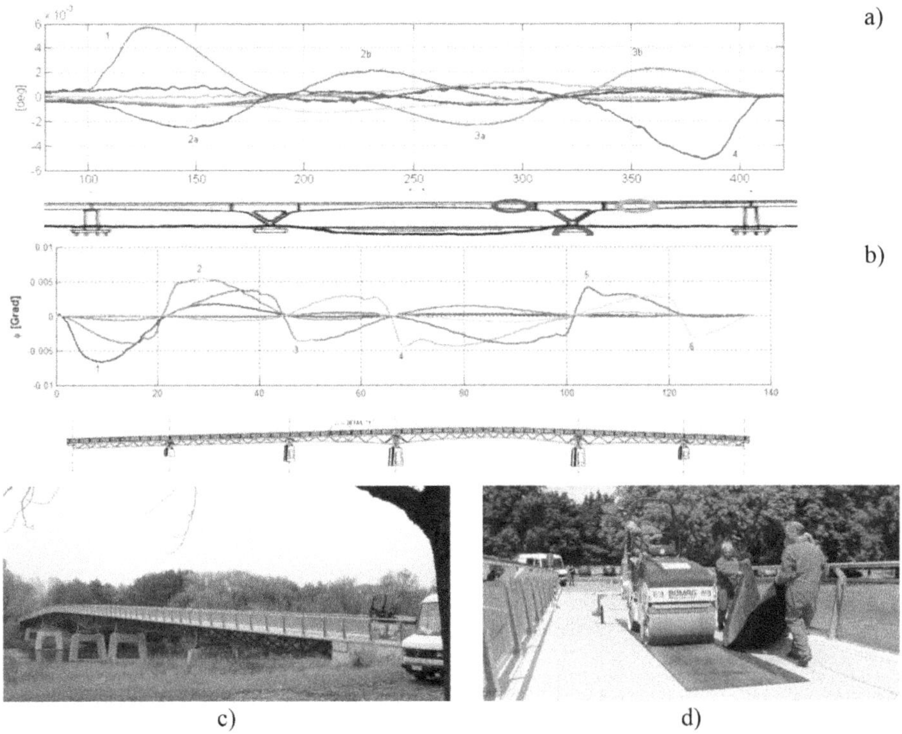

FIGURE 6.8 Influence lines of the inclinations measured at the bridge supports; (a) three-span railway bridge excited by a locomotive, support 1, 2a, 2b, 3a, 3b, 4, (b) six-span footbridge (c) excited by a compaction roller (d), support 1, 2, 3, 4, 5, 6.

Figure 6.9 shows the corresponding influence lines of the modified structures and the difference between the modified and original structures. The difference in the influence lines indicates the modification (or damage), and the maximum difference occurs at the position of the modification, making it possible to detect and localize damage with this moving-load method.

6.7 IDENTIFICATION OF TRACK DAMAGE BY MOVING LOAD RESPONSES AND FREQUENCY RESPONSE FUNCTIONS

The identification of structural damage is easier for the local behavior of a railway track structure than for the global behavior of bridges. On the other hand, the track is highly coupled to the soil, and the strong radiation damping into the soil prevents clear modal behavior. Two situations with damaged slab tracks have been analyzed using the moving load response.

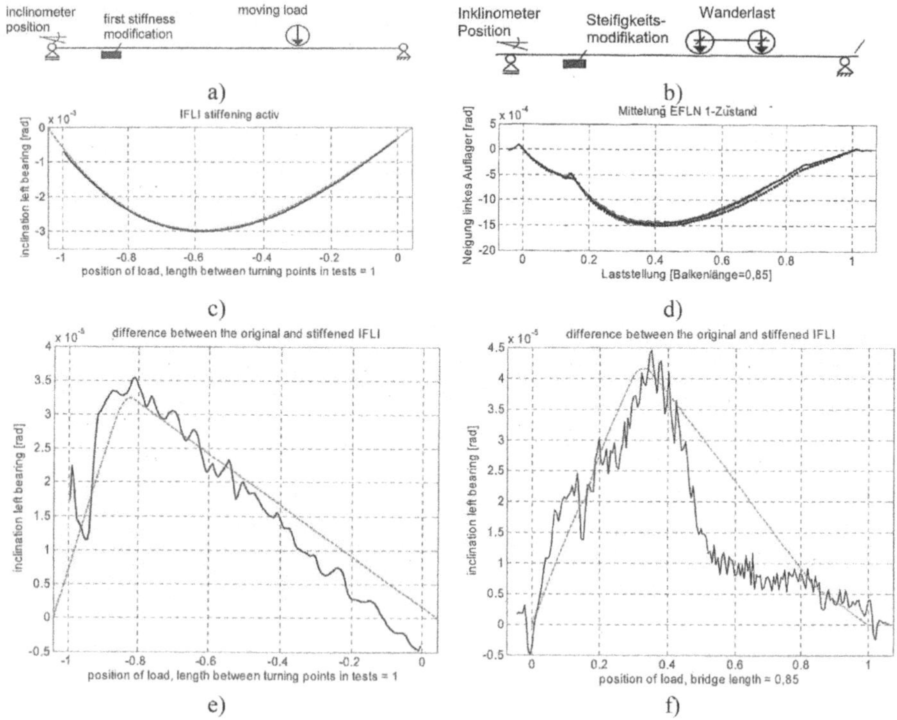

FIGURE 6.9 Influence lines of the inclinations of a laboratory beam under single-axle moving-load device (a, c, e) and a single-span footbridge under a moving truck (b, d, f), measured at the support, (c, d) influence lines of the modified structure, (e, f) difference of the influence lines of the modified and the original structure, measurements in black, finite-element calculation in dashed line.

The first example is a slab track with a gap between the track plate and the base layer, and the second example is a slab track with a loose sleeper. Figure 6.10 shows the responses of the intact and damaged slab tracks to the passage of a high-speed train (Figure 6.10a). The displacements are at 0.4 to 0.5 mm for the track plate of the intact slab tracks (Figure 6.10c,d,g,h). These displacements repeat rather regularly for all 14 cars and 28 bogies. The displacements of the damaged tracks are clearly higher at 1 to 2 mm (Figure 6.10e,f,i,j). These higher amplitudes are the first and clearest indicator of the damage. A second damage indicator is the narrower deflection shape, which can be concluded from the relatively smaller inter-bogie displacements. The more local behavior of the damaged track can also be seen in the presentation of the passage of four neighboring bogies in Figure 6.11. A third damage indicator is the large difference between the track-plate and the base-layer response.

The damaged and intact slab tracks have been modeled using the combined finite-element boundary-element method, where the track elements—rail, sleeper, track plate, and base layer—are modeled by finite elements (Figure 6.10b), and the realistic

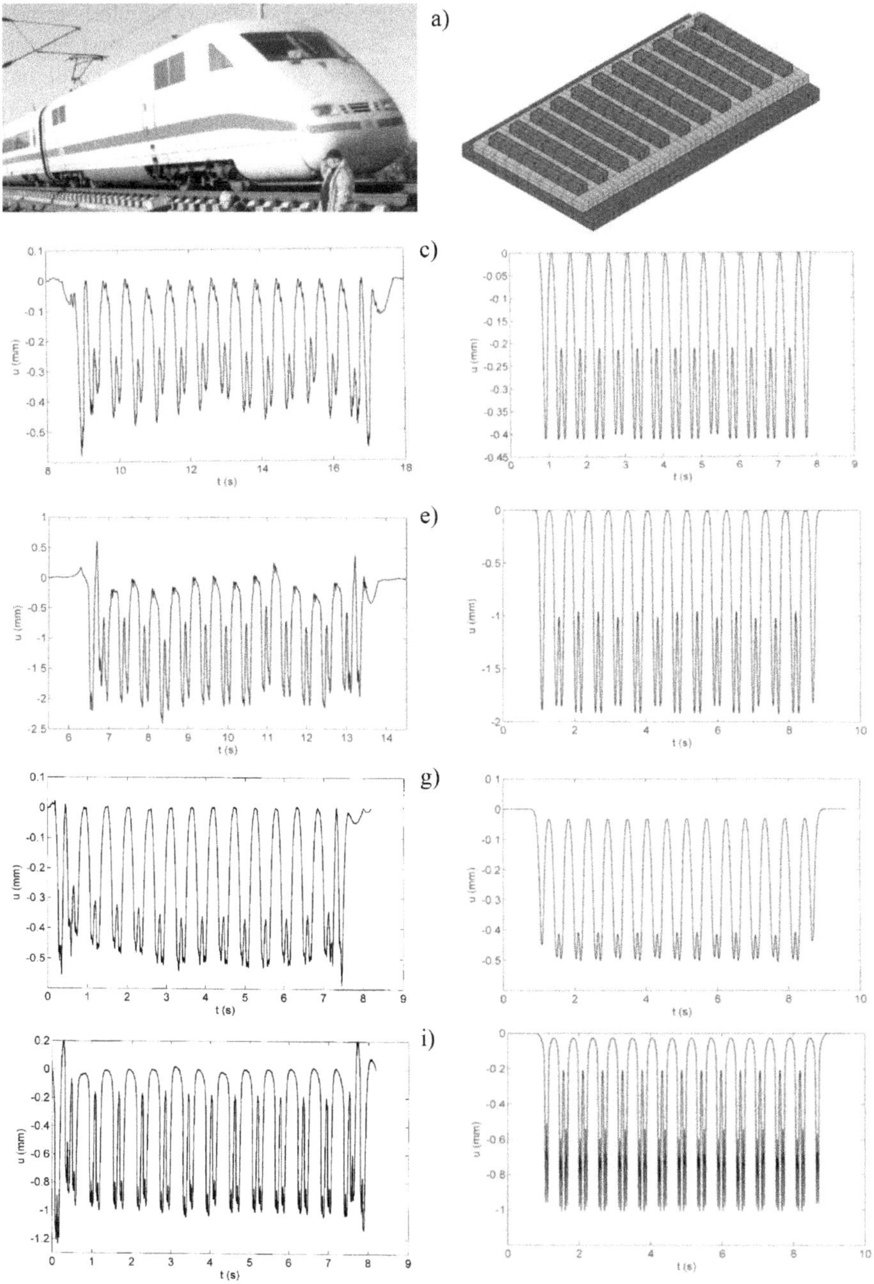

FIGURE 6.10 Passages of an ICE1 train with $v_T \approx 170$ km/h over intact and damaged slab tracks, (a, c, e, g, i) measurements of the track plate and sleeper displacements, (b, d, f, h, j) calculations with the finite-element boundary-element method and updated models, (c, d) intact track (type 1), (e, f) damaged track (type 1, lose track plate), (g, h) intact track (type 2) (i, j) damaged track (type 2, lose sleeper).

behavior of the infinite and continuous soil is represented by a fully coupled boundary matrix [22]. A manual updating of the FEBE model has been done, first for the intact track to get the appropriate soil properties of the measurement site, and second for the damaged track to quantify the extent of the damage. The length of the gap has been varied from 0 m, 0.65 m, 1.95 m, . . . to 7.15 m, corresponding to 0, 1, 3, . . . to 11 sleeper bays. Figure 6.11d shows the calculated results for a damage length of 5.85 m, which shows the best agreement of the maximum amplitudes as well as the most realistic deflection shape.

For the second track example with a loose sleeper, the response of the rail, the sleeper, the track plate, the base plate, and the base layer is shown in Figure 6.11e-h. The best agreement with FEBE calculations has been found for a gap between half the sleeper and the track plate (Figure 6.11h). In theory and measurement, all track components of the intact track have the same amplitudes except for the rail with doubled amplitudes, whereas the amplitudes of the sleeper are higher and almost as high as the rail displacement in the case of the damaged slab track.

The use of hammer tests and frequency response functions is demonstrated for the second track example with a loose sleeper (Figure 6.12). Whereas no resonance can be observed for the intact track (Figure 6.12c,d), a weak resonance can be found at high frequencies for the damaged track (Figure 6.12e,f), which can also be checked by the sudden drop in the phase spectrum. The differences between the different track components are also an indication of the damage. The measured results (Figure 6.12c,e) agree well with the calculations in Figure 6.12d,f.

6.8 UPDATING OF FREQUENCY RESPONSE FUNCTIONS IN CASE OF STRONG SOIL-STRUCTURE INTERACTION

Suitable theoretical models have also been searched for other measured structures, besides the slab tracks of the preceding section, such as ballast tracks, surface and pile foundations, model buildings, and two- to three-story residential buildings. The main problem is to find appropriate soil parameters, whereas the structure is almost known. Because of the time-consuming coupling of all foundation points through the soil, only the manual updating of theoretical models has been carried out, as demonstrated for a box-type model building with a 5.5 m × 4.8 m base area, 2.1 m height, and 0.32 m thick concrete walls (Figure 6.13a,b).

The model building was excited harmonically by an electro-dynamic vibrator which was placed on the soil 15 m away. The measured soil-building transfer functions (Figure 6.13c) show considerable amplification for frequencies below 10 Hz and a reduction for higher frequencies. The whole situation, including wave transmission through the soil, was modeled using the finite-element boundary-element method and calculated for five different wave velocities of the soil (Figure 6.13d-h). The measured characteristics of the building response can be well reproduced. Because of the excitation by passing waves, rocking responses of the building occur—rocking around the building center at a low resonance frequency, and rocking around the end of the building at mid frequencies of 20 to 40 Hz. The vertical eigenfrequency can be observed at 20 Hz but with no absolute amplification compared to the exciting

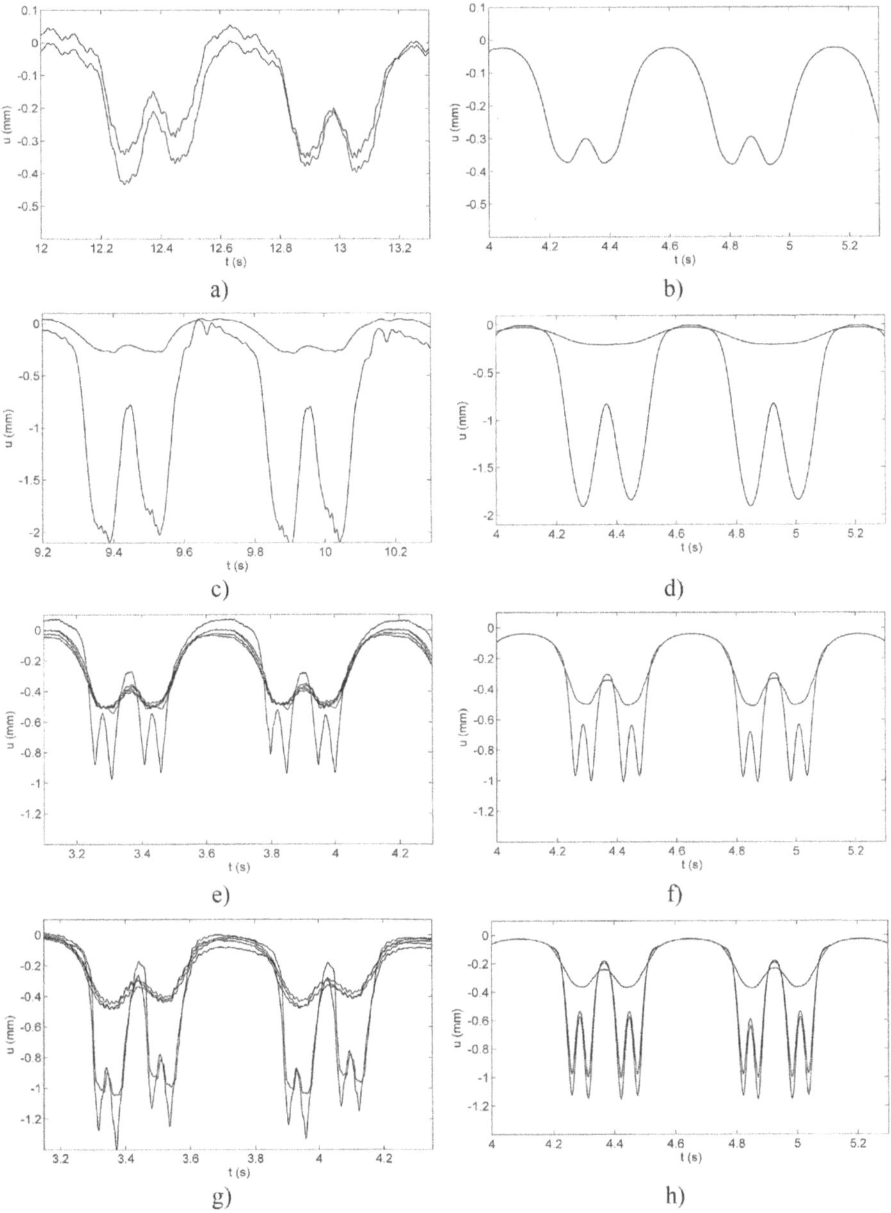

FIGURE 6.11 Passages of four bogies over intact and damaged slab tracks, (a, c) measurements of the track plate and base layer displacements, (e, g) measurements of the rail, sleeper, track plate, base plate and base layer displacements, (b, d, f, h, j) calculations with the finite-element boundary-element method and updated models, (a, b) intact track (type 1), (c, d) damaged track (type 1, lose track plate), (e, f) intact track (type 2), (g, h) damaged track (type 2, lose sleeper).

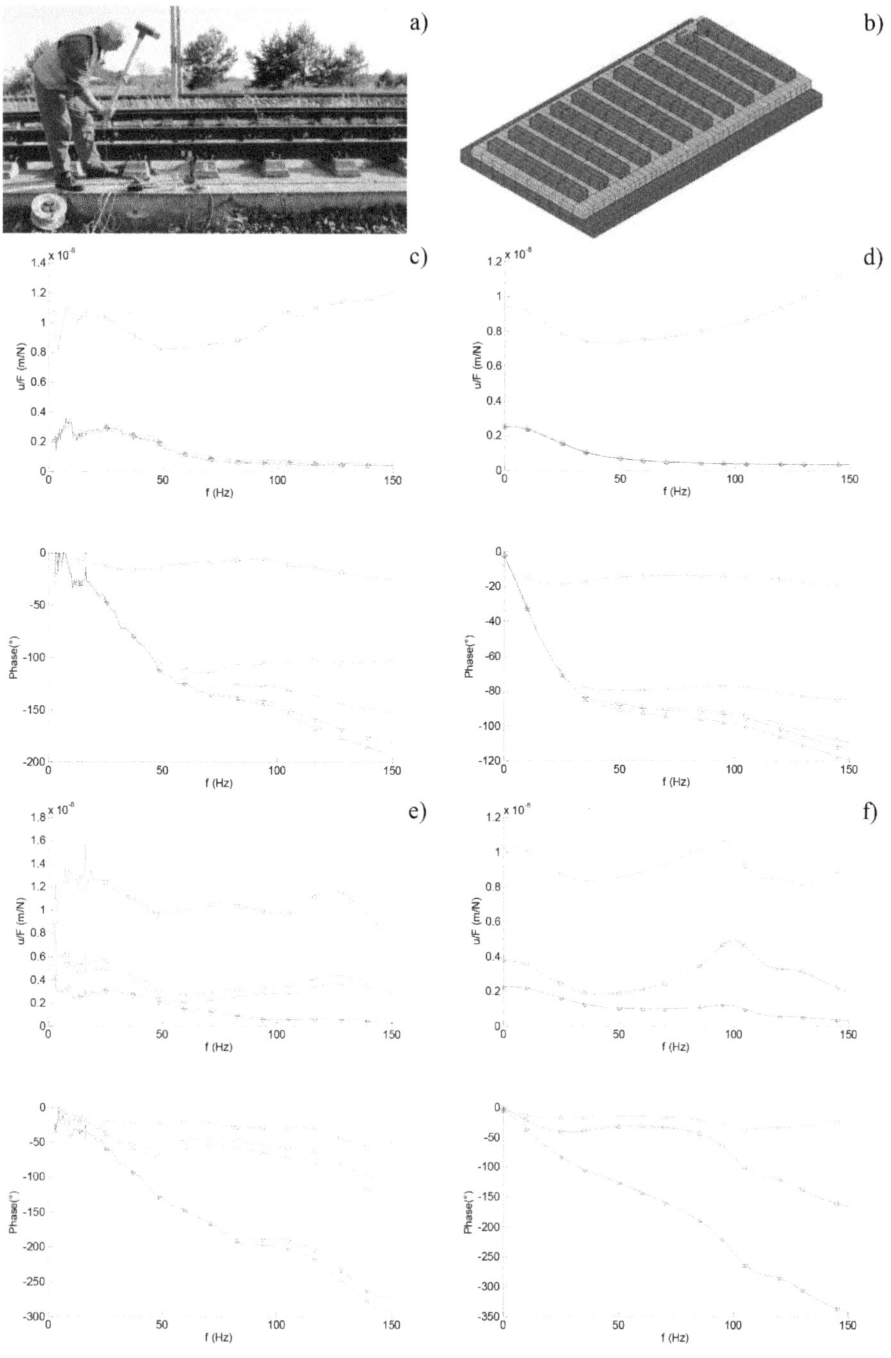

FIGURE 6.12 Compliance functions from hammer impacts (a, c, e) and from calculations with the finite-element boundary-element method (b, d, f), (c, d) intact track (type 2), (e, f) damaged track (type 2, lose sleeper), □ rail, ○ sleeper, △ track plate, + base plate, × base layer.

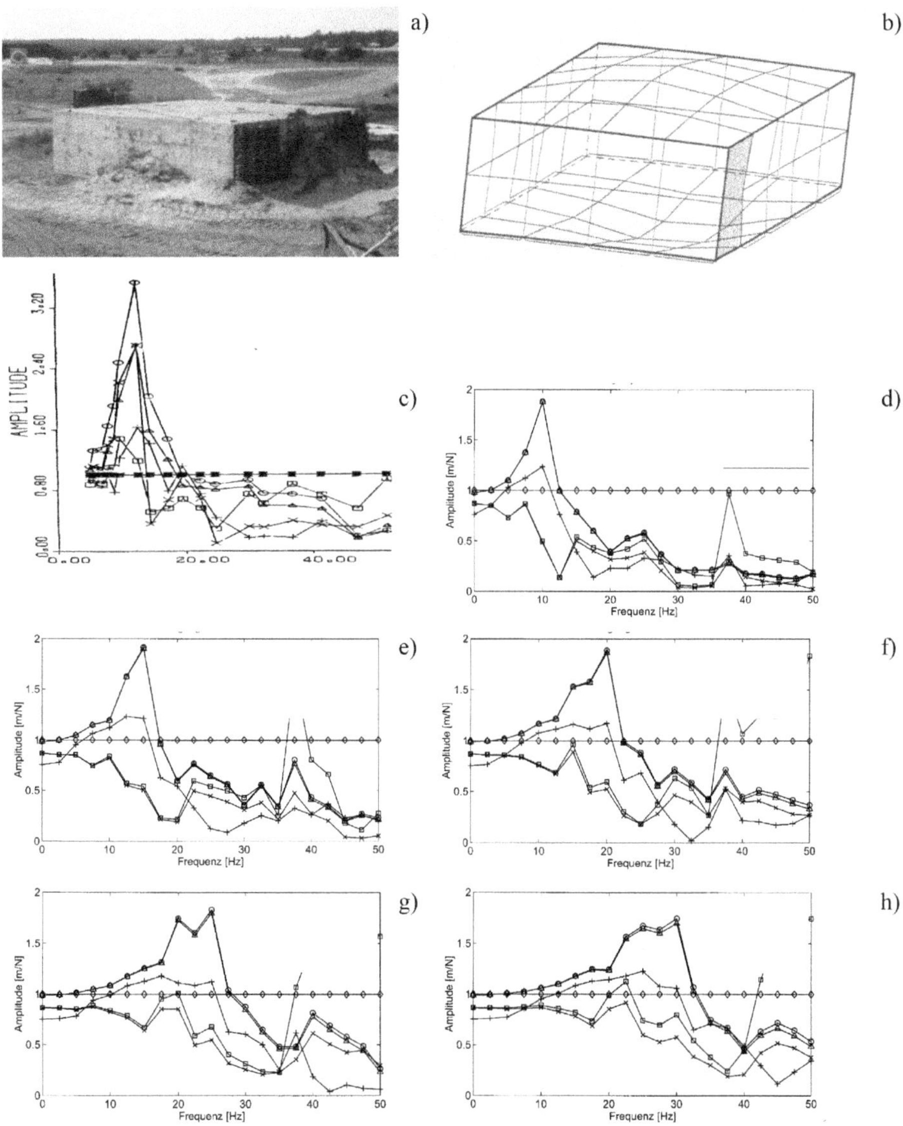

FIGURE 6.13 Soil-building transfer functions of a small box-type building, (a, c) measurements, (b, d, e, f, g, h) calculation with the finite-element boundary-element method and different soils, shear wave velocity (d) 100 m/s, (e) 150 m/s, (f) 200 m/s, (g) 250 m/s, (h) 300 m/s.

wave. At higher frequencies, the soil-building transfer functions are much smaller than 1. The calculations with different soil stiffnesses show the resonance region of vertical and rocking vibrations at 7.5 to 10 Hz for $v_S = 100$ m/s, varying up to 22.5 to 30 Hz for $v_S = 300$ m/s. Looking at the rocking details, amplitudes, and rocking resonance frequency, the calculation with $v_S = 150$ m/s (Figure 6.13e) seems to be the best approximation of the measurements.

6.9 CONCLUSIONS

Some methods for the identification of structures have been presented, which have been developed during research work on railway-induced track, bridge, soil, foundation, building, and floor vibrations. Modal analysis has been applied to bridge and floor vibrations where clear resonances occur. A simple and fast modal analysis based on the approximation of frequency response functions has been successfully applied to several floor structures, but if similar neighboring spans are coupled, the problem of an eigenfrequency cluster appears. This problem has been solved with a special (differential mode shape) method for a large wooden floor in a castle. Other structures, such as tracks, buildings, and foundations, are coupled to the ground, and resonances may be prevented or hidden by strong radiation damping. Therefore, alternative methods to modal analysis have been developed and tested. The interaction of the structure with the infinite and continuous soil is calculated using a combined finite-element boundary-element method. The calculated results, for example, frequency-dependent compliance or admittance functions, are manually updated to find appropriate soil, structure, and damage parameters, which are demonstrated for a small building and two damaged railway tracks. An additional method for the damage identification of railway tracks and bridges is based on the moving load response. The influence lines from inclination measurements are shown for a three-span railway bridge and a six-span footbridge, and the difference in influence lines successfully identified the stiffness modifications of a laboratory beam and a small single-span footbridge. To conclude, new methods for modal analysis and damage detection have been presented, which can be used in difficult situations (eigenfrequency clusters, eigenfrequencies hidden by soil-structure interaction) and during train and vehicle operations.

REFERENCES

[1] Natke, H., 1983, Einführung in Theorie und Praxis der Zeitreihen- und Modalanalyse—Identifikation schwingungsfähiger elastomechanischer Systeme. Vieweg, Braunschweig.

[2] Rohrmann, R., 1985, "Theoretische Ermittlung von Bauwerksschäden mit Hilfe dynamischer Methoden," Proceedings Symposium Zerstörungsfreie Prüfung im Bauwesen, Berlin, 371–382.

[3] Link, M., 1998, "Updating analytical models by using local and global parameters and relaxed optimisation requirements," Mechanical Systems and Signal Processing 12.1, 7–22.

[4] Farrar, C., Worden, K., 2007, "An introduction to structural health monitoring," Philosophical Transactions of the Royal Society A: Mathematical, Physical and Engineering Sciences 365, 303–315.

[5] Brownjohn, J., 2007, "Structural health monitoring of civil infrastructure," Philosophical Transactions of the Royal Society A: Mathematical, Physical and Engineering Sciences 365, 589–622.

[6] Peeters, B., De Roeck, G., 1999, "Reference-based stochastic subspace identification for output-only modal analysis," Mechanical Systems and Signal Processing 13, 55–878.

[7] Basseville, M., Abdelghani, M., Benveniste, A., 2000, "Subspace-based fault detection algorithms for vibration monitoring," Automatica 36, 101–109.

[8] Magalhães, F., 2011, "Operational modal analysis for testing and monitoring of bridges and special structures," PhD Thesis, University Porto.
[9] Döhler, M., Mevel, L., Hille, F., 2014, "Subspace-based damage detection under changes in the ambient excitation statistics," Mechanical Systems and Signal Processing 45, 207–224.
[10] Deraemaeker, A., Reynders, E., De Roeck, G., 2007, "Vibration based structural health monitoring using output-only measurements under changing environment," Mechanical Systems and Signal Processing 22, 34–56.
[11] Rainieri, C., Fabbrocino, G., 2010, "Automated output-only dynamic identification of civil engineering structures," Mechanical Systems and Signal Processing 24, 678–695.
[12] Tomassini, E., Garcia-Macias, E., Reynders, E., Ubertini, F., 2023, "Model-assisted clustering for automated operational modal analysis of partially continuous multi-span bridges," Mechanical Systems and Signal Processing 200, 110587.
[13] Basseville, M., Mevel, L., Goursat, M., 2004, "Statistical model-based damage detection and localization: Subspace-based residuals and damage-to-noise sensitivity ratios," Journal of Sound and Vibration 275, 769–794.
[14] Sohn, H., Worden, K., Farrar, C., 2002, "Statistical damage classification under changing environmental and operational conditions," Intelligent Material Systems and Structures 13, 561–574.
[15] Giglioni, V., Garcia-Macias, E., Venanzi, I., Ierimonti, L., Ubertini, F., 2021, "The use of receiver operating characteristic curves and precision-versus-recall curves as performance metrics in unsupervised structural damage classification under changing environment," Engineering Structures 246, 113029.
[16] Hernández-González, I., Ubertini, F., 2024, "Incorporating digital twins and artificial intelligence for next-generation SHM software," In: Rainieri, C., Gentile, C., Aenlle López, M. (eds.), Proceedings of the 10th International Operational Modal Analysis Conference (IOMAC 2024), Lecture Notes in Civil Engineering, 515. Springer, Cham.
[17] Auersch, L., 1979, "Ermittlung fahrdynamischer Kennwerte aus Fahrzeugmessungen," Diploma Thesis, Technical University Berlin/Daimler Benz AG Stuttgart.
[18] Luong, T., 2018, "Identification of the state of stress in iron and steel truss structures by vibration-based experimental investigations," PhD Thesis, BAM Berlin/BTU Cottbus/Bauhaus-University Weimar.
[19] Auersch, L., 2021, "Resonances of railway bridges analysed in frequency domain by the modal-force-excitation, bridge-transfer and axle-sequence spectra," Engineering Structures 249, 113282.
[20] Rücker, W., Said, S., Schmid, W., Rohrmann, R., 2009, "Zustandsüberwachung der Gärtnerplatzbrücke in Kassel unter Verwendung gemessener Einflusslinien," VDI-Berichte 2063—Baudynamik, Kassel, Germany, 81–91.
[21] Stöhr, S., 2009, "Identifikation von lokalen Steifigkeitsmodifikationen an Brückentragwerken unter Verwendung experimenteller Einflusslinien," PhD Thesis, BAM Berlin/University Kassel.
[22] Auersch, L., 2005, "Dynamics of the railway track and the underlying soil: The boundary-element solution, theoretical results and their experimental verification," Vehicle System Dynamics, 43, 671–695.

7 Temperature compensation in vibration-based structural health monitoring

Static versus dynamic regression models

Soroosh Kamali, Stefano Mariani, and Alessandro Marzani

7.1 INTRODUCTION

Vibration-based structural health monitoring (SHM) systems rely on the continuous monitoring of structural damage-sensitive (SDS) features such as modal frequencies obtained through operational modal analysis (Brincker and Ventura, 2015; Kamali and Hadianfard, 2023; Rainieri and Fabbrocino, 2014) to detect anomalous structural behavior. Initially, SDS features are gathered during an ideal reference period, known as the *baseline*, assuming the structure is healthy. A scalar damage index (DI) is defined. For instance, in this work, the Mahalanobis distance (Mahalanobis, 1936) computed on modal frequencies is considered as the DI. As larger DIs indicate greater deviations from the healthy state of the structure, a threshold is set on the distribution of the DI, and samples that exceed this threshold are used to trigger alarms (Kamali et al., 2024b). A common approach sets the threshold at the mean of baseline DIs plus a specified multiple of the standard deviation, thereby establishing a fixed false positive rate (FPR) or probability of false alarm (PFA) (Falcetelli et al., 2023).

Environmental and operational variations (EOV), particularly temperature, can introduce deviations in the DI not related to anomalies, leading to false alarms—that is, increasing the PFA. Additionally, temperature variations can affect the ability of the system to detect anomalies or damage, influencing the probability of damage detection (POD). Therefore, it is essential to mitigate the impact of temperature variations through a process known as *temperature compensation* (TC) (Anastasopoulos

 DOI: 10.1201/9781003516941-7

et al., 2021; Ente nazionale italiano di unificazione, 2016; Wang et al., 2022; Yan et al., 2005).

An effective TC involves defining a regression model between temperature values at different structural locations (independent variables) and SDS responses, such as modal frequencies (target variables), for the baseline dataset. Among the many, techniques such as multivariate linear regression (MVLR), machine learning (ML), or deep learning methods, including neural networks (NN), can be employed for this purpose. Once the SHM system enters the monitoring phase, the SDS responses (modal frequencies in this work) are compensated for temperature variations. First, the SDS responses are reconstructed at the given temperatures using the regression model. Next, the reconstructed SDS response is subtracted from the original value. Finally, the mean values of the baseline dataset are added to these subtractions. Several studies have demonstrated that TC significantly reduces the variations of the SDS responses, leading to a smaller PFA and an improved POD in undamaged and damaged scenarios, respectively (Kamali et al., 2023, 2025; Reynders et al., 2014; Wang et al., 2022).

A TC strategy can be set up using either a static or a dynamic regression model. *Static* models take only real-time temperatures as independent parameters (Magalhães et al., 2012), i.e., temperature values recorded at the same instant at which SDS responses are also measured or estimated from measured quantities (for instance, in this work, frequencies of vibration are estimated from recorded acceleration time histories). On the other hand, *dynamic* models also incorporate past temperature data as independent parameters (Kamali et al., 2024a; Mariani et al., 2024). This aims at taking into account the time delays that can exist between the measured temperature and structural response variations. These delays depend on the material properties, dimensions, and geometries of the monitored structure, which influence the heat transmission from the environment to the bulk of the structure, thereby producing a nonuniform temperature field at any instant of time. As a result of this phenomenon, temperature values measured at a single instant of time and at a few discrete positions may not be sufficient to characterize the SDS features such as modal frequencies measured at the same temporal instant.

This chapter describes some static and dynamic regression models for TC and discusses their performances in terms of PFA and POD. For this purpose, first, a numerical case of a truss steel bridge, where temperature-dependent pseudo-frequencies are generated for the healthy and damaged scenarios, is considered. Then, the open-access experimental dataset of the KW51 bridge (Maes and Lombaert, 2021) is used for the comparison. For each case study and approach, MVLR and NN regression models are utilized, and optimal time lags for the dynamic method are identified.

7.2 STATIC VERSUS DYNAMIC TEMPERATURE COMPENSATION

Let us assume that L thermocouples and a certain number of accelerometers are installed at various structural locations. The latter are employed to estimate the first M modal frequencies as the SDS responses at a rate Δt, while each thermocouple is set to furnish an average temperature value over the same Δt sampling period. In such a scenario, after a total duration $N \times \Delta t$, the baseline dataset contains $N \times M$

healthy modal frequencies and $N \times L$ corresponding temperature readings. Then, a regression model g can be constructed such that

$$\mathbf{B} \simeq g(\mathcal{T}_S) \tag{7.1}$$

where $\mathbf{B} \in \mathbb{R}^{N \times M}$ is the matrix of the baseline modal frequencies organized as

$$\mathbf{B} = \begin{bmatrix} f_1(1) & f_2(1) & \dots & f_M(1) \\ f_1(2) & f_2(2) & \dots & f_M(2) \\ \vdots & \vdots & & \vdots \\ f_1(N) & f_2(N) & \dots & f_M(N) \end{bmatrix} \tag{7.2}$$

Here, $f_m(n)$ represents the m-th modal frequency at time step n. The block matrix $\mathcal{T}_S = [\mathbf{T}_0, \mathbf{T}_1, \dots, \mathbf{T}_S] \left(\mathcal{T}_S \in \mathbb{R}^{N \times L(S+1)}\right)$ includes the temperature values acquired both at the same temporal instant as the modal frequencies and those recorded at S previous temporal samples, with each block $\mathbf{T}_i \in \mathbb{R}^{N \times L}$ $(i = 0, 1, 2, \dots, S)$ represented as follows:

$$\mathbf{T}_i = \begin{bmatrix} T_1(1-i) & T_2(1-i) & \dots & T_L(1-i) \\ T_1(2-i) & T_2(2-i) & \dots & T_L(2-i) \\ \vdots & \vdots & & \vdots \\ T_1(N-i) & T_2(N-i) & \dots & T_L(N-i) \end{bmatrix} \tag{7.3}$$

Here, $T_l(n-i)$ represents the temperature value of the l-th thermocouple at step $n-i$, where i denotes a time lag with respect to n. Null and negative values for $n-i$ correspond to temperature values acquired prior to the baseline period (practically, they would only be available if the L thermocouples were installed before the start of the baseline period).

Based on these definitions, the regression model is considered static if only real-time data are used for the independent parameters $(S = 0)$, while a time shift $S > 0$ refers to a dynamic model that incorporates historical data up to S time shifts.

Once the regression model g corresponding to the chosen time shift S is formed, at each newly sampled vector of modal frequencies $\mathbf{f} \in \mathbb{R}^{1 \times M}$ and temperature acquisitions $\tau \in \mathbb{R}^{1 \times L}$, the vector of predicted frequencies $\hat{\mathbf{f}}$ is obtained from the set of temperatures $\boldsymbol{\tau}_S \in \mathbb{R}^{1 \times L(S+1)}$ as

$$\hat{\mathbf{f}} = g(\boldsymbol{\tau}_S) \tag{7.4}$$

Finally, the vector of temperature-compensated frequencies $\mathbf{f}^c \in \mathbb{R}^{1 \times M}$ is computed as

$$\mathbf{f}^c = \mathbf{f} - \hat{\mathbf{f}} + \boldsymbol{\mu}_{\mathbf{B}} \tag{7.5}$$

where μ_B is the column-wise mean vector of the baseline frequencies given in Eq. (7.2). If temperature compensation (Eqs. (7.4) and (7.5)) is applied directly to the baseline, the result is the compensated baseline $\mathbf{B}^c$.

7.3 DAMAGE INDEX AND THRESHOLD

A damage index (DI) is calculated for any sample $\mathbf{f}_n^c$ by computing its Mahalanobis distance with respect to the baseline $\mathbf{B}^c$ as follows:

$$\mathrm{DI}_n = \sqrt{\left(\mathbf{f}_n^c - \boldsymbol{\mu}_B^c\right)\mathbf{C}^{-1}\left(\mathbf{f}_n^c - \boldsymbol{\mu}_B^c\right)^T} \tag{7.6}$$

where $\boldsymbol{\mu}_B^c \in \mathbb{R}^{1\times M}$ and $\mathbf{C} \in \mathbb{R}^{M\times M}$ are the column-wise mean vector and the covariance matrix of $\mathbf{B}^c$, respectively.

Figure 7.1 shows an example of DIs that could be obtained from an hourly sampling of structural monitoring data collected over three years, with the first year serving as the baseline and the subsequent two years representing the actual monitoring period. The structure remains healthy until the end of the second year, while some damage is introduced at the beginning of the third year.

An alarm threshold γ needs to be set on the DIs such that DIs higher than γ indicate positive (damaged) samples, while the others indicate negative (healthy) ones. Under this condition, the PFA can be computed on a healthy dataset as follows:

$$\mathrm{PFA} = \frac{\mathrm{FP}}{\mathrm{FP}+\mathrm{TN}} \tag{7.7}$$

where FP and TN are the numbers of false positives and true negatives, respectively. In this work, the alarm threshold γ is determined as the DI corresponding to a certain PFA on the baseline dataset, namely PFA = 0.05. This threshold corresponds to the horizontal dashed line in Figure 7.1.

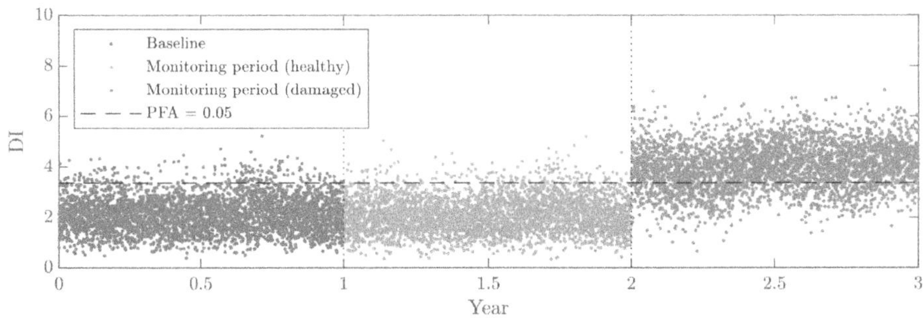

FIGURE 7.1 Damage index (DI) over three years of monitoring (baseline, healthy, and damaged). Damage detection threshold γ (dashed line) defined at PFA = 5% of the baseline DI.

Analogously, after damage occurs, the POD can be defined as follows:

$$\mathrm{POD} = \frac{\mathrm{TP}}{\mathrm{TP} + \mathrm{FN}} \tag{7.8}$$

where TP and FN are the numbers of true positives and false negatives, respectively. Therefore, for a given structural monitoring scenario including both an undamaged and a damaged period, the performance offered by different temperature compensation techniques can be evaluated in terms of PFA and POD, where the goal is to minimize the former and to maximize the latter.

7.4 NUMERICAL CASE STUDY: A TRUSS BRIDGE

An investigation is carried out on a dataset of temperature-dependent pseudo-frequencies obtained from a finite element model (FEM) of a steel truss bridge. In particular, three years of data sampled at hourly intervals are simulated, considering both healthy and damaged scenarios. The computed pseudo-data are exploited to build the regression models, which are used for temperature compensation to show their performance in both the static and dynamic approaches.

7.4.1 Pseudo-data generation

A FEM of the 3D truss structure depicted in Figure 7.2, composed of steel elements with a reference Young's modulus of elasticity of $E_0 = 200$ GPa and various cross-sectional areas, is set up in a MATLAB (The MathWorks Inc., 2022) environment. The green lines at certain nodes in the figure indicate translational restraints along the specified directions. The first six modal frequencies (i.e., $M = 6$) corresponding to a model where the Young's modulus of all elements is set to E_0, are calculated as $f_1 = 7.15$, $f_2 = 8.02$, $f_3 = 8.41$, $f_4 = 11.80$, and $f_5 = 14.68$ and $f_6 = 15.95$ Hz.

Three temperature histories T_1, T_2, and T_3 are hypothesized (i.e., $L = 3$), each of which is obtained as the sum of seasonal $\left(T_l^Y\right)$, daily $\left(T_l^D\right)$, and random $\left(T_l^R\right)$ variations during the three-year period (Figure 7.3a), as follows:

$$T_l = T_l^Y + T_l^D + T_l^R \tag{7.9}$$

where $l = 1, 2$ or 3. Given the assumed hourly sampling rate, 24 temperature readings per day are obtained, resulting in 8,760 samples per year and hence 26,280 in total for each of the three simulated thermocouples. The equations used to produce the three temperature histories are presented in Table 7.1, in which $n = 1$ to 26,280 is the sample number, $N_Y = 8760$ is the number of samples per year, and $N_D = 24$ is the number of samples per day.

T_1, T_2, and T_3 temperature histories are assigned to the red (R), purple (P), and yellow (Y) elements of Figure 7.2, respectively.

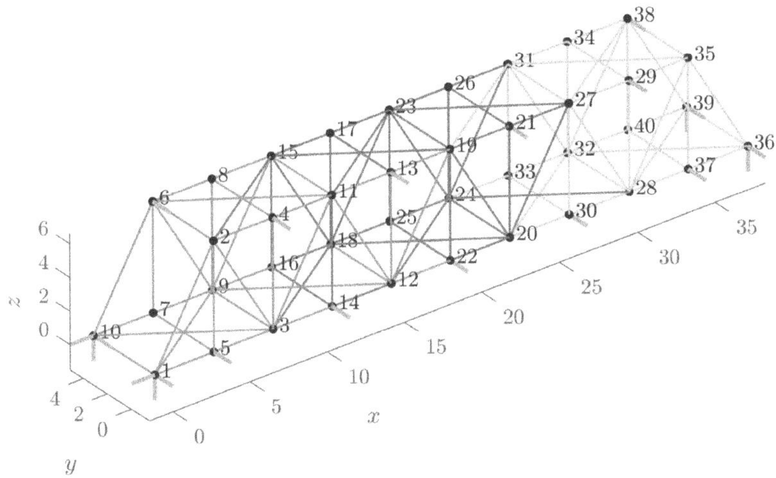

FIGURE 7.2 3D truss structure considered in the numerical example. All measurements are given in meters. The green segments denote constrained displacements.

The effects of varying temperature on the truss structure are incorporated by assigning the following nonlinear temperature-dependent Young's modulus (Seif et al., 2016) to the R, P, and Y elements:

$$E(n) = E_0\left(1 - 0.005 \times \left[\tilde{T}_l(n) - 20\right] e^{-0.01\tilde{T}_l(n)}\right) \tag{7.10}$$

where n is the sample number and $\tilde{T}_l(n)$ is termed the *effective* temperature of the element, which is influenced by both current and past temperature data as follows:

$$\tilde{T}_l(n) = \frac{\sum_{i=0}^{S}\left(1 - \frac{i}{S}\right) \cdot T_l(n-i)}{\sum_{i=0}^{S}\left(1 - \frac{i}{S}\right)} \tag{7.11}$$

Here, $T_l(n-i)$ represents the reading of the l-th thermocouple at sample $n-i$, while S is set to 23 in the present work. The resulting three effective temperature histories are shown in Figure 7.3b.

For the healthy structure, the first six modal frequencies are computed over the first 17,520 temperature samples (two-year duration). The baseline frequency matrix **B** is then constructed using the first year of data (8,760 samples), while the 8,760 samples corresponding to the second year of monitoring are considered the healthy test dataset.

Then, progressive damage is applied over the course of the third year to the elements at elevation $z = 0$ (the bottom truss elements) as a reduction of their cross-sectional area A according to the graph shown in Figure 7.4 in order to mimic the effects of increasing corrosion. Therefore, the 8,760 samples corresponding to the

TABLE 7.1
The equations for seasonal, daily, and random temperature variations used to generate the three thermocouple databases.

	T_1	T_2	T_3
T^Y	$T_1^Y(n) = 20 + 25\sin\left(\frac{2\pi n}{N_Y}\right)$	$T_2^Y(n) = 15 + 20\sin\left(\frac{2\pi n}{N_Y}\right)$	$T_3^Y(n) = 17 + 25\sin\left(\frac{2\pi n}{N_Y} + \frac{2}{3}\right)$
T^D	$T_1^D(n) = 20\sin\left(\frac{2\pi n}{N_D}\right)$	$T_2^D(n) = 16\sin\left(\frac{2\pi n}{N_D}\right)$	$T_3^D(n) = 20\sin\left(\frac{2\pi n}{N_D} + \frac{2\pi}{3}\right)$
T^R	$T_1^R(n) = 5r^*$	$T_2^R(n) = 2r$	$T_3^R(n) = 8r$

$^*r \sim \mathcal{N}(0, 1)$

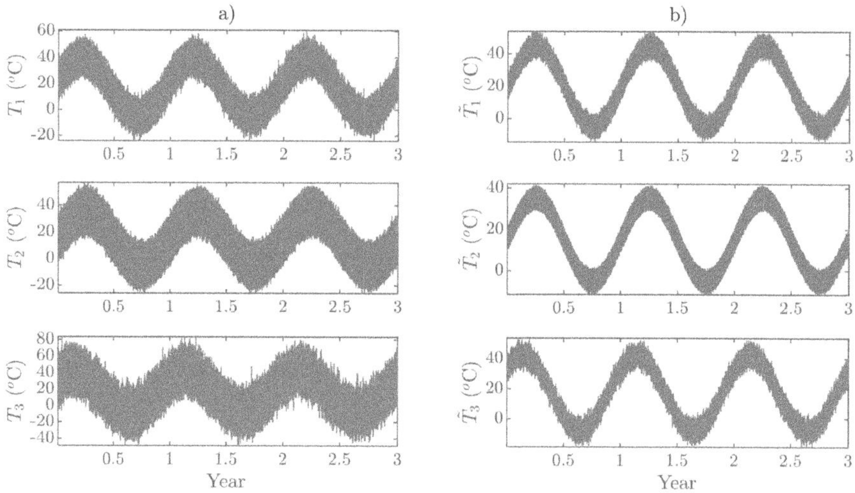

FIGURE 7.3 (a) Temperature variations of the three simulated thermocouples. (b) Effective temperature values used to set the Young's moduli of elasticity of the various elements at each temporal sample.

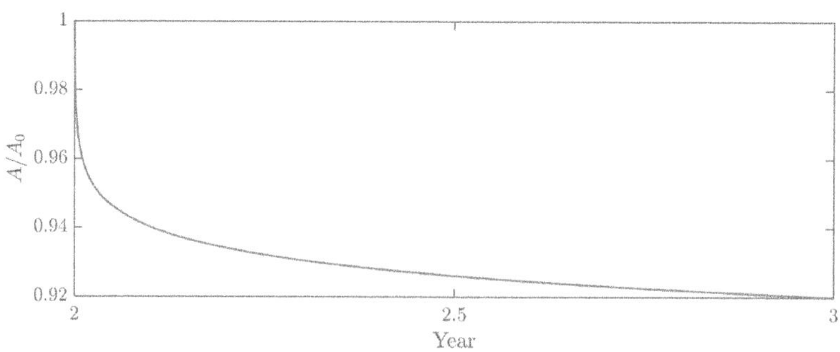

FIGURE 7.4 Damage propagation in the third year is modeled as a reduction of the cross-sectional area A from the initial healthy value A_0 of all truss elements at $z = 0$.

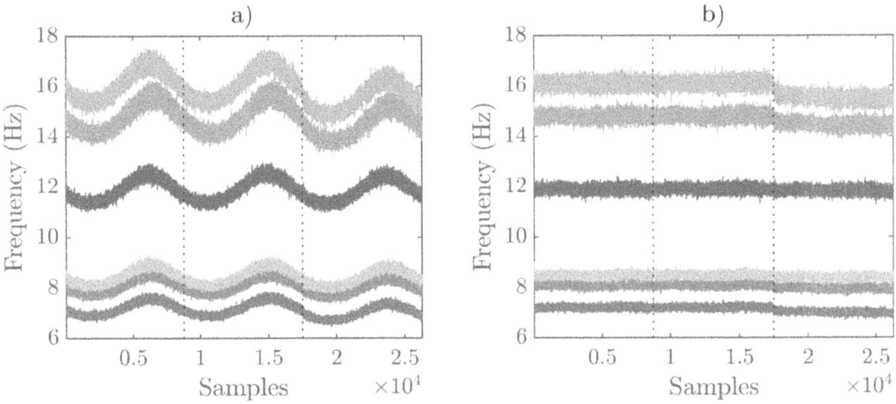

FIGURE 7.5 Modal frequencies: (a) before compensation; (b) after compensation with the NN regression model at $S = 12$.

third year form the damaged test dataset. Finally, the first six modal frequencies (healthy and damaged) are contaminated with Gaussian noise at a 40 dB SNR to simulate typical noise levels of actual SHM systems (Jahangiri et al., 2023) and are shown in Figure 7.5a.

7.4.2 REGRESSION MODELS

The monitoring data acquired in the first year are used as the baseline, i.e., the corresponding temperatures and modal frequencies are used to train two regression algorithms, namely MVLR and NN. The former is a linear regression method that solves a minimization problem to find the matrix of coefficients β such that $\mathbf{B} - \mathcal{T}_S\beta \simeq 0$. On the other hand, NN is a machine learning regression method based on the desired arrangement of input, hidden, and output layers, each of which can be composed of multiple neurons. All neurons in the input layer represent independent variables (in this work, the input layer is a vectorized version of matrix $\mathcal{T}_S$), while each neuron in the following layers is obtained as the weighted sum of the neurons in the previous layer passed through an activation function. The output layer also represents the target variables (in this work, a vectorized version of matrix $\mathbf{B}$). During the *training* phase, the weights are optimized in an iterative manner through procedures denoted as forward and backward propagation in order to minimize a chosen loss function. In this work, two hidden layers are used, and Table 7.2 lists all other relevant hyperparameters.

Both MVLR and NN regression models are trained on the baseline dataset; hence β and a set of optimized weights are obtained for the case of MVLR and NN, respectively, at which point the monitoring data can be fed for prediction.

The trained regression models can thus be used to compensate the modal frequencies for temperature-related variations. For example, Figure 7.5b shows the modal frequencies compensated using the NN regression model at $S = 12$ (note that this plot anticipates the discussion of some results presented in the next section). The

TABLE 7.2
Neural network regression hyperparameters.

Hyperparameter	Value
Number of hidden layers	2
Hidden layers' size	[30, 10]
Training optimization solver	LBFGS*
Iteration limit	10^3
Standardization	On
Activation function	ReLU
Loss function	Mean squared error

* Limited-memory Broyden–Fletcher–Goldfarb–Shanno (Nocedal and Wright, 2006).

comparison with the non-compensated frequencies in Figure 7.5a clearly shows that the temperature compensation procedure greatly reduced the variations in the modal frequencies.

It follows that the DI built on the compensated frequencies is expected to show less variability due to temperature, thus yielding an increased POD for a given PFA.

7.4.3 Results

The DIs computed according to Eq. 7.6 using the non-compensated frequencies in Figure 7.5a are shown in Figure 7.6. The threshold that corresponds to a PFA = 0.05 on the baseline data (first year) is $\gamma = 3.493$. By setting this threshold across the three-year monitoring period, a similar PFA of 0.0454 is obtained on the healthy test dataset of the second year, and a POD of 0.4208 is achieved on the third-year damaged dataset.

Temperature compensation was performed using both MVLR and NN for S increasing from $S = 0$ (static compensation) up to $S = 23$ (dynamic compensation). Some results are displayed in Figure 7.7, where the first and second columns of plots pertain to the use of MVLR and NN, respectively. In particular, Figure 7.7(a,d) shows the DI obtained by static compensation $(S = 0)$, while Figures 7.7(b,e) and (c,f) illustrate the DIs obtained by dynamic compensation for $S = 1$ and 12, respectively. Figure 7.7(a,d) clearly shows that the application of static compensation only yielded very marginal improvements over the case of Figure 7.6, with the POD increasing to 0.4584 and 0.4658 when using MVLR and NN, respectively. On the other hand, the addition of a single past temperature sample $(S = 1)$ was already sufficient to register a sensible increment in the PODs to 0.5864 and 0.7045, respectively. Finally, Figure 7.7(c,f) shows that by using 12 past temperature readings, a significantly improved damage detection performance was obtained, as registered by PODs of 0.8848 and 0.9619 given by MVLR and NN, respectively. As a final remark, the PFAs obtained across all plots essentially remained constant at about 5%.

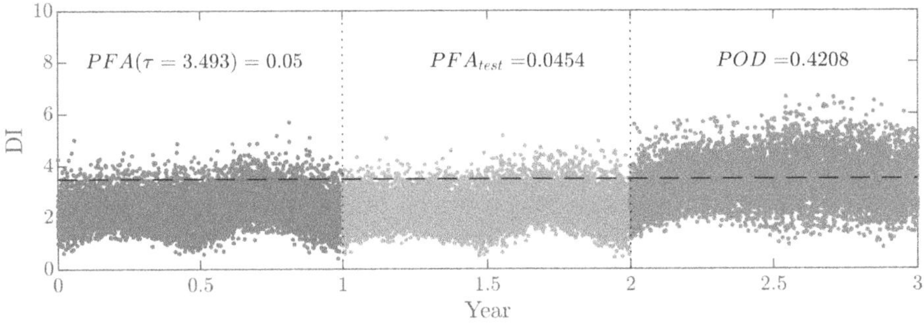

FIGURE 7.6 Damage indexes obtained on non-compensated frequencies.

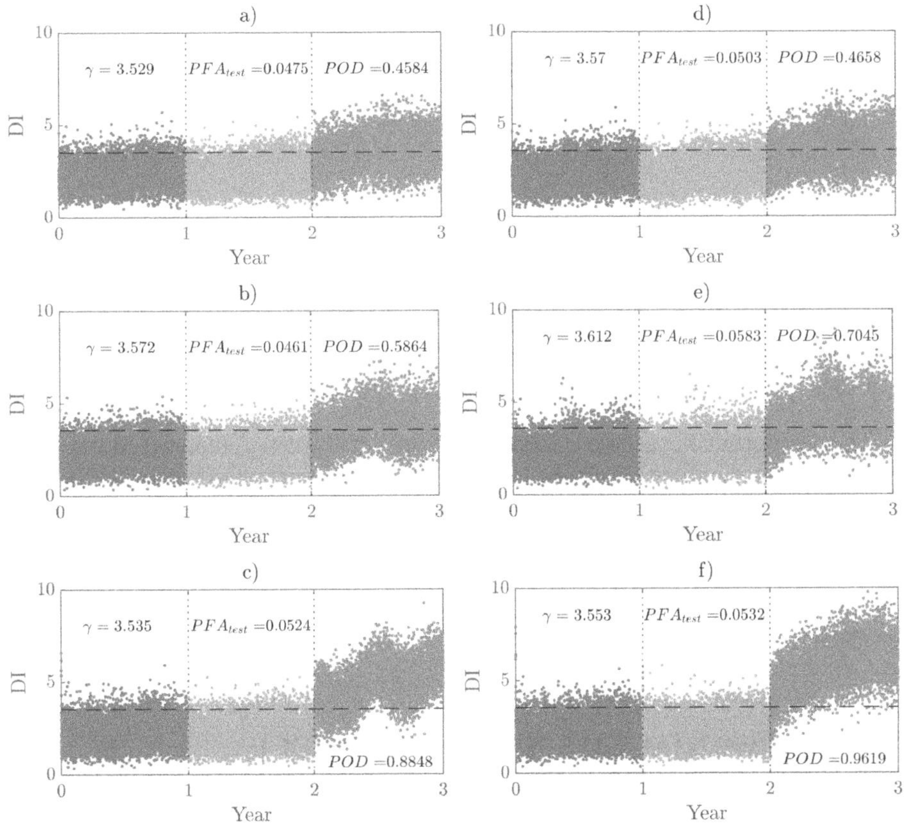

FIGURE 7.7 DIs resulting from temperature compensation using MVLR in the first column and NN in the second column: (a) and (d) $S = 0$; (b) and (e) $S = 1$; and (c) and (f) $S = 12$.

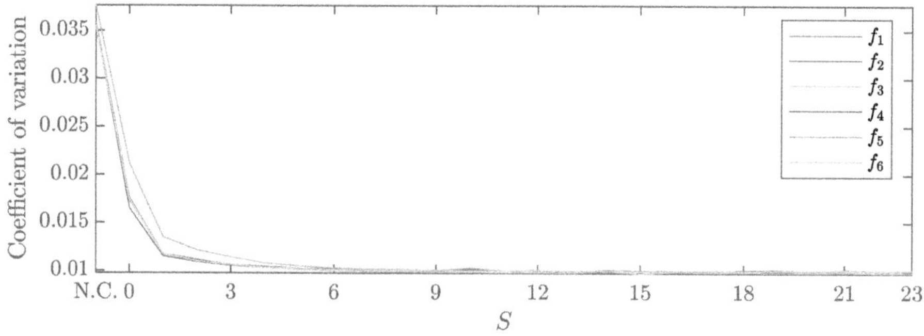

FIGURE 7.8 Coefficients of variation of the healthy modal frequencies for different compensation models using NN (N.C. = no compensation).

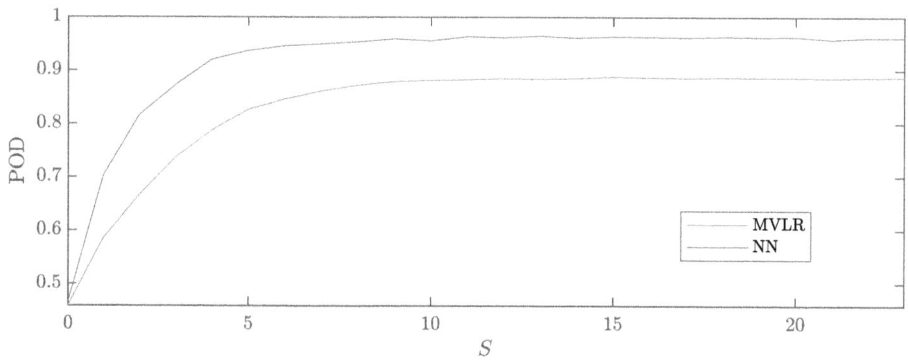

FIGURE 7.9 PODs with respect to different maximum time shifts used in the regression model for temperature compensation.

The coefficients of variation computed on the healthy non-compensated and compensated frequencies as S was varied from 0 to 23 are presented in Figure 7.8. The monotonic reduction clearly suggests an improved performance in terms of POD as S increases, at least until S gets to values higher than $\sim$6, where all curves start to plateau. This expectation is confirmed by the results shown in Figure 7.9, which plot the PODs yielded by the application of MVLR and NN as S varies between 0 and 23. As expected, the PODs of both MVLR and NN increase sharply at the lower values of S until plateauing at about $S = 10$, with NN generally outperforming MVLR.

7.5 EXPERIMENTAL CASE STUDY: KW51 BRIDGE

The open-access monitoring data from the KW51 Bridge in Leuven, Belgium (Maes and Lombaert, 2021), are utilized to compare the different temperature compensation models. The bridge was monitored in three phases: before retrofitting (October 2, 2018–May 15, 2019), during retrofitting (May 15–September 27, 2019), and after retrofitting (September 27, 2019–January 15, 2020). Ambient vibration data were recorded, and 14 modal frequencies were estimated using the reference-based covariance-driven stochastic subspace identification (SSI-cov/ref) method (Peeters and De Roeck, 1999) on an hourly basis. The modal frequency traces are plotted in Figure 7.10a. One temperature history $(L = 1)$ under the bridge deck was also recorded at the same rate.

As the data contain some missing values, only the 3rd, 5th, 6th, 9th, 10th, 11th, 12th, and 13th modal frequencies are used as damage-sensitive parameters $(M = 8)$ since they have fewer missing entries. All gaps in these frequencies are filled using the nearest non-missing values, while missing temperature measurements are addressed using a k-nearest neighbor approach with $k = 1$.

In this example, only the dataset before retrofitting is used for analysis. The dataset during and after retrofitting (red region in Figure 7.10a) is, in fact, excluded from the analysis, since it is characterized by changes in modal frequencies with respect to the pre-retrofitting phase, yielding a POD of unity even without temperature compensation, thus precluding an effective comparison between the discussed compensation models.

The pre-retrofitting dataset was divided into three subsets, namely (D1) training (70% of the data, also serving as the baseline), (D2) validation (15%), and (D3) testing (15%). The training set, as well as the considered temperature history, is illustrated in Figure 7.10b. To simulate an anomaly, a deviation of +0.004 of the mean modal frequencies was added to the testing dataset, creating the anomalous testing dataset (D4).

The training of MVLR models was performed on the baseline dataset (D1), while that of the NN models relied on both the baseline (D1) and validation (D2) datasets. Training began with the static model $(S = 0)$ and was then repeated by incrementally incorporating past temperature values, with maximum included lags $S = 150$.

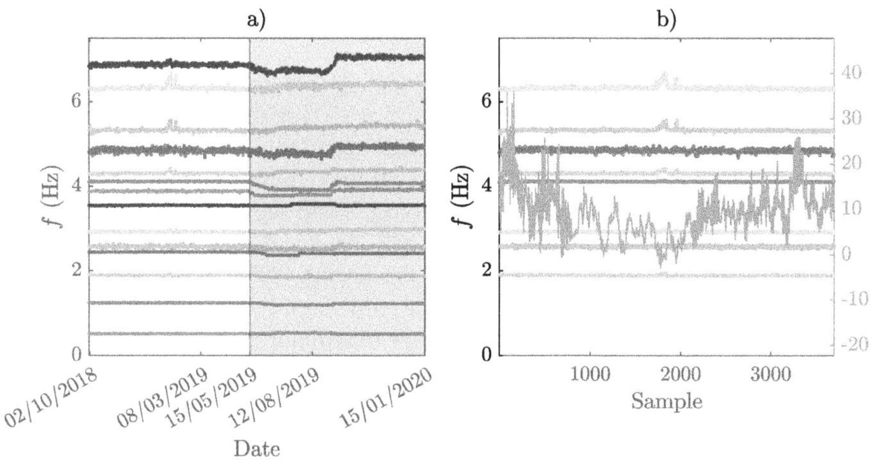

FIGURE 7.10 (a) Modal frequencies of the KW51 Bridge, with the red region indicating the retrofitting and post-retrofitting periods. (b) Training modal frequencies (left axis) and temperature (right axis) with imputed missing data from the pre-retrofitting dataset.

TABLE 7.3
Neural network hyperparameters for the KW51 bridge example.

Hyperparameter	Value	Hyperparameter	Value
Optimizer	Adam	Learning rate drop factor	0.1
Maximum epochs	500	Learning rate drop period	10
Initial learning rate	0.001	Validation frequency	5
Learning rate schedule	Piecewise	Validation patience	30

Before training or testing in each iteration, all independent and dependent datasets were standardized with respect to the training set. The training hyperparameters for NN are summarized in Table 7.3.

Each network begins with an input layer sized to match the number of variables in the temperature matrix ($L\times[S+1]=1$ up to 151). The architecture includes six fully connected hidden layers, gradually reducing the neuron count from 512 to 32, each followed by a ReLU activation function. The output layer is a fully connected layer sized to match the target output (eight modal frequencies).

Once the MVLR and NN models for various lags are trained, the training dataset (D1) is compensated for each model, and DIs with thresholds corresponding to PFA = 5% are determined. The models are then applied to compensate for the testing (D3) and anomalous testing (D4) datasets to evaluate PFA and POD, respectively.

For instance, Figure 7.11 depicts the DIs of the training (blue dots), testing (green dots), and anomalous testing (red dots) datasets for (a) non-compensated data and data compensated with $S=12$ using (b) MVLR and (c) NN. The non-compensated dataset achieves a POD of 0.3856. By incorporating 12 lags, the POD increases to 0.5913 with MVLR and 0.6991 with NN. Notably, the testing PFA remains close to 5%.

The performance of the MVLR (blue lines) and NN (red lines) models with respect to the different maximum time lags ($S=0$ to $S=150$) is shown in Figure 7.12. In this figure, the dashed lines and the right y-axis represent the testing PFAs of the (D3) dataset, which remain close to 5% across all maximum time lags. This consistency indicates the robustness of the trained models and suggests that overfitting to the training dataset has been avoided.

The left y-axis and solid lines illustrate the testing PODs of the (D4) anomalous dataset for different models and maximum time lags. For the NN model, $S=0$ achieves a POD of 0.608, whereas increasing the maximum lag to $S=150$ raises the POD to 0.936. In comparison, the MVLR model achieves PODs of 0.568 and 0.660 for the same maximum lags.

These results indicate that incorporating additional past temperature samples significantly boosts the POD, especially for the NN model. The smaller improvement for MVLR demonstrates its limitation in representing the nonlinear relationship between temperature and modal frequencies, which the NN model captures more effectively.

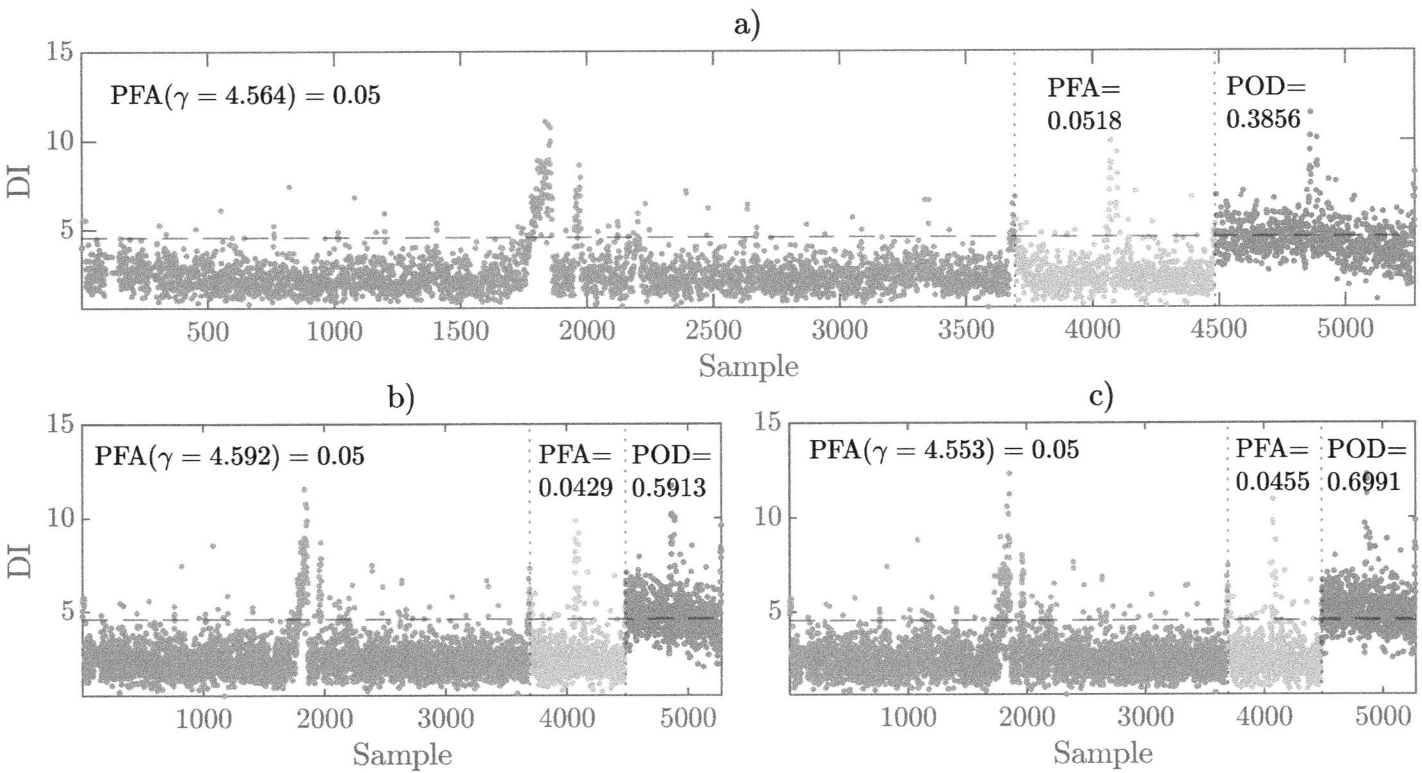

FIGURE 7.11 DIs for the training (grey dots), testing (light grey dots), and anomalous testing (dark grey dots) datasets: (a) non-compensated data, (b) data compensated using MVLR with $S = 12$, and (c) data compensated using NN with $S = 12$. The horizontal dashed lines represent the thresholds (γ).

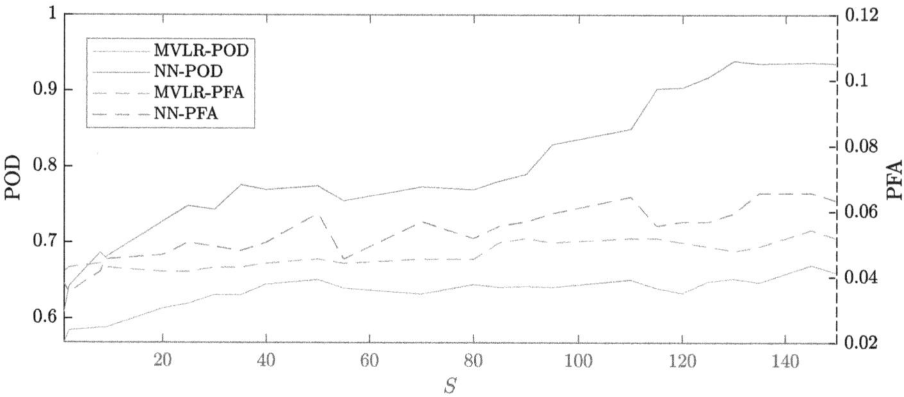

FIGURE 7.12 PODs and PFAs with respect to different maximum time shifts S used in the regression models for temperature compensation.

CONCLUSIONS

This chapter introduced the fundamentals of static and dynamic regression models and their application to temperature compensation in vibration-based data of structural health monitoring systems. The performance of these models was evaluated in compensating modal frequencies using both pseudo-data obtained from an FEM numerical model of a truss bridge and experimental data from the KW51 bridge. Performance evaluation was based on the probability of damage detection (POD) and the probability of false alarms (PFA) for different numbers of past temperature data. Multivariate linear regression (MVLR) and neural network (NN) models were employed for this purpose. For both regression models, the results highlighted the superiority of the dynamic approach, emphasizing the significant role of incorporating past temperature data in developing effective regression models for mitigating temperature-related variations in the structural damage-sensitive responses. Finally, the results demonstrate the superiority of neural networks in capturing the nonlinear relationship between modal frequencies and temperature, leading to more effective compensation and higher PODs compared to MVLR.

REFERENCES

Anastasopoulos, D., De Roeck, G., and Reynders, E. P. (2021). One-year operational modal analysis of a steel bridge from high-resolution macrostrain monitoring: Influence of temperature vs. retrofitting. *Mechanical Systems and Signal Processing*, 161:107951.

Brincker, R. and Ventura, C. (2015). *Introduction to operational modal analysis*. John Wiley & Sons.

Ente nazionale italiano di unificazione (2016). *UNI/TR 11634: 2016: *Linee guida per il monitoraggio strutturale*. UNI.

Falcetelli, F., Yue, N., Rossi, L., Bolognini, G., Bastianini, F., Zarouchas, D., and Di Sante, R. (2023). A model-assisted probability of detection framework for optical fiber sensors. *Sensors*, 23(10):4813.

Jahangiri, M., Palermo, A., Kamali, S., Hadianfard, M. A., and Marzani, A. (2023). A procedure to estimate the minimum observable damage in truss structures using vibration-based structural health monitoring systems. *Probabilistic Engineering Mechanics*, 73:103451.

Kamali, S. and Hadianfard, M. A. (2023). Spectral optimization-based modal identification: A novel operational modal analysis technique. *Mechanical Systems and Signal Processing*, 198:110445.

Kamali, S., Kalantari, A., Mariani, S., Mennuti, C., Augugliaro, G., and Marzani, A. (2024a). Comparing static and dynamic regression models for temperature compensation in vibration-based SHM systems. In *International Operational Modal Analysis Conference*, pages 56–67. Springer.

Kamali, S., Marzani, A., Sciullo, L., Di Felice, M., Augugliaro, G., and Mennuti, C. (2023). Temperature compensation in vibration-based structural health monitoring using neural network regression. In *2023 7th International Conference on System Reliability and Safety (ICSRS)*, Bologna, Italy. IEEE.

Kamali, S., Palermo, A., and Marzani, A. (2025). Virtual baseline to improve anomaly detection of SHM systems with non-stationary data. *Mechanical Systems and Signal Processing*, 224:111968.

Kamali, S., Quqa, S., Palermo, A., and Marzani, A. (2024b). Reducing false alarms in structural health monitoring systems by exploiting time information via binomial distribution classifier. *Mechanical Systems and Signal Processing*, 207:110938.

Maes, K. and Lombaert, G. (2021). Monitoring railway bridge kw51 before, during, and after retrofitting. *Journal of Bridge Engineering*, 26(3):04721001.

Magalhães, F., Cunha, Á., and Caetano, E. (2012). Vibration based structural health monitoring of an arch bridge: From automated OMA to damage detection. *Mechanical Systems and Signal Processing*, 28:212–228.

Mahalanobis, P. C. (1936). On the generalised distance in statistics. In *Proceedings of the National Institute of Science of India*, volume 12, pages 49–55.

Mariani, S., Kalantari, A., Kromanis, R., and Marzani, A. (2024). Data-driven modeling of long temperature time-series to capture the thermal behavior of bridges for SHM purposes. *Mechanical Systems and Signal Processing*, 206:110934.

Nocedal, J. and Wright, S. J. (2006). *Quasi-newton methods*, pages 135–163. Springer.

Peeters, B. and De Roeck, G. (1999). Reference-based stochastic subspace identification for output-only modal analysis. *Mechanical Systems and Signal Processing*, 13(6):855–878.

Rainieri, C. and Fabbrocino, G. (2014). *Operational modal analysis of civil engineering structures*, volume 142, page 143. Springer.

Reynders, E., Wursten, G., and De Roeck, G. (2014). Output-only structural health monitoring in changing environmental conditions by means of nonlinear system identification. *Structural Health Monitoring*, 13(1):82–93.

Seif, M., Choe, L., Gross, J., Luecke, W. E., Main, J. A., McColskey, D., Sadek, F., Weigand, J. M., and Zhang, C. (2016). *Temperature-dependent material modeling for structural steels: Formulation and application.* US Department of Commerce, National Institute of Standards and Technology.

The MathWorks Inc. (2022). *Matlab version: R2022a.* Natick, MA, US.

Wang, Z., Yang, D.-H., Yi, T.-H., Zhang, G.-H., and Han, J.-G. (2022). Eliminating environmental and operational effects on structural modal frequency: A comprehensive review. *Structural Control and Health Monitoring*, 29(11):e3073.

Yan, A.-M., Kerschen, G., De Boe, P., and Golinval, J.-C. (2005). Structural damage diagnosis under varying environmental conditions—part i: A linear analysis. *Mechanical Systems and Signal Processing*, 19(4):847–864.

8 Vibration-based structural health monitoring of a steel-concrete composite bridge

Carmelo Gentile and A. Avramova

8.1 INTRODUCTION

Structural health monitoring (SHM) [1] is generally defined as a multidisciplinary process aimed at assessing the health condition of a structural system based on the continuous measurement of response data through arrays of appropriate sensors. The process often involves the extraction of features from measured data, which are representative of the health condition, and the classification of those features [2] to detect any novelty or abnormal changes in the investigated system.

In recent decades, the SHM strategy based on vibration monitoring and operational modal analysis (OMA, i.e., the output-only identification of modal parameters) has received increasing attention, and many dynamic monitoring systems have been installed, especially on bridges [3]. Well-known examples of permanently monitored bridges in different countries include the Ting Kau Bridge in Hong Kong [4, 5], the Infante Don Henrique Bridge in Portugal [3, 6], the Øresund Bridge in Denmark [7], the Tamar Bridge in the UK [8], and the San Michele Bridge in Italy [9].

The rising scientific and practical interest in dynamic monitoring and vibration-based SHM of bridges has many motivations such as (a) the nondestructive and noninvasive application in operational conditions; (b) the ageing of existing bridges combined with the increase of service loads and climate changes; (c) the increasing complexity of new bridges (where the implementation of monitoring systems is definitely convenient because it is possible to amortize the costs within the construction process); (d) the technological (hardware and software) advances, allowing relatively inexpensive installation of monitoring devices, fully computer-based operation, quick transmission, and accurate processing of the collected data. However, most of the permanently instrumented bridges are concrete or steel infrastructures characterized by a complex structural arrangement such as cable-stayed, suspension, or arch bridges [3–9], whereas there is a lack of studies on multi-span composite bridges. On the other hand, steel-concrete composite bridges have become a very

 DOI: 10.1201/9781003516941-8

common solution in modern bridge engineering due to their optimal strength-to-weight ratio, quick erection time, and sustainability.

Within the systematic monitoring activity recently promoted by the Italian government, dynamic monitoring systems have been installed on various steel-concrete composite structures belonging to the Autostrada Pedemontana Lombarda (APL), an important motorway serving several provinces in the Lombardy region. This chapter presents the results of the continuous monitoring of the Olona West Viaduct, a steel-concrete composite bridge identified by APL for evaluation. The bridge features a four-span configuration, with two central spans of 66 meters each and two lateral spans of 55 meters, resulting in a total length of 242 meters. Two abutments and three concrete piers support the composite deck, which crosses the Olona River in the province of Varese (Northern Italy).

After a concise description of the structure and the installed monitoring system, the dynamic characteristics of the bridge—defined during preliminary ambient vibration tests [10] and at the beginning of monitoring—are presented and compared with the predictions obtained using a 3D finite element (FE) model of the structure. Subsequently, details are given on the software tools employed to process the continuously acquired data within a statistical pattern recognition framework, and the evolution over time of natural frequencies and mode shapes is presented and discussed, along with the effects exerted by environmental and operational variability (EOV) on the modal parameters. Finally, the application of two different novelty analysis methods is presented: the widely used control charts based on the Mahalanobis distance (see, e.g., Worden et al. [11]) and the less commonly applied cointegration technique [12], which does not require the removal of environmental and operational effects but instead relies on defining a linear combination of nonstationary monitored features.

8.2 DESCRIPTION OF THE INVESTIGATED BRIDGE

The Olona West Bridge (Figure 8.1) is a steel-concrete composite bridge that crosses the Olona River in the province of Varese (Lombardy region, Northern Italy). The

FIGURE 8.1 View of the Olona West Bridge.

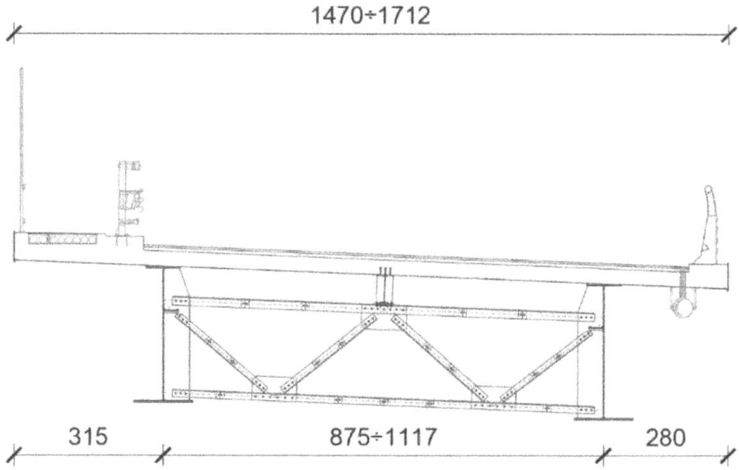

FIGURE 8.2 Typical cross section of the deck (dimensions in cm).

investigated infrastructure, with an overall length of 242 m, consists of four spans of different lengths (55 m + 66 m + 66 m + 55 m) and is characterized by a slightly curvilinear plan.

The bridge deck (Figure 8.2) includes (a) two main longitudinal steel girders, with each girder consisting of a welded I cross section of constant depth (2.80 m) and exhibiting varying girder spacing (between 8.75 m and 11.17 m) and (b) a reinforced concrete slab, 31 cm thick, which is composite with the girders through Nelson-type shear connectors. The roadway platform, whose width ranges from 14.70 m to 17.125 m, is characterized by the presence of a pedestrian curb of 2.70 m.

The steel-concrete composite deck is supported by two concrete abutments (S1 on the Solbiate side and S2 on the Mozzate side) and three piers (P1–P3). All the bearing devices are elastomeric seismic isolators characterized by (a) appropriate dissipative capacity, (b) low horizontal stiffness to guarantee the decoupling between the horizontal motion of the deck and piers/abutments, and (c) high vertical rigidity to support vertical loads.

8.3 DYNAMIC CHARACTERISTICS OF THE BRIDGE, DATA ANALYSIS, AND SHM STRATEGY

To properly design a dynamic monitoring system, prior knowledge of the modal parameters of the structure is generally required. Hence, the investigated bridge underwent preliminary ambient vibration tests (AVTs), which were performed on March 7, 2023 [10]. The response of the bridge was measured at 28 selected points (Figure 8.3), which were also permanently instrumented during the subsequent dynamic monitoring.

Time series of 2700 s were recorded in each setup at a sampling rate of 200 Hz, which is much higher than that required for the bridge under study, as the natural frequencies of the dominant modes are below 10 Hz (Figure 8.4). Hence, low-pass

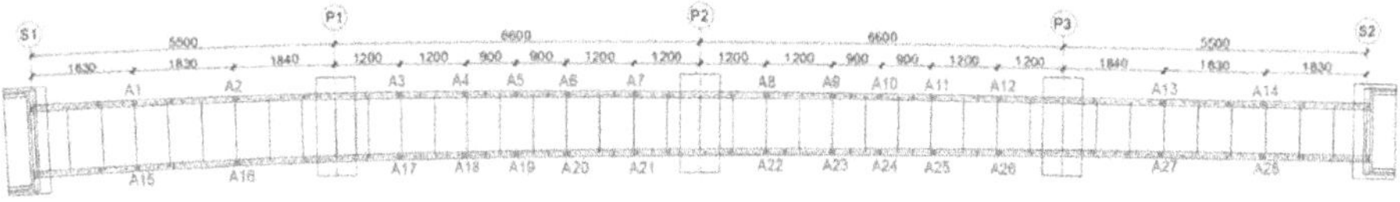

FIGURE 8.3 Sensor layout adopted in both preliminary dynamic tests and continuous dynamic monitoring (dimensions in cm).

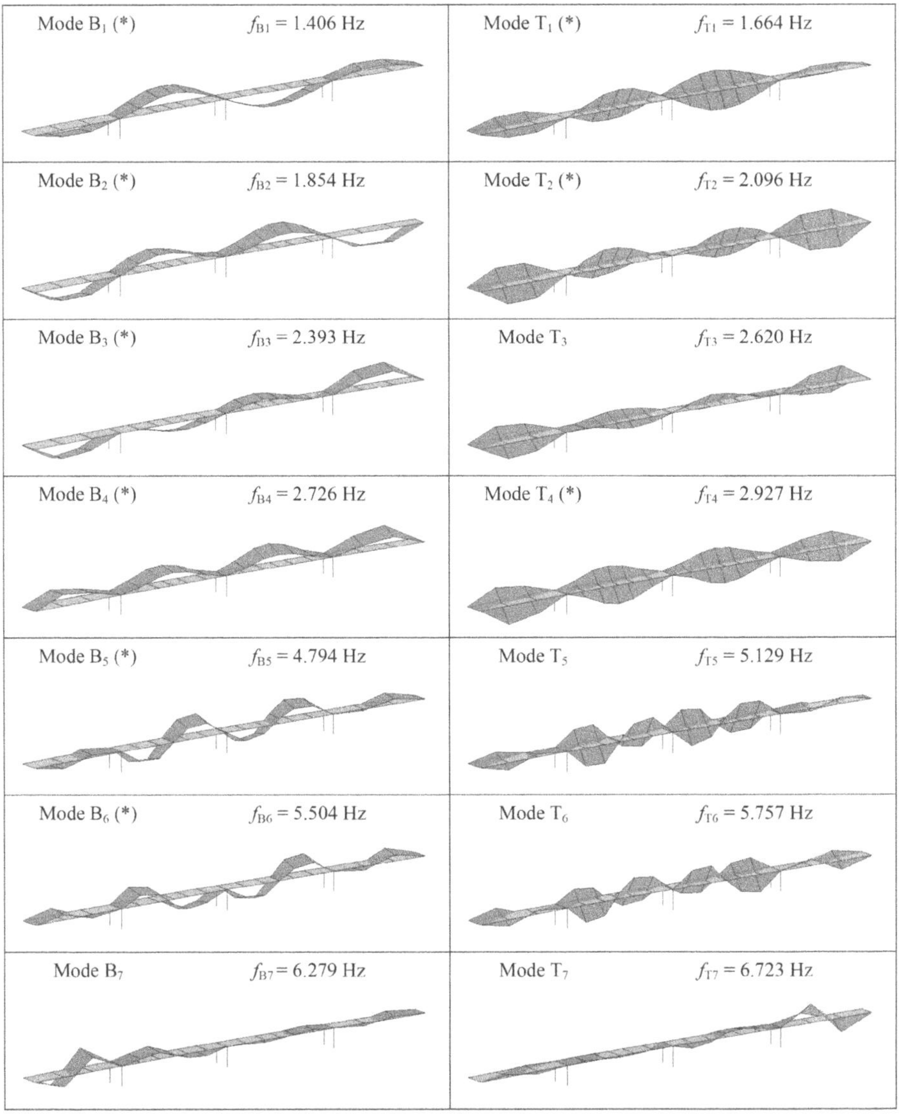

FIGURE 8.4 Reference modes identified at the beginning of the continuous monitoring (SSI-Cov, June 08, 2024, 16:00 h). The (*) modes were also identified in the preliminary tests.

filtering and decimation were applied to the data before the use of the identification tools [13], reducing the sampling rate from 200 Hz to 20 Hz. Nine vibration modes were identified in the preliminary dynamic tests: six bending modes (B_1–B_6 in Figure 8.4) and three torsion modes (T_1, T_2, and T_4 in Figure 8.4). On the other hand, the normal traffic conditions existing at the beginning of continuous monitoring allowed us to successfully identify 14 vibration modes (Figure 8.4) with a good identification rate within the frequency range 0–7 Hz. In more detail, seven vertical bending modes and seven torsion modes were identified. It should be noticed that the lower two series of four bending modes (B_1–B_4 in Figure 8.4) and four torsion modes (T_1–T_4 in Figure 8.3) correspond perfectly to theoretical expectations for a continuous beam with four spans. The mode shapes summarized in Figure 8.4 were obtained by applying the covariance-driven Stochastic Subspace Identification (SSI-Cov) algorithm [14] to the dataset collected on June 08, 2024, from 16:00 to 17:00 and were subsequently used as references in the tracking process.

The monitoring system of the Olona West Bridge has been active since May 20, 2024, and includes 28 MEMS accelerometers (Figure 8.3) and two temperature sensors. The acceleration data are recorded at a sampling rate of 200 Hz, and the length of the windows used for modal identification was set to 3,600 s (1 h). During the monitoring period investigated (i.e., from May 20, 2024 to May 18, 2025), more than 8,700 datasets were collected and processed.

The vibration-based SHM methodology applied to the bridge under study is based on the MATLAB software package called DYMOND (i.e., DYnamic MONitoring and Novelty Detection) [13] and involves the following steps: (a) preprocessing of the raw data (i.e., signal detrending, computation of peak and RMS values, low-pass filtering, and down-sampling); (b) automated modal parameter estimation (MPE) and modal tracking (MT); (c) checking the invariance of mode shapes and examining the dependence of natural frequencies on EOV; (d) removal of the EOV effects from the natural frequencies; (e) novelty detection.

The SSI-Cov algorithm [14] is applied to automatically perform the MPE. The implemented SSI-Cov procedure exploits a stabilization diagram to identify stable physical modes of the structure in terms of natural frequency, mode shape, and damping ratio. In each stabilization diagram (where the poles obtained with increasing model order are represented together), only the poles characterized by realistic values of modal damping ratios (e.g., $0\% < \zeta < 10\%$) and mean phase collinearity (MPC) [15] (i.e., MPC > 0.9) are retained; subsequently, hierarchical clustering is used to group the physical poles representing the same structural mode (based on the sensitivity of frequencies and mode shapes to the increase in model order).

It is further noticed that the automated identification procedure was manually calibrated by tuning the values of a few parameters during the first days of monitoring (time lag equal to 30, model order ranging from 20 to 120).

After each modal estimation, the identified modes that are consistent with the reference modes (Figure 8.4) are tracked over time. In particular, the dynamic properties monitored for SHM purposes are natural frequencies, the modal assurance criterion (MAC) [16]—providing a measure of the correlation between the current mode shapes and the reference ones—and the MPC [15], quantifying the complexity of each mode shape.

After the correlation between EOV and natural frequencies is highlighted, the principal component analysis (PCA) [17] is applied to remove the variance of natural frequencies that is associated with EOV. In more detail, given a matrix $\mathbf{Y} \in \Re^{N\times M}$ collecting M natural frequencies identified for N observations, a PCA regression model is built with a sub-matrix $\mathbf{Y}_{tr} \in \Re^{T\times M}$ (with $T < N$) of features collected during a training period in which the structure is supposed to be in normal condition and under typical EOVs.

Once trained, the PCA model is used to predict the monitored features $\mathbf{Y}_{PCA} \in \Re^{N\times M}$, and the residuals $\mathbf{E}$ obtained as $\mathbf{E} = \mathbf{Y} - \mathbf{Y}_{PCA}$ should contain only the effects of possible structural variations. Hence, reference is made to the following novelty index (NI) [11], based on the Mahalanobis distance:

$$\mathrm{NI} = \sqrt{\boldsymbol{E}^{\mathrm{T}} \boldsymbol{R}^{-1} \boldsymbol{E}} \tag{8.1}$$

$\mathbf{R}^{-1} = (1/N)\mathbf{Y}\mathbf{Y}^{T}$ is the covariance matrix of the identified natural frequencies. The control region of the NI is delimited by an upper control limit (UCL), equal to the m-th percentile of the NI in the training period. When a persistent number of data points lie outside the control zone because they exceed the UCL, a different structural condition concerning the training period can be detected.

As an alternative to the control chart based on PCA and Mahalanobis distance, the cointegration procedure [12] can be adopted. Given M nonstationary time series (vectors) $\mathbf{Y}_j \in \Re^{N}$ (e.g., N observations of M natural frequencies), those vectors are said to be cointegrated if a stationary linear combination ε_k of $\mathbf{Y}_j$ exists:

$$\varepsilon_k = \sum_{j=1}^{M} \beta_j Y_{jk} \tag{8.2}$$

The unknown coefficients β_j ($j = 1, \ldots, M$) in Eq. (8.2) can be found using the Johansen procedure [18] during a training period in which the structure is supposed to be in normal condition under typical EOV. Once the unknown coefficients are determined, the resulting linear combination, denoted as ε_κ, forms a new time series referred to as the cointegration residual. This residual is essentially free from the common trends affecting the natural frequencies, which are primarily associated with EOV, and an anomaly is considered to occur when the cointegration residual departs from the stationary behavior established during the training period.

8.4 FINITE ELEMENT MODELING OF THE BRIDGE

A FE model of the Olona West Bridge (Figure 8.5) was developed using the MIDAS Civil 2024 (v2.1) software [19]. The numerical model includes the four-span steel–concrete composite deck and the concrete intermediate piers (Figure 8.5); as usual, the model was developed based on the “as built” drawings of the bridge.

As shown in Figure 8.5, the steel–concrete composite deck modeling includes (a) the two longitudinal main girders and the central spine beam, (b) the steel bracing

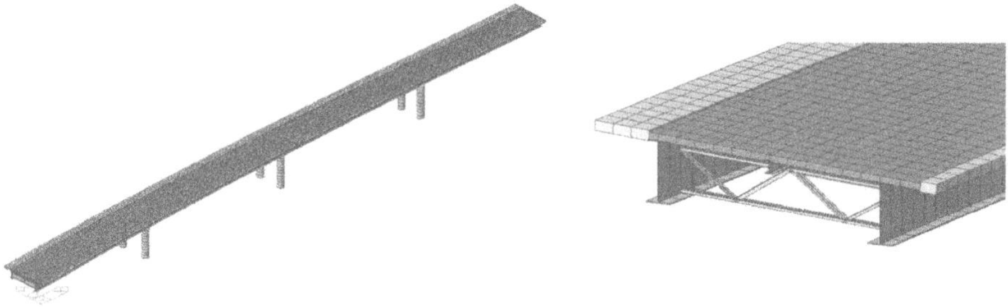

FIGURE 8.5 FE model: general view and detail of deck modeling.

system, and (c) the reinforced concrete deck. The steel members were modeled using beam elements, whereas four-node shell elements were used to model the concrete deck. It is worth mentioning that (a) a 1×1 m mesh was adopted in modeling the concrete slab so that an appropriate mass distribution is attained, as well as a good resolution for stress evaluation; (b) nonstructural elements of the deck, including the 110 mm asphalt layer, were included as distributed mass.

The elastic properties of the aforementioned structural members were assigned according to the S355 steel grade and the C32/40 concrete grade, as provided by the relevant Eurocodes. Consequently, the following values of Young's modulus and Poisson's ratio were assumed for steel and concrete: (a) steel, $E_s = 210.00$ GPa and $\nu_s = 0.30$; (b) concrete, $E_{c1} = 33.35$ GPa and $\nu_{c1} = 0.20$.

As shown in Figure 8.5, the model includes three circular reinforced concrete piers (C25/30, $E_{c2} = 31.48$ GPa, and $\nu_{c2} = 0.20$) with constant 2.5 m diameters and varying heights (10.70 m for P1, 11.55 m for P2, and 12.35 m for P3). Each pier was modeled as a vertical beam element, ideally fixed at the ground level. The top node of each pier is connected to the upper structural system through links that simulate the seismic isolation bearings. The isolation system was accurately represented using two similar modeling approaches: (a) at the abutments, where the substructure was not explicitly modeled, point spring supports with a horizontal stiffness of 3.17 kN/mm in both orthogonal directions (representing FIP SI-H 550/105 isolators), and infinite vertical stiffness were used; (b) at the piers, elastic link elements with 7.04 kN/mm horizontal stiffness (modeling FIP SI-H 800/100 isolators), and, again, infinite vertical stiffness were adopted to model the connection with the deck. Furthermore, the orientation of the links was adjusted to follow the deck's curvature.

As shown in Figures 8.6 and 8.7 and in Table 8.1, there is a very good correspondence between the modal parameters of the FE model and the reference experimental values (Figure 8.4). In more detail, it can be observed that (a) there is a one-to-one correspondence between the identified and numerical modes; (b) the average frequency error (Table 8.1 and Figure 8.7) does not exceed 2%, with the higher discrepancy being equal to 5.89% (mode T_1); (c) the mode shape agreement turns out to be fully satisfactory, with only three modes exhibiting MAC lower than 0.90.

$$\text{Error} = 100 \times (f_{\text{FEM}} - f_{\text{OMA}})/f_{\text{OMA}}.$$

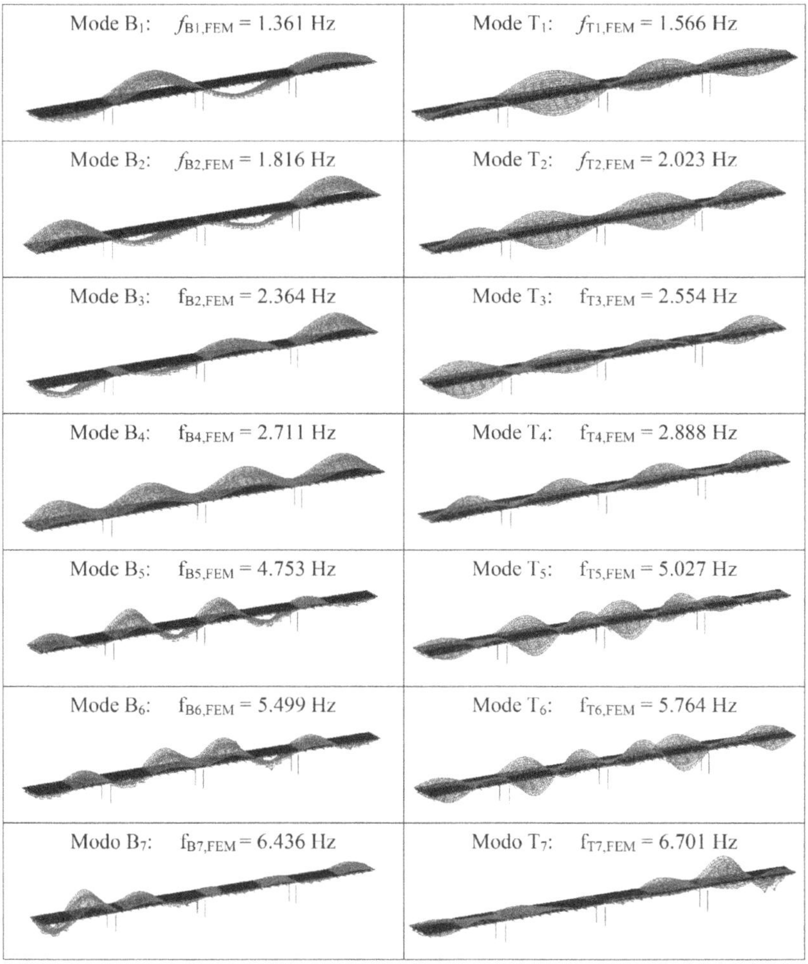

FIGURE 8.6 Selected vibration modes of the FE model.

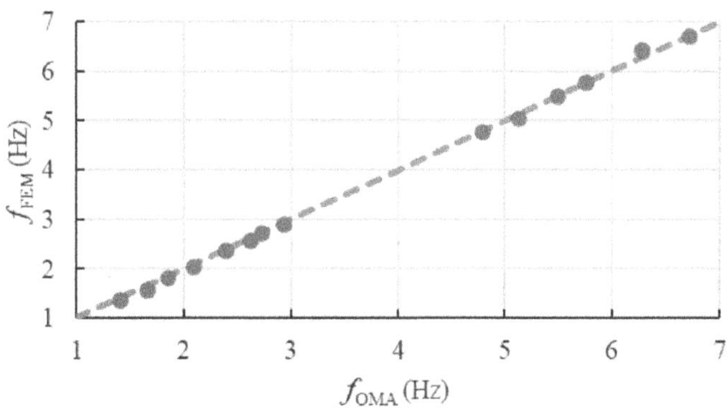

FIGURE 8.7 Correlation between experimental (f_{OMA}) and numerical (f_{FEM}) natural frequencies.

TABLE 8.1
Olona West Bridge: comparison between the modal parameters obtained from the FE model and the experimental ones.

Mode	f_{OMA} (Hz)	f_{FEM} (Hz)	Error (%)	MAC
B_1	1.406	1.361	−3.20	0.973
T_1	1.664	1.566	−5.89	0.998
B_2	1.854	1.816	−2.05	0.957
T_2	2.096	2.023	−3.48	0.998
B_3	2.393	2.364	−1.21	0.937
T_3	2.620	2.554	−2.52	0.987
B_4	2.726	2.711	−0.55	0.898
T_4	2.927	2.888	−1.33	0.990
B_5	4.794	4.753	−0.86	0.914
T_5	5.129	5.027	−1.99	0.969
B_6	5.504	5.499	−0.09	0.823
T_6	5.757	5.764	0.12	0.978
B_7	6.279	6.436	2.50	0.925
T_7	6.723	6.701	−0.04	0.886

Note: Reference modes identified on May 20, 2024, Figure 8.4.

8.5 CONTINUOUS DYNAMIC MONITORING AND RESULTS

The time evolution of natural frequencies during the investigated monitoring period is illustrated in Figure 8.8, whereas the corresponding changes in air temperature (approximately ranging between −3°C and +37°C) are shown in Figure 8.9. The inspection of Figures 8.8 and 8.9 clearly reveals that the automatically identified natural frequencies exhibit clear daily and seasonal trends, both of which can mostly be attributed to temperature changes.

The statistics of the natural frequencies and mode shapes identified during the first year of monitoring are summarized in Table 8.2 in terms of the average value (f_{avg}), standard deviation (σ_f), extreme values (f_{min}, f_{max}) of each frequency, and the coefficient of determination between frequency and temperature ($R^2_{f\text{-}T}$). Furthermore, Table 8.1 also presents the average, standard deviation, and minimum value of MAC.

The values in Table 8.2 show that the natural frequencies exhibit relatively low variability, with standard deviations being generally below 0.06 Hz and fluctuations being limited to a few percentage points of the corresponding mean values. To better understand the dependence of frequencies on temperature, the correlation coefficient R^2 was computed. The results suggest that frequency variations are mainly driven by temperature variations. In addition, the bending and torsion modes turn out to exhibit different sensitivity to thermal effects, with the frequency of the torsion modes being characterized by a stronger correlation with

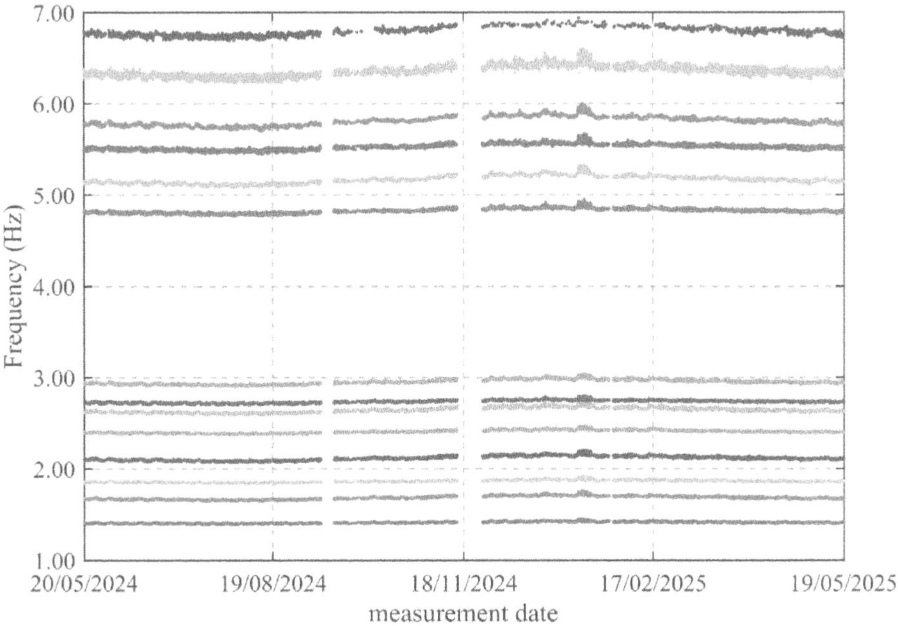

FIGURE 8.8 Time evolution of the automatically identified (SSI-Cov) natural frequencies during the first year of monitoring (from May 20, 2024 to May 18, 2025).

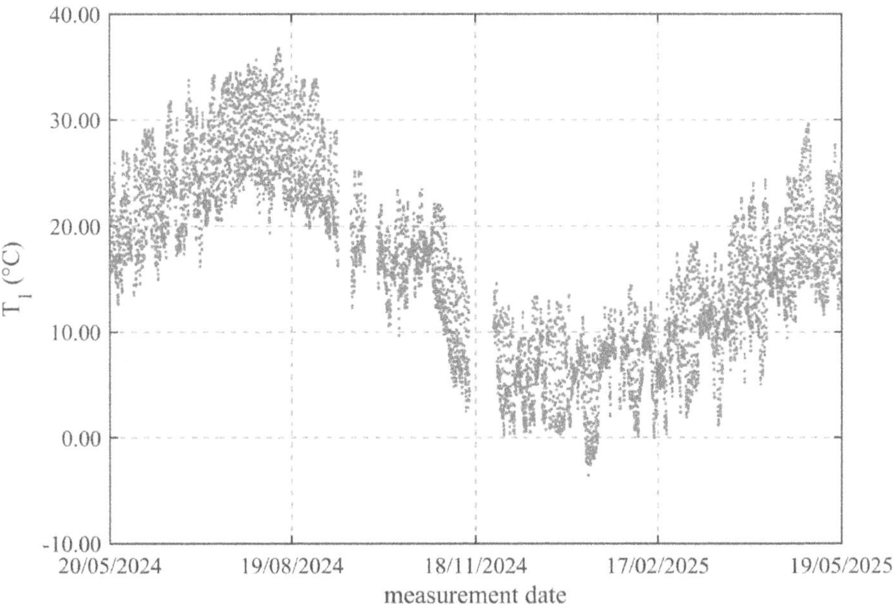

FIGURE 8.9 Time evolution of the measured air temperature during the first year of monitoring (from May 20, 2024 to May 18, 2025).

TABLE 8.2
Olona West Bridge: statistics of the identified modal parameters during the first year of monitoring.

Mode	f_{avg} (Hz)	σ_f (Hz)	f_{min} (Hz)	f_{max} (Hz)	R^2_{f-T}	MAC_{avg}	σ_{MAC}	MAC_{min}
B_1	1.412	0.010	1.390	1.451	0.663	0.999	0.0002	0.990
T_1	1.683	0.019	1.642	1.759	0.906	0.991	0.0123	0.902
B_2	1.863	0.013	1.837	1.916	0.772	0.999	0.0004	0.989
T_2	2.118	0.023	2.073	2.204	0.903	0.994	0.0083	0.902
B_3	2.409	0.017	2.376	2.473	0.878	0.999	0.0005	0.986
T_3	2.643	0.027	2.593	2.740	0.893	0.982	0.0177	0.901
B_4	2.737	0.016	2.701	2.801	0.812	0.998	0.0032	0.928
T_4	2.952	0.025	2.896	3.040	0.888	0.996	0.0076	0.902
B_5	4.825	0.025	4.767	4.948	0.874	0.999	0.0008	0.9800
T_5	5.169	0.041	5.082	5.325	0.873	0.983	0.0150	0.903
B_6	5.528	0.031	5.451	5.671	0.862	0.996	0.0031	0.915
T_6	5.817	0.046	5.701	5.999	0.860	0.976	0.0180	0.900
B_7	6.355	0.057	6.230	6.604	0.850	0.993	0.0062	0.923
T_7	6.779	0.043	6.694	6.941	0.849	0.955	0.0245	0.900

Note: From May 20, 2024 to May 18, 2025.

temperature ($0.849 \le R^2 \le 0.906$) compared to the bending modes ($0.663 \le R^2 \le 0.878$). It is further noticed that among all the identified frequencies, only the R^2 related to the first vertical bending frequency f_{B1} is characterized by a value that is lower than 0.70 (Figure 8.10).

Since a well-distributed grid of accelerometers is available on the bridge, the investigation of mode shape variations might be significant because mode shapes provide both local and global information on the structure. As a matter of fact, even for complex systems, the mode shapes are supposed to be less sensitive than the natural frequencies to environmental changes (which affect almost uniformly the structure) and conceivably more sensitive to local structural changes. In the present case, the mode shape invariance was evaluated using both the MAC and the MPC. Both quantities should ideally remain close to unity (meaning that no increase in mode complexity occurs, with the reference shapes being maintained) without exhibiting any abrupt or irreversible drop.

As shown in Table 8.2 and Figure 8.11, the bending modes exhibit average MAC values very close to unity during the entire observation period, whereas the average MAC is slightly larger for the torsion modes (which are also characterized by higher dispersion than that observed for the bending modes).

As previously stated, the minimization/removal of EOV effects on natural frequencies was first performed by applying the classic strategy based on PCA and specifically involving the following steps: (a) in defining the PCA-based regression,

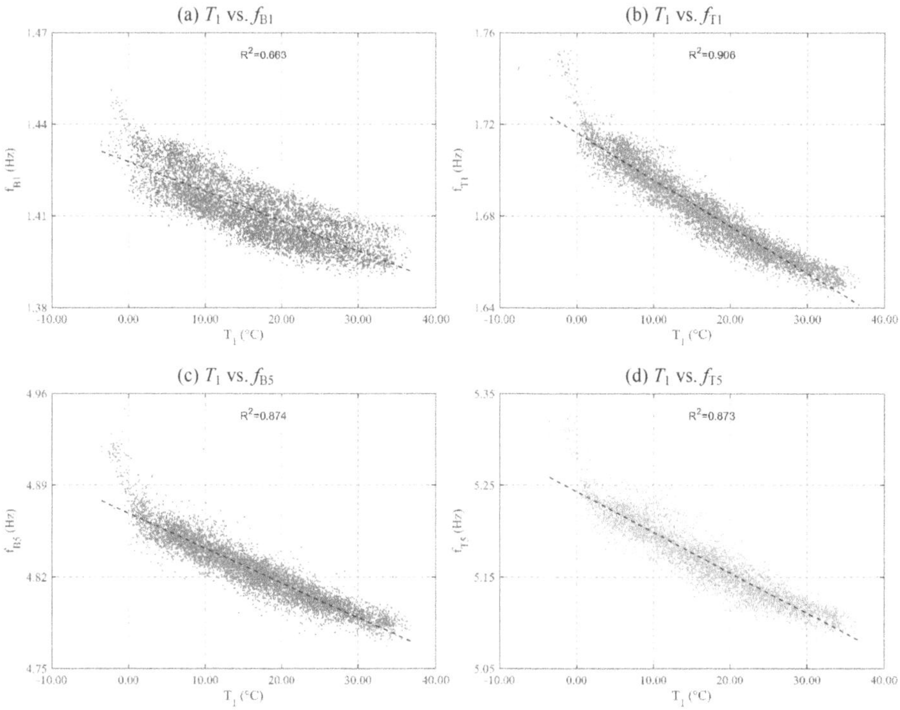

FIGURE 8.10 Selected temperature–frequency correlation (modes B_1, T_1, B_5, and T_5).

the first nine months of monitoring were assumed as the training period; (b) the lower 12 natural frequencies (Figure 8.4), characterizing the vibration modes with higher identification rates, were selected for the application of PCA; (c) only one principal score was retained to define the PCA-based regression. The effectiveness of the minimization/removal of EOV can be proven by comparing the time evolution of the natural frequencies in Figure 8.8 and the cleansed natural frequencies in Figure 8.12.

As expected, the cleansed frequencies (Figure 8.12) still exhibit small dispersion, possibly hiding minor structural changes. To investigate the possibility of occurrence of small changes, two different procedures have been used.

The first one is the control chart based on the Mahalanobis distance (Figure 8.13a), defined using a 99th percentile upper control limit (UCL). According to this approach, when the NI (Eq. 8.1) remains below the UCL, the system is operating normally; if the UCL is consistently exceeded, it may indicate the presence of a structural anomaly. As shown in Figure 8.13a, the control chart reveals no consistent exceedances in the present case; this observation supports the conclusion that no significant structural anomalies have occurred in the investigated infrastructure over the one-year-long monitoring period.

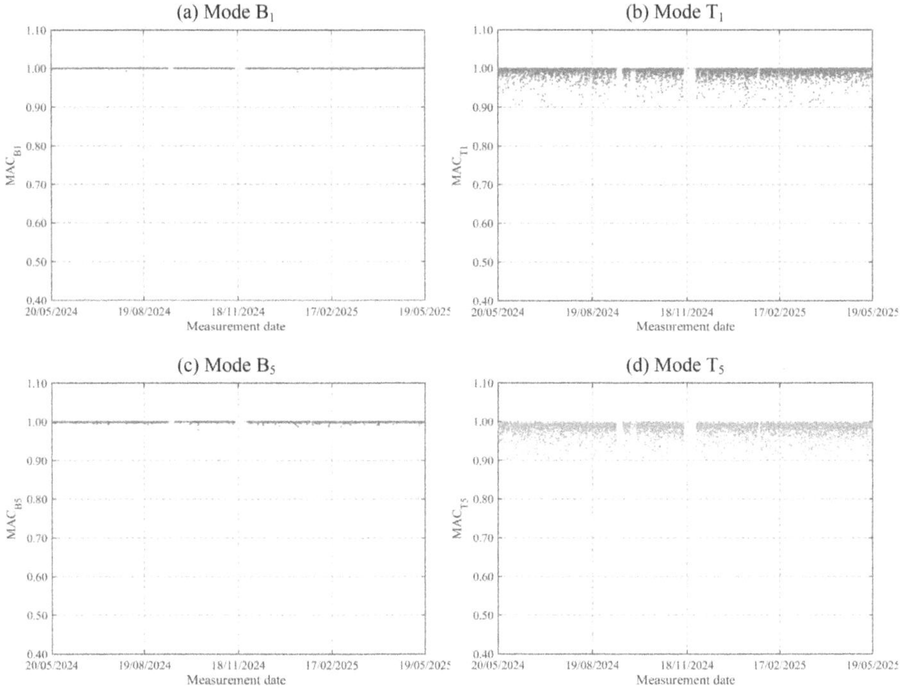

FIGURE 8.11 Variation in time of the MAC of modes B_1, T_1, B_5, and T_5.

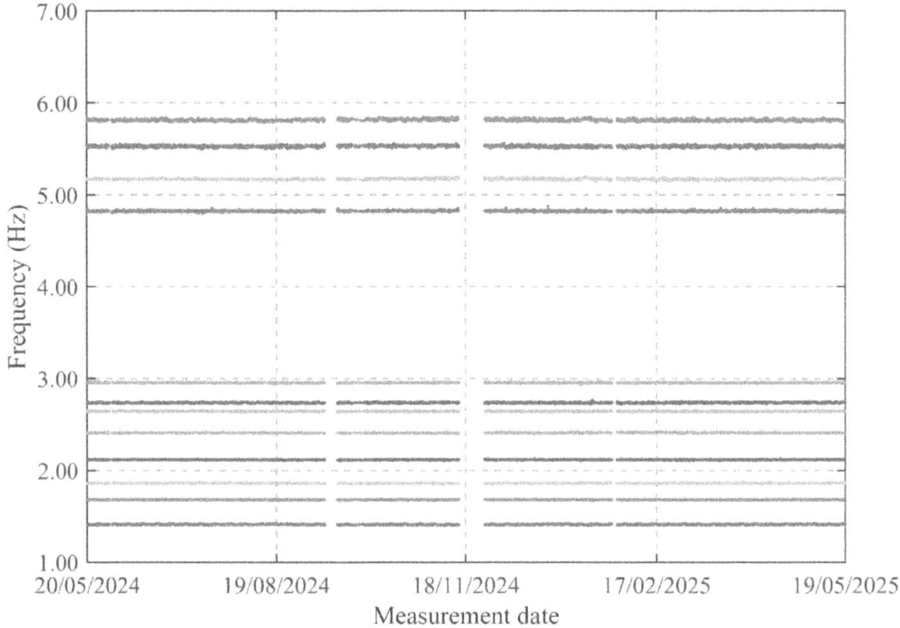

FIGURE 8.12 Time evolution of the cleansed natural frequencies during the first year of monitoring.

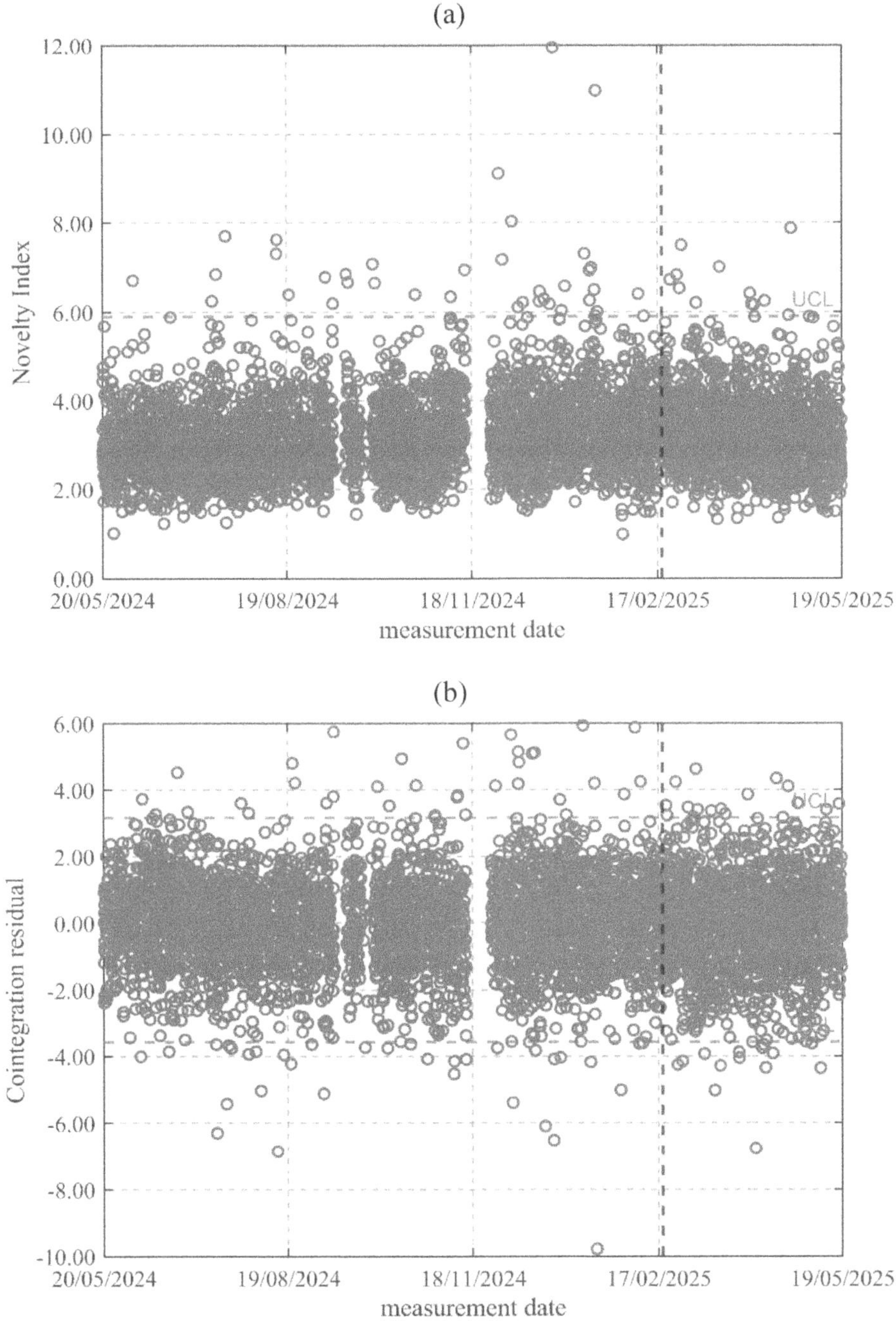

FIGURE 8.13 Control charts based on (a) Mahalanobis distance and (b) cointegration residual.

Similar results have been obtained through the cointegration residual; Figure 8.13b shows the time evolution of the cointegration residual computed from the identified natural frequencies of the bridge, spanning one year from May 2024 to May 2025. The coefficients of the cointegration residual (Eq. 8.2) are estimated by applying the Johansen procedure during a reference period of nine months. As suggested in the literature, the UCL and lower control limit (LCL) in Figure 8.13 are defined as ±3 times the standard deviation (estimated in the reference period).

As expected, Figure 8.13b highlights that the residual remains stationary, with a minimal presence of outliers, once again indicating that no significant anomalies occurred during the investigated monitoring period.

8.6 CONCLUSIONS

The chapter illustrates the OMA-based strategy adopted for the SHM of a multi-span steel–concrete composite bridge, namely the Olona West viaduct, located along the Autostrada Pedemontana Lombarda in Northern Italy. Preliminary investigations, FE modelling, and the results collected during the first year of monitoring allowed the following main conclusions:

1. Within the frequency range of 0–7 Hz, 14 vibration modes were successfully identified during continuous monitoring.
2. A 3D FE model was developed and showed a fairly good correlation with the experimental results.
3. The monitoring results suggest a strong temporal consistency in the identified dynamic characteristics of the bridge, with natural frequencies exhibiting limited variability and mode shapes demonstrating remarkable time invariance, as highlighted by MAC values that are always very close to unity.
4. The air temperature turned out to be a dominant driver of the daily and seasonal fluctuations of the natural frequencies of all modes, with the frequency of torsion modes generally exhibiting a higher correlation with temperature fluctuations than the frequency of bending modes.
5. The analysis of OMA results and control charts based on both the Mahalanobis distance and the cointegration residual indicates the absence of any significant structural anomalies in the Olona West viaduct throughout the one-year monitoring period.

8.7 ACKNOWLEDGMENTS

The support of Autostrada Pedemontana Lombarda (APL) and the cooperation of APL Technical Director, MSc Sabatino Fusco, in managing the collected data are gratefully acknowledged.

The collaboration of Valeria Pepe, MSc, PhD, in the development of the numerical model is gratefully acknowledged.

REFERENCES

[1] Boller C, Chang FK, Fujino Y (2009). *Encyclopedia of Structural Health Monitoring*. John Wiley & Sons, New York.

[2] Figueiredo E, Brownjohn J (2022). Three decades of statistical pattern recognition paradigm for SHM of bridges, *Structural Health Monitoring* 21(6), 3018–3054. https://doi.org/10.1177/14759217221075241

[3] Cunha A, Caetano E, Magalhães F, Moutinho C (2013). Recent perspectives in dynamic testing and monitoring of bridges, *Structural Control and Health Monitoring* 20, 853–877. https://doi.org/10.1002/stc.1516

[4] Wong KY (2004). Instrumentation and health monitoring of cable-supported bridges, *Structural Control and Health Monitoring* 11, 91–124. https://doi.org/10.1002/stc.33

[5] Ni YQ, Zhou HF, Chan KC, Ko JM (2008). Modal flexibility analysis of cable-stayed Ting Kau bridge for damage identification, *Computer-Aided Civil and Infrastructure Engineering* 23, 223–236.

[6] Magalhães F, Cunha Á, Caetano E (2012) Vibration based structural health monitoring of an arch bridge: From automated OMA to damage detection, *Mechanical Systems and Signal Processing* 28, 212–228.

[7] Peeters B, Couvreur G, Razinkov O, Kündig C, van der Auweraer H, de Roeck G. (2009). Continuous monitoring of the Øresund bridge: System and data analysis, *Structure Infrastructure Engineering* 5, 395–405. https://doi.org/10.1080/15732470701478362

[8] Cross EJ, Koo KY, Brownjohn J, Worden K (2013). Long-term monitoring and data analysis of the Tamar bridge, *Mechanical Systems and Signal Processing* 35, 16–34. https://doi.org/10.1016/j.ymssp.2012.08.026

[9] Gentile C, Saisi A (2015). Continuous dynamic monitoring of a centenary iron bridge for structural modification assessment, *Frontiers of Structural and Civil Engineering* 9, 26–41.

[10] Garcia-Fernandez N, Aenlle M, Gentile C (2024). Vibration testing and finite element modelling of a steel-concrete composite bridge, in: Rainieri C, Gentile C, Aenlle López M (eds.), *Proceedings of the 10th International Operational Modal Analysis Conference (IOMAC 2024).* Lecture Notes in Civil Engineering. Springer, Cham. https://doi.org/10.1007/978-3-031-61425-5_14

[11] Worden K, Manson G, Feller NRJ (2000). Damage detection using outlier analysis, *Journal of Sound and Vibration* 229(3), 647–667.

[12] Cross E, Worden K, Chen Q (2011). Cointegration: A novel approach for the removal of environmental trends in structural health monitoring data. *Proceedings of the Royal Society A: Mathematical, Physical and Engineering Sciences* 467(2133), 2712–2732.

[13] Gentile C, Pirrò M, Borlenghi P (2024). DYMOND: A Matlab toolbox for the dynamic monitoring of bridges according to the Lombardia regional guidelines, in: *Lecture Notes in Civil Engineering*. Springer, 476–487. https://doi.org/10.1007/978-3-031-61421-747

[14] Peeters B, de Roeck G. (1999). Reference-based stochastic subspace identification for output-only modal analysis, *Mechanical Systems and Signal Processing* 13, 855–878. https://doi.org/10.1006/mssp.1999.1249

[15] Pappa RS, Elliott KB, Schenk A (1992). A consistent-mode indicator for the Eigen system realization algorithm, in: *NASA Technical Memorandum 107607.* NASA Langley Research Center, Hampton.

[16] Allemang RJ, Brown DL (1982). A correlation coefficient for modal vector analysis, in: *Proceedings of the 1st International Modal Analysis Conference*, 110–116.

[17] Jolliffe IT (2002). *Principal Component Analysis*. Springer, New York.
[18] Johansen S (1988). Statistical analysis of cointegration vectors, *Journal of Economic Dynamics and Control* 12 (2–3), 231–254.
[19] MIDAS Civil 2024 (v2.1), MIDAS Information Technology Ltd., Korea.

9 Deep learning-based structural health monitoring of historical towers

M. Girardi, G. Gurioli, N. Messina, C. Padovani, and D. Pellegrini

9.1 INTRODUCTION

Artificial intelligence is transforming the traditional approach to science, technology, and industry, and deep learning (DL) techniques are increasingly used in various applications. The possibility of storing, analyzing, and processing big data recorded by sensor networks allows scientists and technicians to train data-driven models that reproduce the complexity of real-world phenomena.

The complexity of heritage structures, stemming from their nonlinear mechanical behavior, various geometric schemes, constituent materials, and building techniques, is heightened by the difficulty in getting accurate information on the materials' mechanical properties due to the scarcity of documentation and the need to limit destructive in-situ tests. For these reasons, structural health monitoring (SHM) has become increasingly attractive for safeguarding architectural heritage [1–8]. Continuous, long-term SHM based on ambient vibration tests can effectively contribute to detecting possible anomalies in the structures' dynamics. Within the framework of artificial intelligence, the large datasets collected during the monitoring campaigns, containing the velocities, accelerations, and environmental parameters recorded by the sensor networks installed on a building, can be processed, used as training data, and exploited to extract relevant features, predict the dynamic properties at future time instants, and detect anomalies. Time series forecasting and anomaly detection are based on DL techniques, whose applications to the SHM of heritage structures are still relatively recent and limited [9–19].

The present chapter shows the application of deep neural networks to the analysis and prediction of the ambient vibrations of a medieval masonry tower. The study relies on an extensive dataset recorded in the historic center of Lucca on the Saint Frediano bell tower [20] to train a temporal fusion transformer (TFT) network [21], predict the dynamical behavior of the building, and perform anomaly detection.

DOI: 10.1201/9781003516941-9

9.2 TIME SERIES FORECASTING AND ANOMALY DETECTION

In this section, we recall the main features of TFT [21]. TFT is a relatively new DL technique whose applications to the SHM of heritage buildings are quite limited. For this reason, we decided to compare the performance of TFT with the outcome of principal component analysis (PCA) [22], a well-established machine learning (ML) approach widely used for eliminating environmental and operational influences on damage-sensitive features such as natural frequencies [23] and for performing anomaly detection [20, 24, 25]. Unlike TFT, which uses environmental parameters like temperature and wind to forecast the targets, PCA is an output-only method that does not require any information on the environmental conditions. The TFT and PCA approaches are applied to the dynamics of the Saint Frediano bell tower, and their performances in anomaly detection are compared.

9.2.1 Temporal fusion transformers

The TFT approach proposed in Lim et al. [21] is devoted to multi-horizon time-series forecasting and unsupervised anomaly detection: A network is trained to understand the normality patterns in the data, and, at inference time, the more significant deviations from the learned normality patterns are regarded as anomalies.

In this work, we use a TFT network within the framework of SHM for learning the structure's dynamics from the available data, consisting of natural frequencies (calculated from the velocities recorded by the sensor network via dynamic identification techniques) and environmental parameters such as temperature, wind speed, rain, and humidity. The TFT approach has already been applied in Carrara et al., Falchi et al., Girardi et al. [26, 10, 27] to the Saint Frediano bell tower and the Guinigi Tower in Lucca, discussing how the TFT can predict the structures' dynamics while characterizing its uncertainty and how this output can be used for anomaly detection.

The algorithm's core is constituted by the network, which predicts the structure's dynamic evolution given the data from the recent past and estimates the uncertainty of the prediction. This estimation is crucial during the anomaly detection phase: It is possible to know whenever the actual observed values fall inside the predicted confidence interval and, if not, to easily quantify the deviation from the interval.

Input data are split into observed inputs $\mathbf{z}_t$, measured at each $t \in [0, T]$ by the sensors, and inputs $\mathbf{x}_t$ that are known without measurements (known inputs such as the date at a prescribed time t). Input data are used to train the TFT model, which is then employed to predict the scalar targets $y_{i,t}$ (e.g., the i-th natural frequency, as in Carrara et al., Falchi et al., Girardi et al. [26, 10, 27], and in this chapter) from t onward. TFT can simultaneously predict various percentiles (e.g., the 1st, 50th, and 99th) of the targets at each future time step of interest. Each quantile forecast can be written as

$$\hat{y}_i(q, t, \tau) = f_q(\tau, y_{i,t-k:t}, \mathbf{z}_{t-k:t}, \mathbf{x}_{t-k:t+\tau}; \omega). \tag{9.1}$$

In Eq. (9.1), $\hat{y}_i(q, t, \tau)$ is the predicted qth quantile (e.g., $q = 0.01, 0.5, 0.99$) of the i-th target, referred to the τ-step ahead of the starting time t; $f_q(.; \omega)$ is the model induced by the TFT architecture. Such a model depends on the parameter ω, the data $y_{i,t-k:t}$ and $\mathbf{z}_{t-k:t}$, representing the past target values and the observed inputs from the starting time t up to k time steps before, and the known inputs $\mathbf{x}_{t-k:t+\tau}$ across the entire range. The TFT network computes the parameter vector ω by solving a minimization problem on the domain of the input training data. Commonly, the input data are divided into a training set, a validation set reserved for further model refinement, and a test set on which the network test is conducted. We refer to the original work [21] for more details on TFTs and related issues and to Falchi et al. [10] for its application to the SHM domain.

Exploiting its prediction capability, TFT can detect possible anomalies or unexpected events by inspecting how much the observed targets deviate from the predicted ones. Once the TFT has been trained on a set of non-anomalous data, the prediction $\hat{y}_i(0.5, t, \tau)$ of the i-th target value, along with the 1st and the 99th percentiles, namely $\hat{y}_i\ (0.01, t, \tau)$ and $\hat{y}_i\ (0.99, t, \tau)$, is obtained. An anomaly occurs at time $t + \tau$ if the observed value $y_{i,t+\tau}$ lies outside the confidence interval $[\hat{y}_i(0.01, t, \tau), \hat{y}_i(0.99, t, \tau)]$ predicted by the network.

9.2.2 Principal component analysis

Let $\mathbf{y}_h = (y_{1,h}, y_{2,h}, \ldots, y_{d,h})^T$, $h = 1, \ldots, ns$ be the vectors of centered experimental data, where d is the number of targets and ns the number of samplings.

We consider a set of training data $\mathbf{y}_1, \mathbf{y}_2, \ldots, \mathbf{y}_{nt}$, with $nt<ns$ and the matrix

$$Y = \frac{1}{\sqrt{nt}}\left[\mathbf{y}_1\ \mathbf{y}_2 \ldots \mathbf{y}_{nt}\right] \in \mathbb{R}^{d\times nt} \text{ with } d < nt. \tag{9.2}$$

We can estimate the data outside the training set by looking for some $\mathbf{x}_k \in \mathbb{R}^{nt}$ such that

$$\mathbf{y}_h = Y\mathbf{x}_h,\ h = nt+1, \ldots, ns. \tag{9.3}$$

System (9.3) is underdetermined since $d < nt$, and we assume that the rank of Y is d. Let us consider the singular value decomposition of Y:

$$Y = U\Sigma V^T = \sum_{j=1}^{d} \sigma_j \mathbf{u}_j \mathbf{v}_j^T, \tag{9.4}$$

with $U = \left[\mathbf{u}_1\ \mathbf{u}_2 \ldots \mathbf{u}_d\right] \in \mathbb{R}^{d\times d}$ and $V = \left[\mathbf{v}_1\ \mathbf{v}_2 \ldots \mathbf{v}_{nt}\right] \in \mathbb{R}^{nt\times nt}$ as orthogonal matrices and $\Sigma \in \mathbb{R}^{d\times nt}$ as a diagonal matrix with nonnegative diagonal elements $\sigma_1 \geq \sigma_2 \geq \cdots \geq \sigma_d$, called the singular values of Y. The columns of U and $\sigma_1^2, \sigma_2^2, \ldots, \sigma_d^2$ are the eigenvectors and eigenvalues of the covariance matrix $YY^T \in \mathbb{R}^{d\times d}$.

If, in Eq. (9.4), we consider only the m largest singular values of Y, with $m<d$, we then obtain an approximation of $\mathbf{y}_h$

$$\tilde{\mathbf{y}}_h = \sum_{j=1}^{m} \sigma_j \mathbf{u}_j \mathbf{v}_j^T \mathbf{x}_h,\quad h = 1, \ldots, ns; \tag{9.5}$$

$\tilde{\mathbf{y}}_h$ is the principal component approximation of $\mathbf{y}_h$ and belongs to the subspace $\mathcal{U}_m$ of $\mathbb{R}^d$ spanned by the vectors $\mathbf{u}_1, \mathbf{u}_2, \ldots, \mathbf{u}_m$.

If an anomaly appears at time $h > nt$, the experimental data $\mathbf{y}_h$ are expected not to belong to $\mathcal{U}_m$. As suggested by Reynders et al. and Diamantaras and Kung [24, 22], we can define the reconstruction error $\mathbf{r}_k$ as the projection of the error $\mathbf{y}_h - \hat{\mathbf{y}}_h$ onto the subspace of $\mathbb{R}^d$ orthogonal to $\mathcal{U}_m$, and the misfit (or novelty index) as the norm of the reconstruction error,

$$r_h = \left(I - U_m U_m^T\right)\left(y_h - \tilde{y}_h\right) = y_h - U_m U_m^T y_h, \tag{9.6}$$

where I is the identity matrix and $U_m = \left[\mathbf{u}_1\ \mathbf{u}_2 \ldots \mathbf{u}_m\right] \in \mathbb{R}^{d \times m}$.

9.3 THE SAINT FREDIANO BELL TOWER IN LUCCA

The Basilica of Saint Frediano, shown in Figure 9.1, is one of the most fascinating monuments in Lucca. The church's bell tower, 52 m high, was recently restored along with its bells (Figure 9.1) and opened to visitors in 2023.

From October 2015 to November 2017, in collaboration with the Seismic Observatory of Arezzo (INGV), the tower was instrumented with four SARA

FIGURE 9.1 The Saint Frediano Basilica and its bell tower.

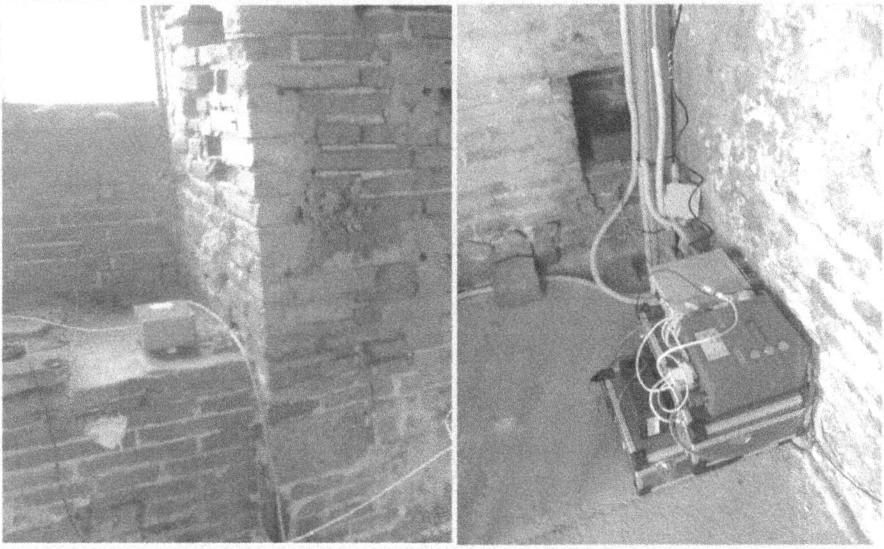

FIGURE 9.2 The SARA seismic stations (provided by the Seismic Observatory of Arezzo–INGV) installed in the bell chamber (left) and at the base of the bell tower (right).

triaxial seismometric stations (Figure 9.2), each composed of a seismometer (electrodynamic velocity transducer) coupled with an SL06 24-bit digitizer and arranged in different layouts to identify the tower's mode shapes. In all the experiments, the sampling frequency was set to 100 Hz. The recorded velocities (https://zenodo.org/records/7849700) were split into hourly slots and processed using the stochastic subspace identification (SSI) method [28] to extract the first five vibration frequencies of the tower [20].

9.3.1 The experimental campaign

Different long-term dynamic monitoring campaigns on historic towers have shown that the natural frequencies depend on environmental parameters such as temperature and humidity [28, 29]. The daily variations of the first two natural frequencies of the Saint Frediano bell tower from 6 to 8 June 2016 are shown in Figure 9.3a, along with the temperature plot. The curves exhibit a daily periodicity, with the frequencies increasing during the first half of the day, when temperatures rise, and decreasing in the latter half. The frequencies appear to be shifted by about 1–2 hours compared to the temperature trend. This phenomenon is likely due to the tower's thermal inertia and guarantees that the reported effects on the frequencies are not attributable to the experimental apparatus. The daily variations observed in the frequency values are on the order of 2–3%. Figure 9.3b,c shows the dependence of the first two frequencies on temperature. The relative differences of the frequencies over the entire monitoring period (evaluated at the 1st and 99th percentile) are on the order of 5–6% [20]. The first two frequencies depend almost linearly

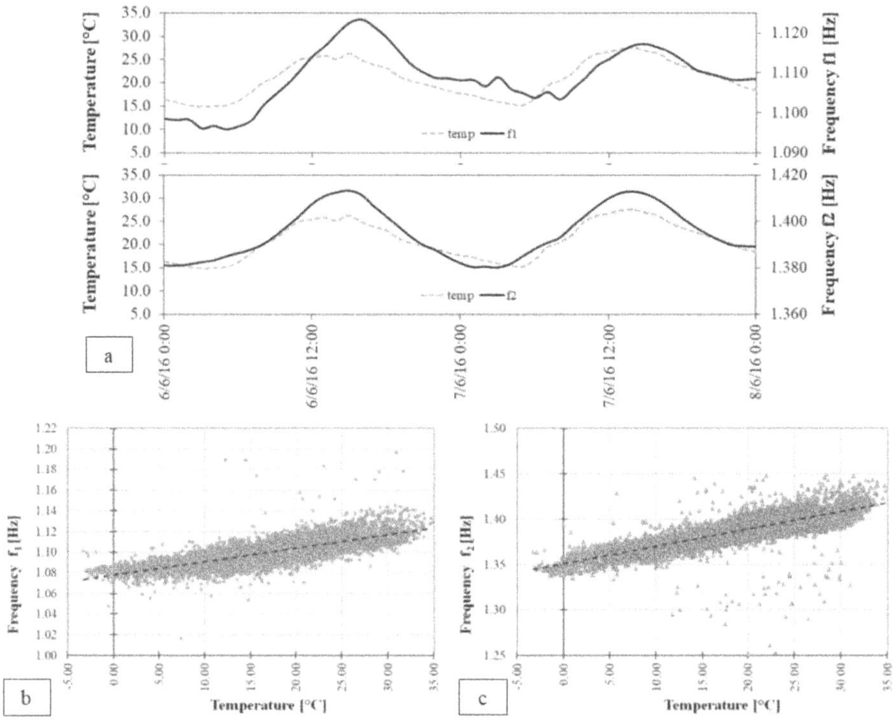

FIGURE 9.3 The first two natural frequencies (black line) and temperature (grey dashed line) versus time (a); first (b) and second (c) natural frequencies versus temperature.

on temperature. As observed in other long-term vibration monitoring studies of ancient buildings [1, 30, 31], the frequencies tend to increase with temperature. The thermal expansion caused by positive temperature variations induces a reduction in the micro-cracks inside the masonry and can thus temporarily increase the global stiffness of the structure.

The principal vibration sources for the tower during the monitoring period were the wind, the swinging of the bells, the passage of people nearby, and seismic events. Figure 9.4 shows the maximum absolute values of the velocities per hour recorded at the top of the bell tower (+42 m) from October 29 to December 31, 2015. The most relevant values correspond to the swinging of the bells on Saturday and Sunday, with the maximum velocity values in the direction of the bell's swinging (y in the current reference) on the order of 1–2 mm/s. Unusual amplification of the tower's movements, up to 3 mm/s, was recorded during the periods October 29 to November 1, 2015 (Lucca Comics & Games festival), November 17–18, 2015 (celebrations of Saint Frediano), and December 24–25 (Christmas). All these events (except Lucca Comics & Games, whose effects on the tower's vibrations were due to the crowd in the square) corresponded to the full swinging motion of the Saint Frediano bells. The wind affected the tower to a lesser extent. The vibrations reached relevant values

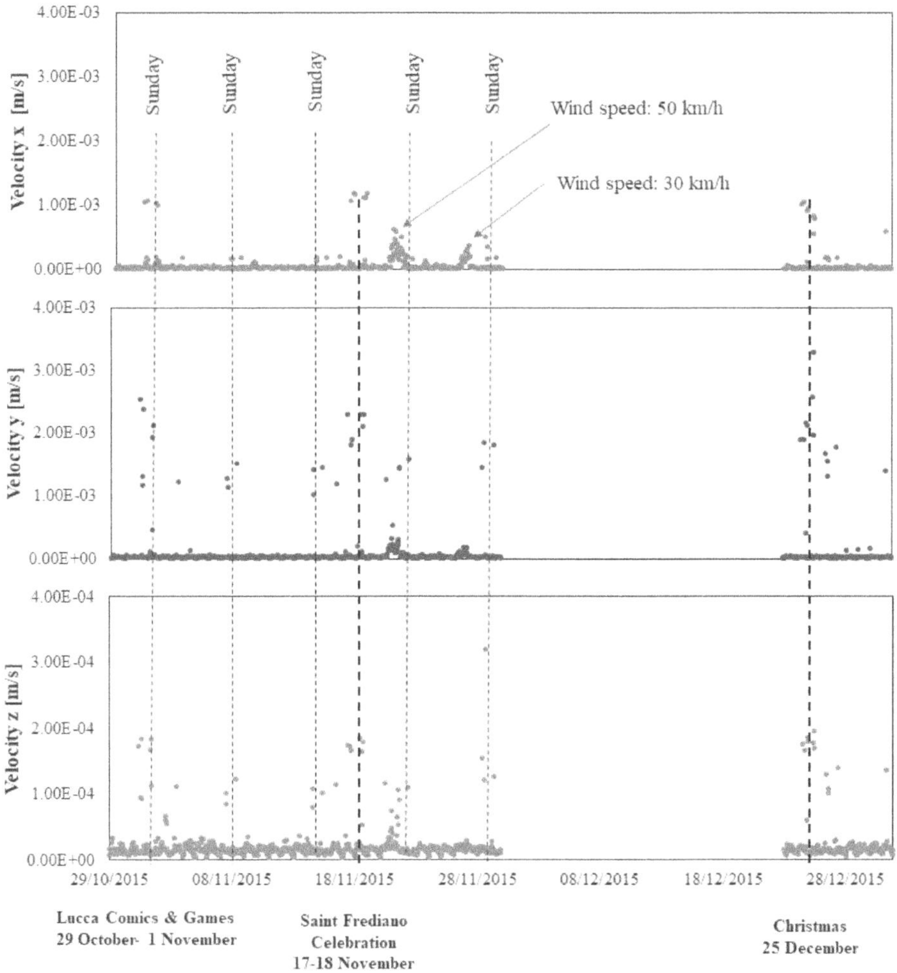

FIGURE 9.4 Maximum absolute values per hour of the velocities recorded by the seismic station at the top (+42 m) in the x, y, and z directions from October 29 to December 31, 2015.

only at relatively high wind velocities, over 50 km/h (for example, on November 21, 2015, the maximum recorded velocity on the tower was on the order of $5{\cdot}10^{-4}$ m/s, corresponding to a maximum wind velocity of 50 km/h). The velocities in the z direction are an order of magnitude lower than those in the horizontal direction. The effects of the swinging bells and the wind are also recognizable in the vertical signals.

Figure 9.5 plots the daily maximum velocities (absolute values) measured at the top along the x (grey line) and y (black line) directions, along with the daily maximum wind speeds (grey dashed line). The figure shows a correlation between the tower's velocities and the wind velocity, whose peaks were reached on November 21 and 27.

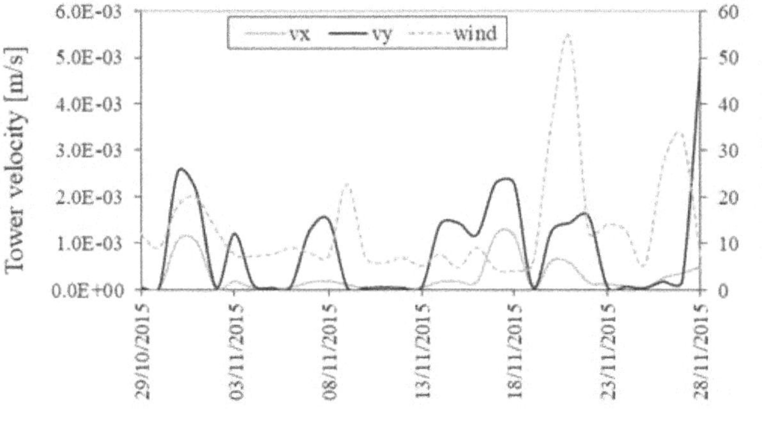

FIGURE 9.5 Daily maximum of the maximum hourly velocities measured at the top along the x (grey line) and y (black line) directions, and the daily maximum of the maximum hourly wind speeds (grey dashed line).

9.3.2 TFT and PCA: Analysis and Comparison of the Results

A preliminary application of the TFT network to the Saint Frediano bell tower is described in the work of Carrara et al. and Girardi [26, 27], where the TFT algorithm was trained, validated, and tested on the input database constituted by the frequencies and environmental parameters (such as temperature and humidity) measured on the tower or in its proximity.

The database used in this chapter includes data from October 28, 2015, to October 16, 2016, split into training, validation, and test sets. The frequencies f_1, f_2, f_3, f_4, f_5 are considered damage indicators. The known, observed, and target values are

$$\mathbf{x} = [date,\ hour\ slot,\ we], \tag{9.7}$$

$$\mathbf{z} = [temp,\ rain,\ hum,\ wind_v,\ wind_d,\ f_1, f_2, f_3, f_4, f_5], \tag{9.8}$$

$$\mathbf{y} = [f_1, f_2, f_3, f_4, f_5], \tag{9.9}$$

where *date* is the reference date, *hour slot* the reference hour slot, *we* a Boolean variable marking weekend days, f_1, f_2, f_3, f_4, f_5 the first five natural frequencies [Hz] ordered in ascending order, *temp* and *hum* the temperature [°C] and humidity (%) values registered by the thermo-hygrometers on the tower, *rain* [mm] the rainfall measured in the Botanic Garden in Lucca, and $wind_v$ and $wind_d$ the wind speed [m/s] and direction (degrees from the North, clockwise positive direction) measured in Pieve di Compito (environmental data available at www.sir.toscana.it).

Concerning the parameters k and τ in Eq. (9.1), a time window of $k = 96$ has been considered to identify the past steps used to predict the targets at $t + \tau$, with τ equal to 12 at training time and 1 at inference time.

Some events measured on the tower have been considered for testing the algorithm's sensitivity to potential anomalies: the Amatrice earthquake (M 6.0, about 400 km from Lucca), which occurred on August 24, 2016, at 1:36 UTC; the ringing of the

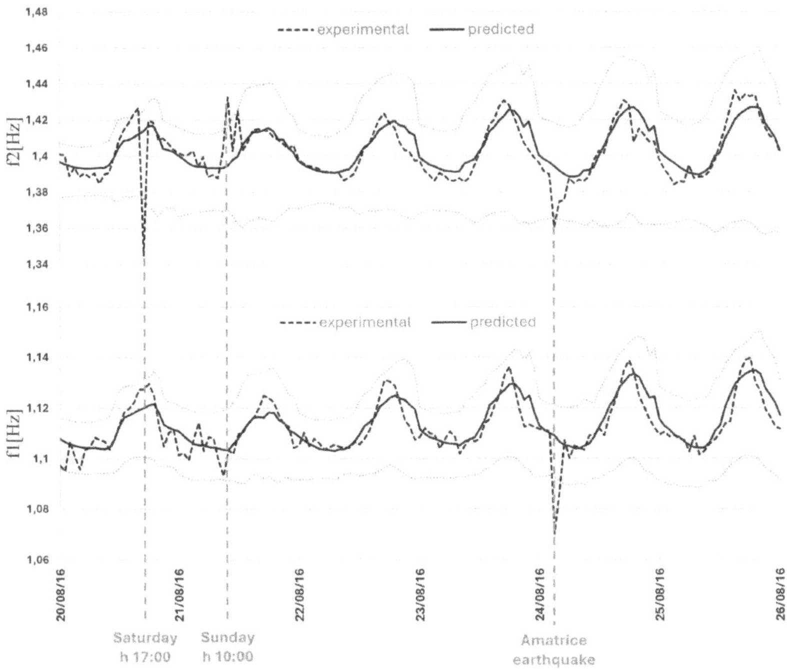

FIGURE 9.6 First two predicted (solid line) and experimental frequencies (dashed line). The dashed lines represent the 1st and 99th percentiles predicted by the model (testing period: August 20–26, 2016).

bells, which usually occurs on Saturday, Sunday, and Christmas; and wind velocity peaks. Figure 9.6 shows the anomaly detection results when the testing of the TFT network is conducted over a period containing the earthquake, focusing on the first two frequencies. The red line represents the measured (experimental) frequency, the continuous blue line represents the 50th percentile predicted by TFT, and the two blue dashed lines represent the predicted confidence interval between the 1st (lower) and 99th (upper) percentiles. According to the damage detection criterion introduced at the end of Section 9.2, an anomaly occurs when at least one experimental frequency exceeds the predicted interval. The network can capture the seismic event as an anomaly affecting the first two frequencies. The algorithm can also detect the bell tower vibrations induced by the swinging bells. This anomaly involves the second frequency (related to the swinging direction) and corresponds to the main religious ceremonies held at the Cathedral, particularly on Saturday (August 20 at 17:00) and Sunday (August 21 at 10:00). It is worth noting that the frequencies of the Saint Frediano bell tower exhibit a very marked oscillatory behavior over the day; this behavior is due to the daily temperature variations and is well predicted by the TFT network.

Figure 9.7 summarizes the results of the TFT network trained using the samples from the time intervals January 1, 2016–April 10, 2016, and May 10, 2016–17 August 2016. The anomaly plot focuses on the test set from November 1 to December 2, 2015. It shows the five observed frequencies (orange line) and the ones predicted by

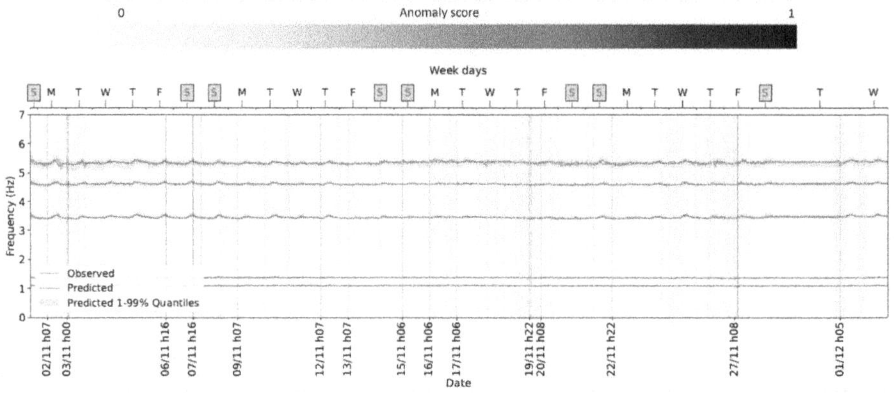

FIGURE 9.7 Anomaly plots for the TFT model, testing period November 1–December 2, 2015, observed (dashed line) and predicted (solid line) frequencies (Hz). The confidence intervals are represented by the grey areas between the 5th and 95th percentiles (top) and between the 1st and 99th percentiles (bottom).

the algorithm (corresponding to the 50th percentile, blue line). The grey areas represent the confidence intervals between the 1st and 99th percentiles at every prediction time step for each predicted frequency. The vertical bars show the time locations where at least one predicted frequency does not fall within the estimated confidence interval, thus indicating anomalies. In [20], a suitable anomaly score was defined to visualize the magnitude of the anomalies via the color map reported in the figure. The anomaly score, ranging from 0 to 1, quantifies how much the observed target falls outside the confidence interval predicted by the network and is set to 1 when the observed value overshoots the estimated interval size of a percentage greater than or equal to 100%. The windiest days (November 20–21 and 26–27) and the bells swinging on Sunday and Saturday and on the Saint Frediano celebration (November 17) are recognizable in the plot. These anomalies systematically arise when a drop in the frequencies of the tower is associated with the higher velocity values caused by the wind and the bells (Figure 9.4). These frequency drops as the amplitude of the velocity increases are typical of softening systems and can be attributed to the masonry's inability to withstand tensile stress [32].

Figure 9.8 shows the anomalies detected by the algorithm for a simulated damage scenario in which the frequency values are permanently reduced to simulate some structural damage, as suggested by the experimental results shown in [30], where a decrease was observed in the natural frequencies of the Gabbia Tower (Mantua) after a seismic event. In this case, the training period goes from October 28, 2015, to August 22, 2016. During the testing period, from August 23, 2016, to October 16, 2016, the five experimental frequencies were reduced by 4%, 2%, 1%, 1%, and 1%, starting from the Amatrice earthquake to simulate damage. To quantify the anomalies detected by the TFT network, we introduce a damage index B_{test}, defined as the ratio between the number of vertical bars in the anomaly plot for a given testing period and the number of hours in the same period. The vertical bars after the

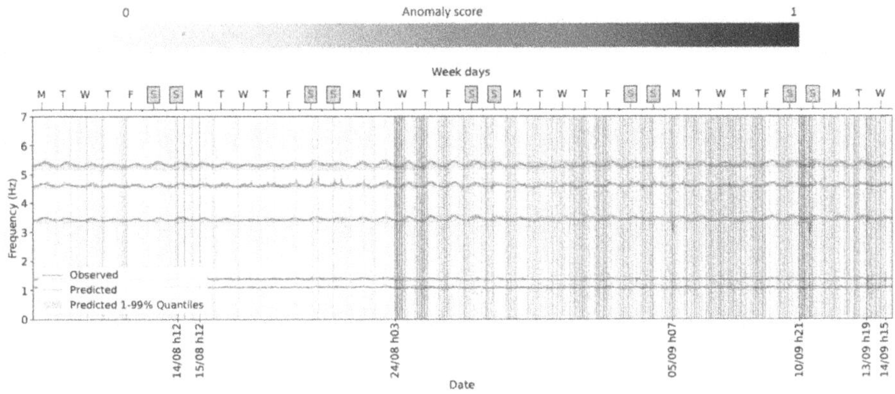

FIGURE 9.8 Simulated damage scenario, anomaly plots for the TFT model, testing period from August 23, 2016, to October 16, 2016, observed (dashed line) and predicted (solid line) frequencies (Hz). The confidence intervals are represented by the grey areas between the 1st and 99th percentiles.

earthquake indicate the presence of severe damage, with B_{test} = 93.8% in the damage scenario versus B_{test} = 15.6% in the absence of damage for the 1–99 percentile configuration.

A preliminary application of the PCA to the Saint Frediano bell tower is reported by Azzara et al. [20]. In particular, the performance of the PCA algorithm in detecting the Amatrice earthquake and the anomalies associated with simulated damage scenarios was checked by considering a training set containing the first three frequencies measured before the Amatrice earthquake.

In this chapter, we investigate the results of PCA conducted using the first five experimental frequencies. In Figure 9.9, the dashed horizontal line corresponds to the 99th percentile of the misfits—the norm of the reconstruction errors defined in Eq. (9.6)—for the training period (from October 28, 2015, to August 22, 2016). Then, considering data measured from August 23, 2016, to October 16, 2016, thus including the Amatrice earthquake, the percentage N_{test} of points exceeding the threshold of the 99th percentile was evaluated. As expected, the value of N_{test} = 2.25% indicates that no damage arose in the tower due to the earthquake.

Figure 9.10 illustrates the application of PCA to the simulated damage scenario considered earlier. The training period goes from October 28, 2015, to August 22, 2016 (blue dots). During the testing period, from August 23, 2016, to October 16, 2016, the five experimental frequencies were reduced by 4%, 2%, 1%, 1%, and 1% (black dots), starting from the Amatrice earthquake, consistent with the damage scenario used in the TFT tests. The results were obtained by taking four principal components, corresponding to an explained variance of 96.72%. The dashed horizontal line represents the 99th percentile of the misfit during the entire training period. The percentage of points N_{test} exceeding the threshold in the test set increases from 2.25% in the absence of damage to 17.90% for this artificial damage scenario, thus indicating a permanent change in the dynamic behavior of the tower.

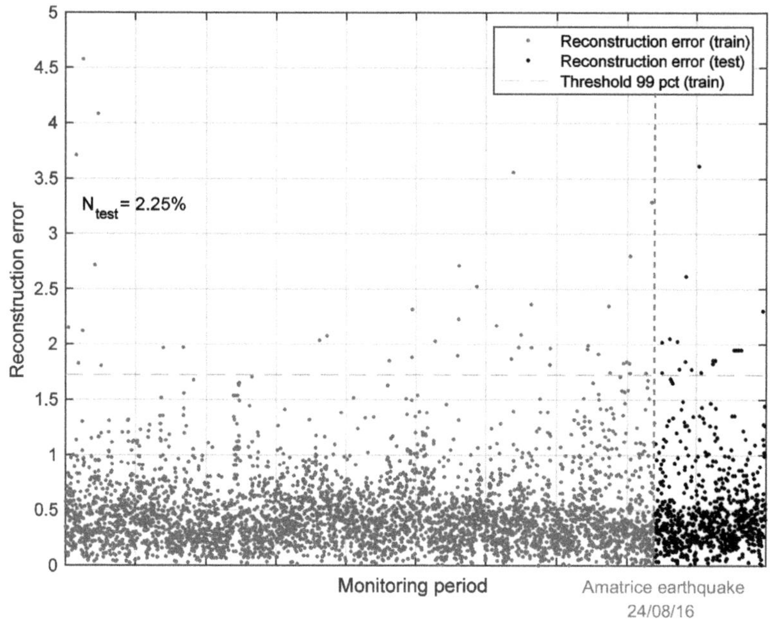

FIGURE 9.9 Misfit of the linear PCA using five frequencies and three principal components for the training period (grey dots) and testing period (black dots), which includes the Amatrice earthquake.

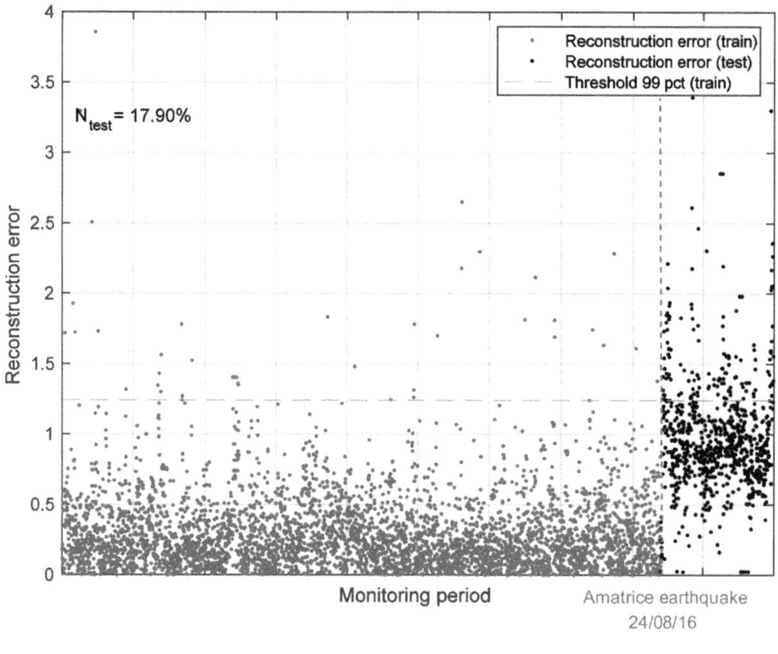

FIGURE 9.10 Misfit of the linear PCA using five frequencies and four principal components for the training period (grey dots) and the testing period with simulated damage scenario (black dots), characterized by frequency reductions of 4%, 2%, 1%, 1%, and 1%.

9.4 CONCLUSIONS

This chapter describes the TFT method and its application to the SHM of the Saint Frediano bell tower in Lucca. The algorithm is trained on the first five natural frequencies of the tower, extracted from the velocity data recorded by high-sensitivity seismic stations and other relevant input variables describing the environment (temperature and humidity) and the loads acting on the structure (wind and swinging bells).

The TFT network has been employed to predict the vibrational features (natural frequencies) and to detect possible anomalies or unexpected events by inspecting how much the actual frequencies deviate from the predicted ones. TFT has been tested on the Amatrice earthquake measured on the tower during the monitoring period and on a simulated damage scenario in which the frequencies of the bell tower are permanently reduced after the earthquake. The network has shown good accuracy in predicting the tower's frequencies and sensitivity in anomaly detection. A comparison with the well-established PCA approach has been conducted. PCA is an output-only method and gives good results in anomaly detection, but it is outperformed by the TFT approach, which can also predict the target quantities and their dependence on environmental parameters.

9.5 ACKNOWLEDGMENTS

This research has been partially supported by the Italian National Research Council (REVOLUTION Project, Progetti di Ricerca @CNR, 2022–2025) and by PNRR—M4C2—Investimento 1.3, Partenariato Esteso PE00000013—"FAIR—Future Artificial Intelligence Research"—Spoke 1 "Human-centered AI", funded by the European Union—NextGenerationEU. These supports are gratefully acknowledged.

REFERENCES

1. Ramos, L.F., Marques, L., Lourenco, P.B., De Roeck, G., Campos-Costa, A., Roque, J. Monitoring historical masonry structures with operational modal analysis: Two case studies. Mech. Syst. Sig. Process. 24, 1291–1305 (2010).
2. Magalhaes, F., Cunha, A., Caetano, E. Vibration based structural health monitoring of an arch bridge: From automated OMA to damage detection. Mech. Syst. Sig. Process. 28, 212–228 (2012).
3. De Stefano, A., Matta, E., Clemente, P. Structural health monitoring of historical heritage in Italy: Some relevant experiences. J. Civ. Struct. Health Monit. 6, 83–106 (2016).
4. Ubertini, F., Comanducci, G., Cavalagli, N. Vibration-based structural health monitoring of a historic bell tower using output-only measurements and multivariate statistical analysis. Struct. Health Monit. 15(4), 438–457 (2016).
5. Gentile, C., Ruccolo, A., Canali, F. Continuous monitoring of the Milan Cathedral: Dynamic characteristics and vibration-based SHM. J. Civ. Struct. Health Monit. 9, 671–688 (2019).
6. Baraccani, S., Azzara, R.M., Palermo, M., Gasparini, G., Trombetti, T. Long-term seismometric monitoring of the two towers of Bologna (Italy): Modal frequencies identification and effects due to traffic induced vibrations. Front. Built Environ. 6, 85 (2020).

7. Barsocchi, P., Bartoli, G., Betti, M., Girardi, M., Mammolito, S., Pellegrini, D., Zini, G. Wireless sensor networks for continuous structural health monitoring of historic masonry towers. Int. J. Archit. Herit. 15(1), 22–44 (2021).
8. Sun, Q., Rainieri, C., Ren, W.X., Yan, W.J., Fabbrocino, G. Automated operational modal analysis of bell towers subjected to narrowband input. Structures 54, 78–88 (2023).
9. Rosso, M.M., Cucuzza, R., Marano, G.C., Aloisio, A., Cirrincione, G. Review on deep learning in structural health monitoring. In: Bridge Safety, Maintenance, Management, Life-Cycle, Resilience and Sustainability, pp. 309–315, CRC Press (2022).
10. Falchi, F., Girardi, M., Gurioli, G., Messina, N., Padovani, C., Pellegrini, D. Deep learning and structural health monitoring: Temporal fusion transformers for anomaly detection in masonry towers. Mech. Syst. Sig. Process. 215, 111382 (2024).
11. Dang, H.V., Tran-Ngoc, H., Nguyen, T.V., Bui-Tien, T., De Roeck, G., Nguyen, H.X. Data-driven structural health monitoring using feature fusion and hybrid deep learning. IEEE Trans. Autom. Sci. Eng. 18(4), 2087–2103 (2020).
12. Fernandez-Navamuel, A., Magalhaes, F., Zamora-Sanchez, D., Omella, A.J., Garcia-Sanchez, D., Pardo, D. Deep learning enhanced principal component analysis for structural health monitoring. Struct. Health Monit. 21(4), 1710–1722 (2022).
13. Li, J., Chen, W., Fan, G. Structural health monitoring data anomaly detection by transformer enhanced densely connected neural networks. Smart Struct. Syst. 30(6), 613–626 (2022).
14. Rosafalco, L., Torzoni, M., Manzoni, A., Mariani, S., Corigliano, A. Online structural health monitoring by model order reduction and deep learning algorithms. Comput. Struct. 255, 106604 (2021).
15. Seventekidis, P., Giagopoulos, D., Arailopoulos, A., Markogiannaki, O. Structural health monitoring using deep learning with optimal finite element model generated data. Mech. Syst. Sig. Process. 145, 106972 (2021).
16. Sun, H., Song, L., Yu, Z. A deep learning-based bridge damage detection and localization method. Mech. Syst. Sig. Process. 193, 110277 (2023).
17. Giglioni, V., Venanzi, I., Poggioni, V., Milani, A., Ubertini, F. Autoencoders for unsupervised real-time bridge health assessment. Comput. Aided Civ. Infrastruct. Eng. 38(8), 959–974 (2023).
18. Cimino, M.G., Galatolo, F.A., Parola, M., Perilli, N., Squeglia, N. Deep learning of structural changes in historical buildings: The case study of the Pisa tower. In: Proceedings of the 14th International Joint Conference on Computational Intelligence (IJCCI 2022), vol. 1, pp. 396–403. Scitepress (2022).
19. Bui-Tien, T., Nguyen-Chi, T., Le-Xuan, T., Tran-Ngoc, H. Enhancing bridge damage assessment: Adaptive cell and deep learning approaches in time-series analysis. Constr. Build. Mater. 439, 137240 (2024).
20. Azzara, R.M., De Roeck, G., Girardi, M., Padovani, C., Pellegrini, D., Reynders, E. The influence of environmental parameters on the dynamic behaviour of the San Frediano bell tower in Lucca. Eng. Struct. 156, 175–187 (2018).
21. Lim, B., Arik, S.Ö, Loeff, N., Pfister, T. Temporal fusion transformers for interpretable multi-horizon time series forecasting. Int. J. Forecast. 37(4), 1748–1764 (2021).
22. Diamantaras, K.I., Kung, S.Y. Principal Component Neural Networks: Theory and Applications. John Wiley & Sons, Inc., New York, NY, USA (1996).
23. Saisi, A., Gentile, C., Guidobaldi, M. Post-earthquake continuous dynamic monitoring of the Gabbia tower in Mantua, Italy. Constr. Build. Mater. 81, 101–112 (2015).
24. Reynders, E., Wursten, G., De Roeck, G. Output-only structural health monitoring in changing environmental conditions by means of nonlinear system identification. Struct. Health Monit. 13(1), 82–93 (2014).

25. García-Macías, E., Ruccolo, A., Zanini, M.A., Pellegrino, C., Gentile, C., Ubertini, F., Mannella, P. P3P: A software suite for autonomous SHM of bridge networks. J. Civ. Struct. Health Monit. 13(8), 1577–1594 (2023).
26. Carrara, F., Falchi, F., Girardi, M., Messina, N., Padovani, C., Pellegrini, D. Deep learning for structural health monitoring: An application to heritage structures. Mater. Res. Proceed. 2023(26), 581–586 (2022).
27. Girardi, M., Gurioli, G., Messina, N., Padovani, C., Pellegrini, D. Vibration monitoring of historical towers: New contributions from data science. In International Operational Modal Analysis Conference, pp. 15–24. Cham: Springer Nature Switzerland (2024).
28. Peeters, B., De Roeck, G. Reference-based stochastic subspace identification for output-only modal analysis. Mech. Syst. Sig. Process. 13(6), 855–878 (1999).
29. Cabboi, A., Gentile, C., Saisi, A. From continuous vibration monitoring to FEM-based damage assessment: Application on a stone-masonry tower. Constr. Build. Mater. 156, 252–265 (2017).
30. Gentile, C., Guidobaldi, M., Saisi, A. One-year dynamic monitoring of a historic tower: Damage detection under changing environment. Meccanica 51(11), 2873–2889 (2016).
31. Ubertini, F., Comanducci, G., Cavalagli, N., Pisello, A.L., Materazzi, A.L., Cotana, F. Environmental effects on natural frequencies of the San Pietro bell tower in Perugia, Italy, and their removal for structural performance assessment. Mech. Syst. Signal Process. 82 (1), 307–322 (2017).
32. Girardi, M. On the dynamic behaviour of masonry beam-columns: An analytical approach. Eur. J. Mech. A/Solids 45, 174–184 (2014).

10 Assessment of bridges using dynamic monitoring and sparse autoencoders

M. Pirrò and Carmelo Gentile

10.1 INTRODUCTION

Dynamic monitoring of bridges and structures, complemented by the automation of operational modal analysis (OMA), represents a well-known strategy in structural health monitoring (SHM) [1], mainly based on the correlation between modal and structural parameters. However, modal parameters are sensitive not only to mass and stiffness changes but also to environmental and operational variability (EOV) [2–4]. As environmental and operational factors can significantly influence modal parameters, it is crucial to mitigate EOV effects to avoid false positives when detecting structural anomalies and changes.

While OMA-based approaches have proven effective for early anomaly detection, there is increasing interest in applying deep learning (DL) frameworks for SHM [5]. This interest is driven by the advanced processing power of DL networks, which can model essential data features using layers of artificial neurons. Among DL algorithms, the sparse autoencoder (SAE) [6] has shown promise for SHM because SAE networks (a) can discover hidden features by reducing the dimensionality of input data and (b) are believed to inherently handle EOV, potentially removing the need for additional steps to mitigate its impact [7, 8].

Unlike traditional methods, which train separate networks for each channel of data [7], the approach discussed herein trains a single network using all available channels from the structure simultaneously [9, 10]. During training, the structure is assumed to be undamaged and subjected to normal variations of environmental and operational conditions. Once trained, the SAE can be fed with new data and reconstruct the original signals. If the structure's condition and EOV do not change after training, the SAE should accurately reconstruct the input data. Hence, the structural health is evaluated by calculating the mean absolute error (MAE) between the input and reconstructed data, with deviations indicating potential damage. In addition, higher MAE values are expected in damaged areas, assisting in identifying the presence of critical regions.

The application of the proposed SAE-based methodology is demonstrated using data collected from two different benchmark bridges [11–17].

The first case study is related to the ADA bridge [11], a steel truss infrastructure, where dynamic tests were performed in undamaged and damaged conditions

 DOI: 10.1201/9781003516941-10

to identify and localize the inflicted damage scenarios. However, since the dynamic measurements were taken over just two days and under consistent temperature conditions, the influence of EOV on the data was minimal.

The spectral analysis of the accelerations measured on the ADA bridge [12] provided evidence of the damage occurrence, but no information was given on the damage location. In addition, the variation of the mode shapes is investigated by Chang and Kim [13], and minor changes were observed only when the damage was applied asymmetrically. The previous results [12, 13] are herein improved by detecting the introduced damages directly from the raw signals using an SAE network, without the need for spectral functions or modal parameters. Moreover, unlike prior studies [12, 13], the present approach successfully localizes the inflicted damages.

The second application is based on the results obtained during the continuous dynamic monitoring of the historical San Michele iron bridge (1889) [14, 15]. In previous studies [16, 17], automated OMA and especially the information provided by the mode shapes allowed the detection of structural performance anomalies and changes in the dynamic characteristics of the bridge, conceivably related to the progress of the damage due to corrosion.

The SAE-based methodology is therefore applied to identify structural degradation during the monitoring period and possibly locate the phenomenon along the bridge, relying solely on the recorded signals without requiring any modal identification.

10.2 SAE NETWORK

An AE network is aimed at reconstructing the input time series using three types of neural layers (Figure 10.1): encoder layer, hidden layer, and decoder layer. The encoder compresses the input data by reducing its dimensionality and capturing hidden patterns to create a more condensed representation. The decoder then uses the features learned by the hidden layer to reconstruct the original input data at the output layer.

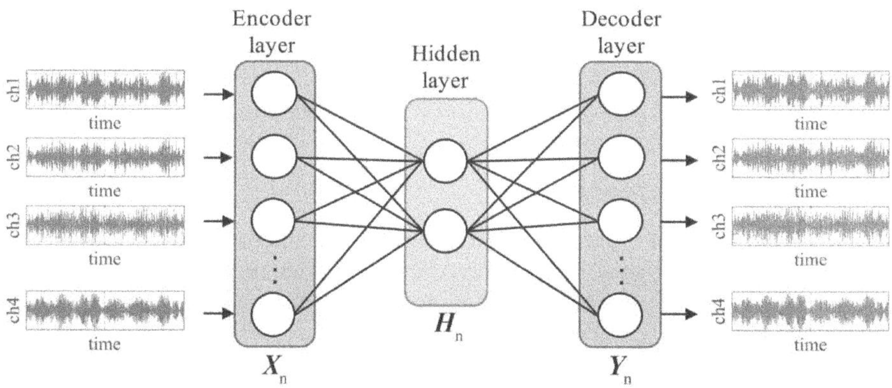

FIGURE 10.1 Basic architecture of an AE network.

Given the time series $\boldsymbol{x}_n \in \Re^I$, the encoding procedure can be formally expressed as

$$\boldsymbol{h}_n = f(\boldsymbol{W}\boldsymbol{x}_n + \boldsymbol{b}_1) \tag{10.1}$$

where $\boldsymbol{h}_n \in \Re^J$ (J < I) is the vector collecting the hidden patterns learned by the network, $\boldsymbol{W} \in \Re^{J\times I}$ is the unknown weight matrix, and $\boldsymbol{b}_1 \in \Re^J$ is the bias term for the input layer.

Subsequently, the decoder operation maps the hidden patterns $\boldsymbol{h}_n$ by means of:

$$\boldsymbol{y}_n = g(\boldsymbol{W}^T \boldsymbol{h}_n + \boldsymbol{b}_2) \tag{10.2}$$

where $\boldsymbol{y}_n \in \Re^I$ is the reconstructed time series, $\boldsymbol{W}^T \in \Re^{I\times J}$ is the unknown transpose weight matrix, and $\boldsymbol{b}_2 \in \Re^I$ is the bias term for the hidden layer. Typical functions f and g can be chosen as sigmoid, linear, or hyperbolic tangent functions [7].

The unknown parameters $\theta = (\boldsymbol{W}, \boldsymbol{b}_1, \boldsymbol{b}_2)$ can be estimated by minimizing a loss function $Z(\boldsymbol{x}_n, \boldsymbol{y}_n)$, i.e., the error between the original input x_n and the reconstructed output y_n.

To avoid the network from merely copying the input directly to the output during training, thus missing meaningful hidden patterns, a SAE is employed [18]. Specifically, a sparsity penalty is applied to the hidden layer, encouraging the network to keep most of its neurons inactive during training, thereby focusing on learning more complex structures from the input.

10.3 THE VIBRATION-BASED ANOMALY DETECTION APPROACH

This section presents the vibration-based anomaly detection procedure based on an SAE, distinguishing three main steps: (a) organization of the input, (b) training of the SAE, and (c) anomaly detection using the SAE.

10.3.1 Organization of Database of Signals

Unlike conventional approaches, which train individual networks for each data channel, the method developed herein trains a single network to process input data $\boldsymbol{X}_n \in \Re^{M\times I}$ from all available M channels simultaneously (Figure 10.1) [9]. The input signals must be sufficiently long to capture the dynamic behavior of the investigated structure.

10.3.2 Training of the Network

The training signals are organized into $M \times I$ matrices, which are collected when the structure is assumed to be undamaged and operating under normal EOV. It should be noticed that 75% of those signals are used for training, whereas the remaining 25% provides the basis for validation.

During training, several parameters should be properly determined, such as (a) the activation encoder-decoder functions f and g, (b) the number of nodes in the hidden layer (i.e., the compression rate), and (c) the mini-batch size.

TABLE 10.1
List of SAE parameter ranges.

Parameter	Input range
Number of encoder/decoder layers	1
Number of hidden neurons	10–50% of the signal length
Encoder/decoder function	Sigmoid/linear
Max. epochs	[500–5000]
Mini-batch size	[4; 16; 32; 64]

It should be noticed that, among the previous parameters, the number of hidden neurons—determining the compression rate of the input data—is crucial. Training parameters are determined via a type of grid search method [19], with the optimal estimate of the input parameters (Table 10.1) corresponding to minimum reconstruction errors within the validation set. By doing so, the network will be able to correctly reproduce the time series collected under normal structural conditions and EOV. The range of training parameters—inspired by [20] and also based on the authors' experience—is summarized in Table 10.1.

The ADAM algorithm [21] is adopted to minimize the loss function $Z(x_n, y_n)$, and the input data are normalized [22] to prevent numerical issues during the optimization process.

10.3.3 Testing of the network and anomaly detection

Once the training process has been completed, new time series collected from unknown scenarios are fed into the network. For each channel, the reconstruction error between the original input x_n and the reconstructed output y_n is quantified by the MAE:

$$MAE_n^{(i)} = \frac{1}{I}\sum_{k=1}^{I}\left|x_{n,k}^{(i)} - y_{n,k}^{(i)}\right| \tag{10.3}$$

where

- I is the length of the signals;
- $x_{n,k}^{(i)}$ and $y_{n,k}^{(i)}$—referring to the original input and the reconstructed output of channel i, respectively—are the n-th samples in time.

When test sequences refer to structural conditions that are similar to those observed during training, the network should reconstruct the input with high precision. Conversely, if changes in the structural conditions occur, the SAE may fail to accurately reproduce the input, leading to higher MAEs. A test sequence is considered an outlier if its reconstruction error exceeds the m-th percentile (e.g., 95th–99th) of the error values from the training data. Frequent occurrences of such outliers may indicate the presence of abnormal structural conditions.

In the case of available labeled datasets, a common way to compare the experimental results with a set of true classifications is achieved by defining the number of (a) non-anomalous samples correctly predicted (true negatives, TN), (b) non-anomalous samples classified as anomalous (false positives, FP), (c) anomalous samples correctly predicted (true positives, TP), and (d) anomalous samples classified as non-anomalous (false negatives, FN). From these values, it may be useful to define the following metrics:

$$\mathrm{ACC} = (\mathrm{TP}+\mathrm{TN})/(\mathrm{TP}+\mathrm{FP}+\mathrm{TN}+\mathrm{FN}) \tag{10.4}$$

$$\mathrm{DR} = \mathrm{TP}/(\mathrm{TP}+\mathrm{FN}) \tag{10.5}$$

$$\mathrm{FAR} = \mathrm{FP}/(\mathrm{FP}+\mathrm{TN}) \tag{10.6}$$

where (a) the accuracy (ACC) is the percentage of correct predictions among all samples, (b) the detection rate (DR) is the percentage of predicted anomalous samples among all the anomalous samples, and (c) the false alarm rate (FAR) is the ratio of non-anomalous samples classified as anomalous.

10.4 THE OLD ADA BRIDGE

10.4.1 The Bridge and the Test Scenarios

Vehicle-induced and ambient vibration measurements were acquired on the steel truss bridge, named the Old ADA Bridge, located in Japan [11]. As shown in Figure 10.2, the simply supported bridge is 59.2 m long, and the deck is 3.6 m wide.

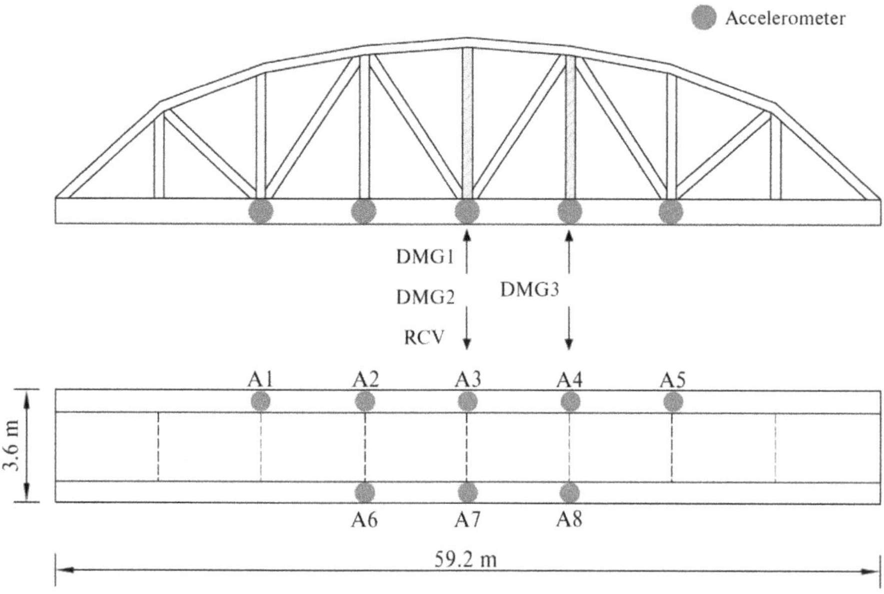

FIGURE 10.2 Old ADA Bridge: sensor layout and damage locations (adapted from Kim et al. [11]).

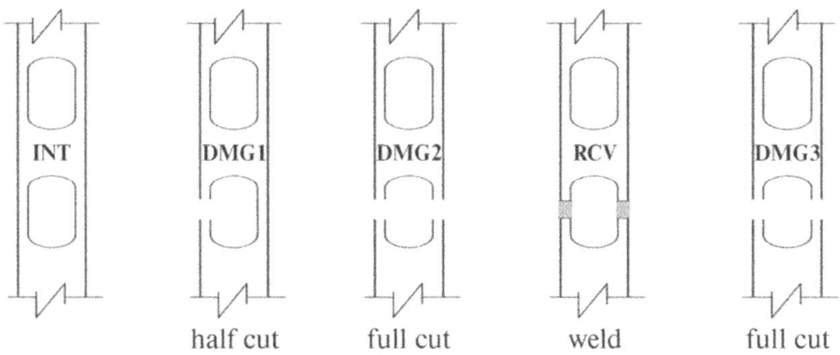

FIGURE 10.3 Old ADA Bridge: damage scenarios applied to the tension members (adapted from Kim et al. [11]).

After more than 50 years of service, the infrastructure was scheduled for demolition in 2012. Before the bridge removal, dynamic tests were carried out (March 1–2, 2012, from 10:00 to 17:00), with artificial damages being introduced during the tests.

Acceleration responses from both vehicle-induced and ambient sources were recorded at a sampling rate of 200 Hz, using eight uniaxial accelerometers placed vertically on the deck (Figure 10.2); however, only vehicle-induced data are considered in this study. The vehicle used for the tests was a two-axle vehicle (Serena model; Nissan Motor Co. Ltd.) with a total weight of 21 kN. It is further noticed that no other vehicles were allowed to use the bridge during the tests. For more details on the experimental measurements, the interested reader is referred to Kim et al. [11]. As shown in Figure 10.3, five damage scenarios were chosen for the vibration tests:

- INT, where no artificial damage was applied;
- DMG1, involving a half-cut to a vertical truss member at the midspan of the bridge, on the same side as accelerometer A3;
- DMG2, featuring a full cut to the same vertical truss member at midspan, in the vicinity of accelerometer A3;
- RCV (recovery state), wherein the full cut of the midspan member was repaired;
- DMG3, where a full cut was made to a vertical truss member at 5/8 of the span length, on the same side of accelerometer A4.

It should be noted that (a) the cuts were made to simulate damage patterns caused by corrosion or overloading, and (b) for the RCV state, a jack was utilized to minimize the gap in the cut member before welding steel components, though full restoration to the original intact state of the bridge was not guaranteed [11].

Each scenario involved multiple test runs. In the INT scenario [11], the speed of the moving vehicle varied from 30 to 50 km/h. Since the signal lengths differed slightly across runs, the first 40 seconds of data were selected for analysis in each

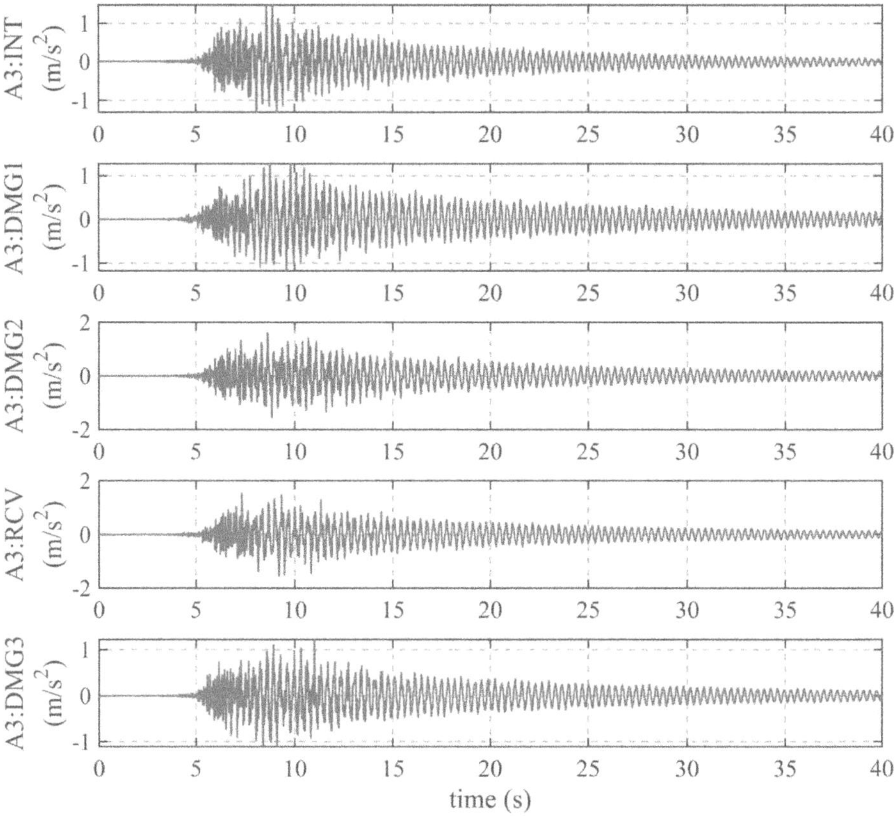

FIGURE 10.4 Vehicle-induced accelerations in different scenarios (sensor A3).

case (Figure 10.4). It is worth mentioning that, as shown in Figure 10.4, distinguishing between intact and damaged scenarios based on the time history alone was challenging.

In the work of Al-Ghalib [12], linear discriminant analysis is applied to power spectral density functions to detect the inflicted damages. This method, not implying any system identification, does not provide any insight into the damage location; the classification results do not allow to distinguishing between points that are close to or far from the damaged regions. In the work of Chang and Kim [13], a stabilization diagram-aided multivariate autoregressive analysis is employed to determine the bridge's modal parameters. Natural frequencies turn out to be effective damage indicators when a sufficient number of modes are considered and combined with a multivariate control chart. On the other hand, mode shapes exhibit detectable variations only when the damage is applied asymmetrically.

It is worth mentioning that the tests on the ADA Bridge were conducted over two consecutive days under consistent temperature conditions (approximately 15°C), ensuring minimal environmental effects on the bridge's dynamic characteristics.

Hence, the measured accelerations were first used to assess the effectiveness of the SAE in detecting and localizing anomalies, with negligible influence of the EOV.

10.4.2 Training of the SAE

A total number of 26 (8,000 × 8) matrices, collected in the INT scenario, are used for training and testing. In more detail, 19 datasets (approximately corresponding to 75% of the total) are used for training, and the remaining 7 (about 25%) for validation purposes.

The optimal values of the SAE parameters, estimated by applying the grid search method [19], are (a) encoder and decoder functions as sigmoid and linear, respectively, (b) mini-batch size equal to 4, (c) epochs equal to 500, and (d) compression rate equal to 30% of the signal length.

It should be noticed that the time series are rescaled to [–1, +1] to avoid numerical issues in the convergence of the ADAM optimization method [21]. The computational time required during the training procedure was about 20 minutes, using a PC with a 16 GB RAM and a 2.7 GHz Intel Core i5 dual-core processor.

After estimating the SAE parameters in the INT scenario, the matrices of accelerations from the other scenarios (i.e., DMG1, DMG2, RCV, and DMG3) are fed into the network. As previously stated, the indicator adopted to evaluate the quality of the data reconstruction is the MAE between the input and the output sequences. It is expected that—for data associated with DMG1, DMG2, and DMG3—the network is not fully capable of reconstructing the acceleration sequences, resulting in MAE values higher than those obtained for the INT scenario. On the other hand, it is expected that the MAE values associated with the RCV state should be comparable to those of the INT scenario.

By assuming that the MAE values follow a normal distribution and that most of the values lie within 99% of the curve, the 99th percentile of the MAE values computed during training is selected as the threshold for outlier occurrence. Figure 10.5 shows the reconstructed sequences, respectively from the INT (validation set) and DMG2 scenarios, confirming the capability of the trained network to reconstruct the original time series when collected from a normal intact condition. On the other hand, the network fails to reconstruct the signal if it comes from a damaged scenario.

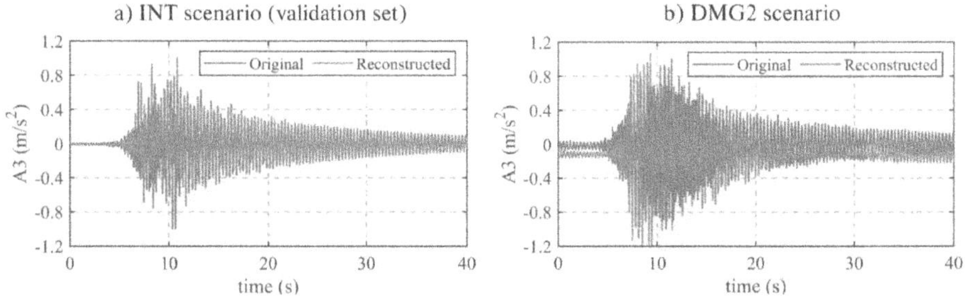

FIGURE 10.5 Sensor A3: comparison between original and reconstructed sequences.

10.4.3 Testing of the SAE and discussions

Figure 10.6 shows the evolution of the MAE over the test datasets and for each channel. By assuming the 99th percentile of the MAE values computed with the training data as the threshold to classify the data reconstruction as normal or anomalous, it is possible to conclude that (a) almost all the time series collected in the damaged scenarios (i.e., DMG1, DMG2, and DMG3) are correctly classified; (b) within the same damage scenario, higher MAE values correspond to points that are closer to the damaged strut; (c) the MAE values increase with increased damage severity;

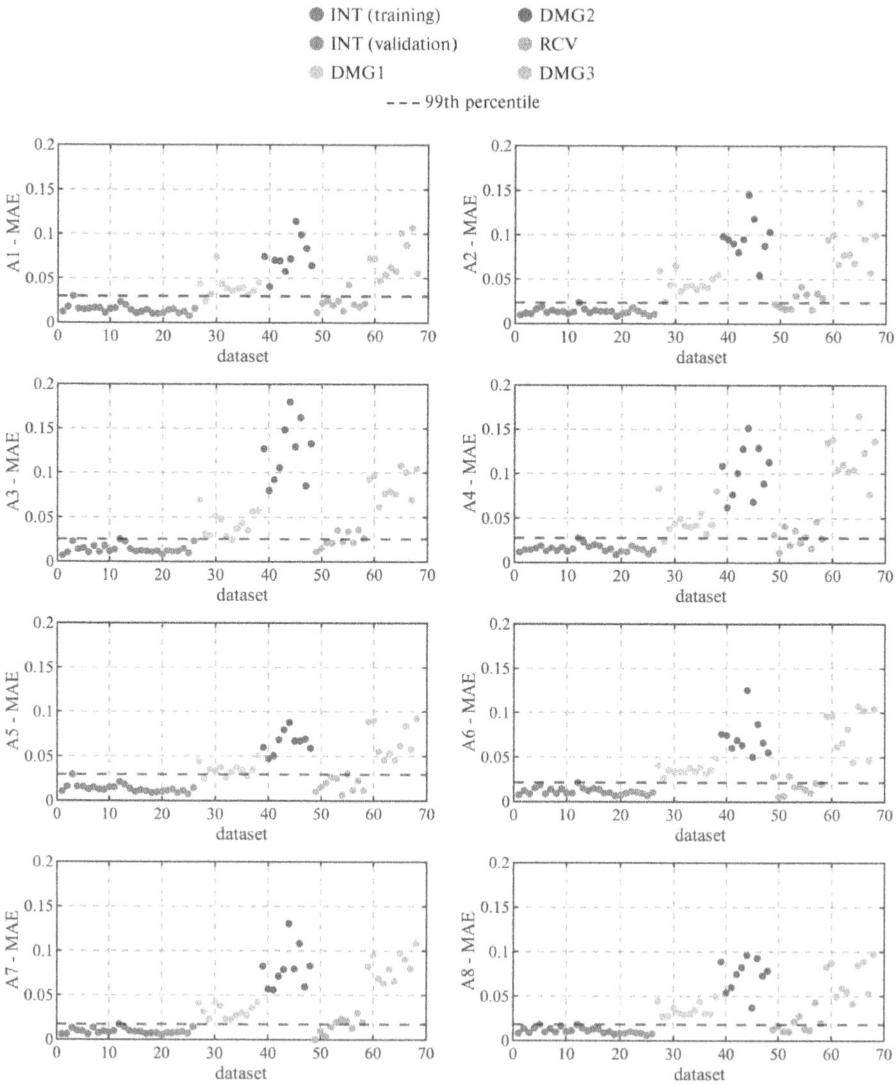

FIGURE 10.6 Mean absolute error (MAE) over the datasets.

TABLE 10.2
Performance indicators for the damage detection method computed with (a) and without (b) the datasets collected in the RCV scenario.

%		A1	A2	A3	A4	A5	A6	A7	A8
ACC	(a)	95.9	87.8	91.8	85.7	95.9	93.9	89.8	91.8
	(b)	97.4	100	100	97.4	97.4	100	100	100
DR	(a)	96.9	100	100	96.9	96.9	100	100	100
	(b)	96.9	100	100	96.9	96.9	100	100	100
FAR	(a)	5.9	35.3	23.5	35.3	5.9	17.6	29.4	23.5
	(b)	0	0	0	0	0	0	0	0

TABLE 10.3
Average ($\overline{\text{MAE}}$) and standard deviation (σ_{MAE}) of MAE for each test scenario.

Scenario	Index	A1	A2	A3	A4	A5	A6	A7	A8
INT	$\overline{\text{MAE}}$	0.015	0.014	0.014	0.016	0.014	0.012	0.009	0.011
	σ_{MAE}	0.004	0.003	0.005	0.004	0.004	0.003	0.002	0.003
DMG1	$\overline{\text{MAE}}$	0.037	0.045	0.044	0.046	0.036	0.036	0.032	0.034
	σ_{MAE}	0.010	0.010	0.016	0.017	0.008	0.005	0.007	0.007
DMG2	$\overline{\text{MAE}}$	0.074	0.097	0.123	0.099	0.066	0.073	0.080	0.074
	σ_{MAE}	0.020	0.023	0.033	0.026	0.012	0.021	0.023	0.018
RCV	$\overline{\text{MAE}}$	0.021	0.025	0.024	0.027	0.017	0.017	0.014	0.018
	σ_{MAE}	0.008	0.009	0.008	0.011	0.008	0.008	0.009	0.010
DMG3	$\overline{\text{MAE}}$	0.071	0.087	0.086	0.119	0.067	0.080	0.083	0.069
	σ_{MAE}	0.020	0.022	0.016	0.024	0.019	0.024	0.014	0.020

(d) the RCV scenario leads to ambiguities in the classification because—as stated by Chang and Kim [11]—the complete restoration of the intact state was not attained [11].

To better demonstrate the performance of the anomaly detection method, Table 10.2 summarizes the values of ACC, DR, and FAR that were computed including (a) and removing (b) the datasets collected in the RCV scenarios, respectively. It should be noticed that the trained network always shows great capabilities of ACC and DR, with values higher than 85% and 96%, respectively. On the other hand, the FAR performance—which is not negligible when computed including the RCV scenarios—decreases to 0% for all channels after removing the "uncertain" datasets.

Furthermore, the average ($\overline{\text{MAE}}$) and standard deviation (σ_{MAE}) of the MAE values, computed for each channel and damage scenario and reported in Table 10.3, also allow the unambiguous identification of the damaged portion of the bridge in the different scenarios. For instance (see Table 10.3), channels A2–A3–A4 exhibit higher $\overline{\text{MAE}}$ and σ_{MAE} and correctly indicate the location of damage DMG1 (which is applied in the vicinity of sensor A3).

As the severity of the damage increases, the inspection of the values reported in Table 10.3 reveals that the damage localization capability increases as well, since DMG2 scenario affects mostly the zone closer to channel A3, and the DMG3 scenario affects the zone closer to channel A4, confirming the damage localization performance.

10.5 THE HISTORICAL SAN MICHELE BRIDGE (1889)

10.5.1 THE BRIDGE AND THE MONITORING SYSTEM

The San Michele Bridge is an iron arch structure that connects the small towns of Paderno and Calusco d'Adda, about 50 km from Milan [14]. Constructed in 1889 to complete one of Italy's earliest railway lines, the bridge is still in use, serving both road and rail traffic. As illustrated in Figures 10.7 and 10.8, the bridge's key features include a parabolic iron arch, an upper truss box metal girder, and supporting piers. The arch consists of two ribs, with a span of 150 m and a rise of 37.5 m; as shown in Figures 10.7 and 10.8, the two ribs tilt inward, with the distance ranging between 16

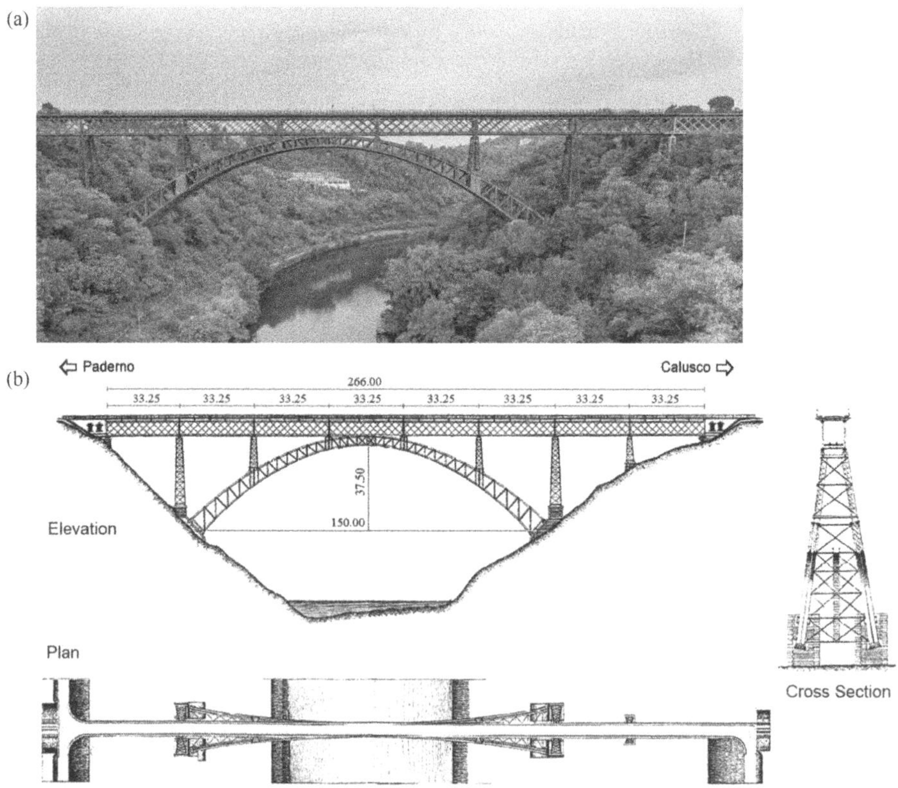

FIGURE 10.7 (a) View of the San Michele Bridge (1889); (b) elevation, plan, and cross section of the bridge (dimensions in m, adapted from Società Nazionale delle Officine di Savigliano [14]).

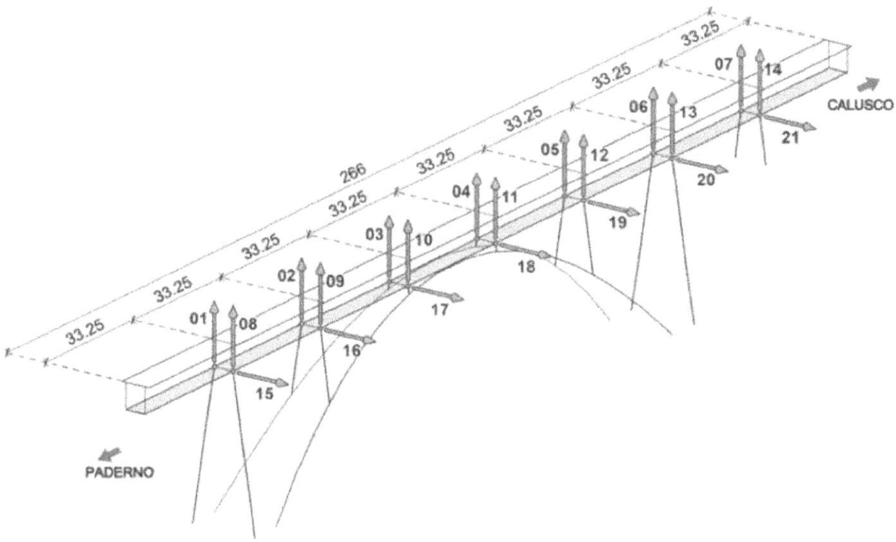

FIGURE 10.8 Schematic of the continuous dynamic monitoring system installed on the San Michele Bridge [16].

m (at the base) and 5 m (at the top). The upper girder is 266 m long and is supported by nine equally spaced bearings. The girder supports two levels: the upper deck accommodates alternating road traffic and pedestrians, whereas the lower deck is dedicated to a single railway line.

The bridge's condition has deteriorated over time due to challenges in routine maintenance, particularly on the arches, with many structural components suffering from significant corrosion. As part of an infrastructure assessment program conducted by the Province of Lecco, ambient vibration tests were performed on both the roadway and railway decks between June 2009 and June 2011 [15]. The initial findings revealed a correlation between the dynamic characteristics of the structure and the amplitude of the excitation and response [15].

In collaboration with the Italian Railway Authority (RFI), the main institutional owner of the bridge, it was decided to install a dynamic monitoring system on the railway deck, as maintenance activities could be more efficiently managed there with support from RFI technical staff. The dynamic monitoring system has been fully operational since November 28, 2011 [16, 17], and the data collected during the first 18 months of monitoring are herein considered.

The monitoring system included 21 MEMS accelerometers, seven data acquisition (DAQ) units, two thermocouples, two Ethernet switches, and one industrial PC. Seven cross sections of the bridge, corresponding to the truss-box girder bearings between the abutments (Figure 10.8), were instrumented with three accelerometers: two sensors measured the vertical accelerations on the upstream and downstream sides, whereas the third one measured the lateral response. The temperature sensors were installed on the second and fifth instrumented cross sections (starting

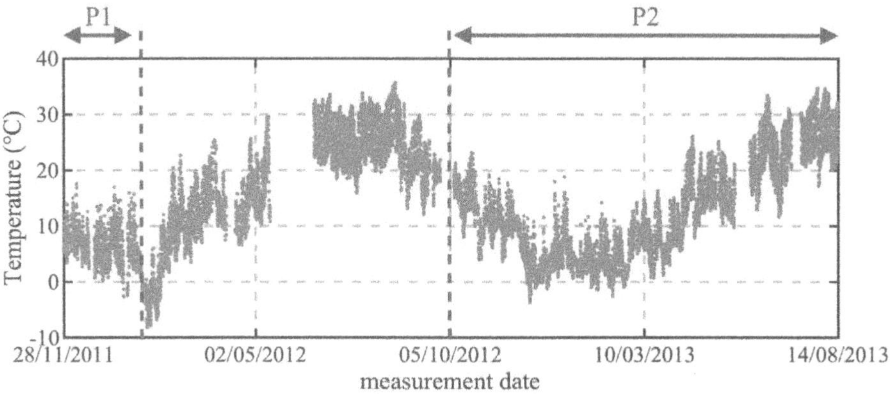

FIGURE 10.9 Average temperature between downstream and upstream records from November 28, 2011 to August 14, 2013.

from the Paderno side, Figure 10.8). The two Ethernet switch devices aggregate data from each group of channels and transmit the digitized information to the on-site industrial PC. Every hour, a new binary file containing 21 acceleration time series (sampled at 200 Hz) and temperature data was created, stored locally on the PC, and transmitted to the Politecnico di Milano for further analysis.

In previous studies [16, 17], automated OMA was used to identify the modal parameters and track their evolution over the monitoring period. Mode shape variations in terms of the modal assurance criterion (MAC) and mean phase deviation (MPD) [16, 17] showed a gradual variation in the bridge behavior, especially in the region corresponding to the crown of the arch, with this region exhibiting a worse state of preservation and greater corrosion than the neighboring areas.

In the present application, seven vertical accelerations measured on the downstream side (corresponding to sensors 1–7 in Figure 10.8) were used to test the SAE-based methodology. Figure 10.9 shows the temperature record obtained by averaging the records on the upstream and downstream sides of the bridge from November 28, 2011 to August 14, 2013: several distinct periods were affected by snowfalls and low temperatures, such as during the first weeks of February 2012 and, more frequently, between December 2012 and February 2013.

10.5.2 Training of the SAE

From each 1-h dataset, a one-minute segment is extracted, including the seven vertical accelerations measured on the downstream side; hence, the signals are organized into 7 × 2,400 matrices (seven channels, each with 2,400 time samples).

Two different SAE networks were defined, corresponding to two different training periods, P1 and P2. More specifically:

1. The training period P1 spans from November 28, 2011, to February 01, 2012; 1,274 signal matrices are selected for training purposes. During training,

the average temperature (Figure 10.9) varies from –2.6°C to 17.6°C, but the first weeks of February 2012—affected by the lowest temperatures—are not included in the training period.

2. The training period P2 spans from October 05, 2012, to August 14, 2013; 9,860 signal matrices are adopted during the training phase. The average temperature (Figure 10.9) varies over a wider range compared to case P1, i.e., from –3.6°C to 34.7°C.

In both cases, about 75% of the matrices collected during training are used to train the SAE model, while the remaining 25% is set aside for validation purposes. Moreover, the time series are normalized within [–1, +1].

The training process, performed through the grid search method [19], involved the following parameters: (a) the encoder and decoder functions are set as sigmoid and linear, respectively; (b) mini-batch size equal to 16; (c) 1,000 and 3,000 epochs are assumed for P1 and P2, respectively; (d) the adopted compression rates are equal to 15% and 30% of the signal length for P1 and P2, respectively.

The training took approximately 20 min and 50 min for P1 and P2, respectively, on a PC with 16 GB of RAM and a 2.7 GHz Intel Core i5 dual-core processor.

10.5.3 Testing of the SAE and discussions

Figure 10.10 exemplifies the reconstruction capability of the network; the figure refers to the P1 training procedure and exemplifies reconstructed signals from (a) validation set and (b) testing set (Figure 10.10a,b). Figure 10.10b clearly highlights that the trained network is not able to accurately reconstruct the testing time series, suggesting that the testing data include conditions not present during training.

To better understand this result, Figures 10.11 and 10.12 show the evolution of the MAE for four selected channels (i.e., channels 1, 3, 4, and 6),

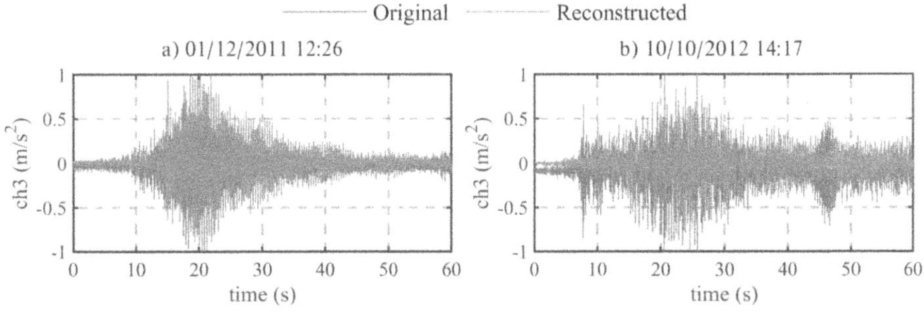

FIGURE 10.10 Channel 3, comparison between original and reconstructed sequences: (a) validation set; (b) testing set.

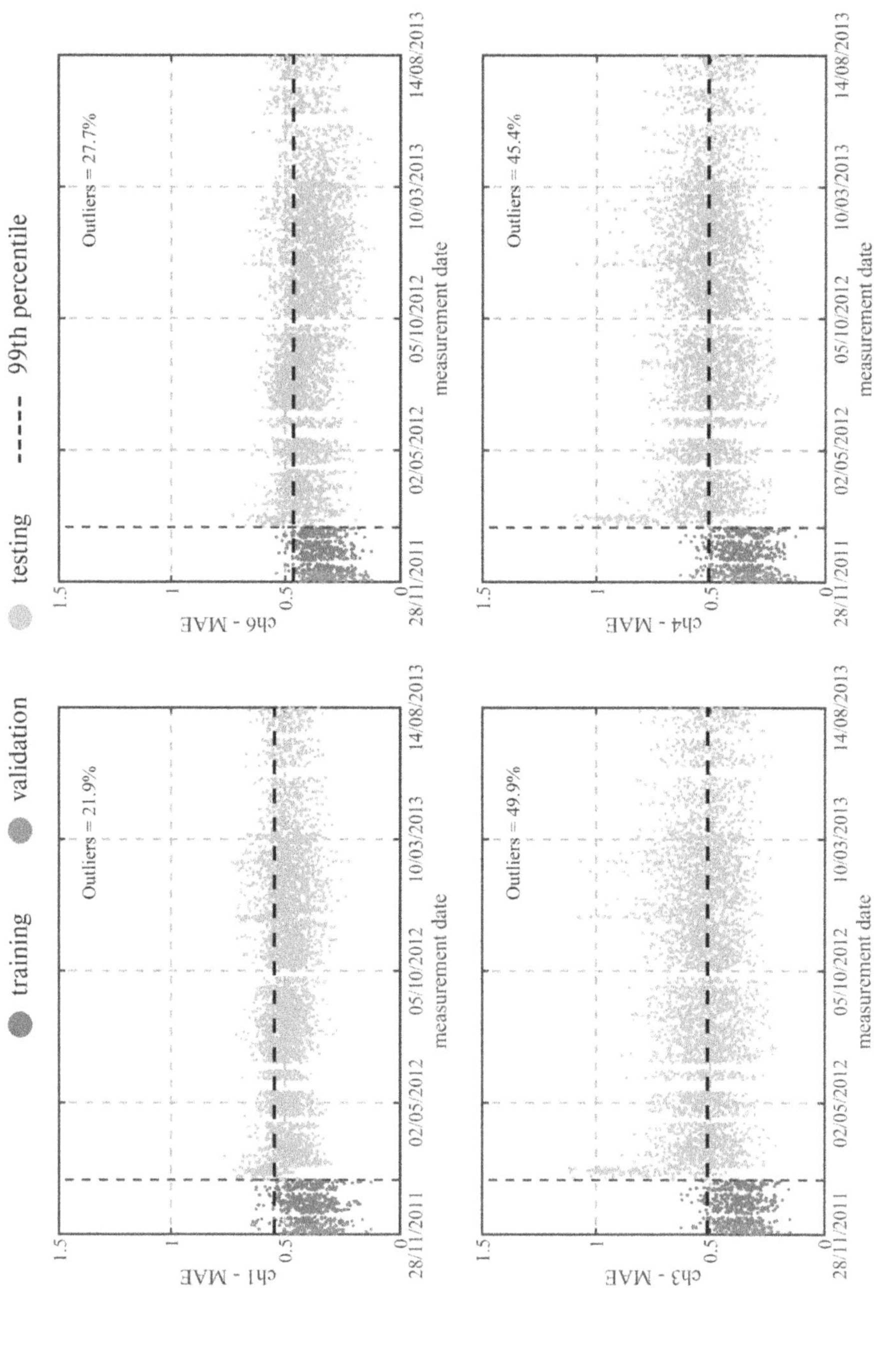

FIGURE 10.11 Mean absolute error (MAE) for the P1 training procedure.

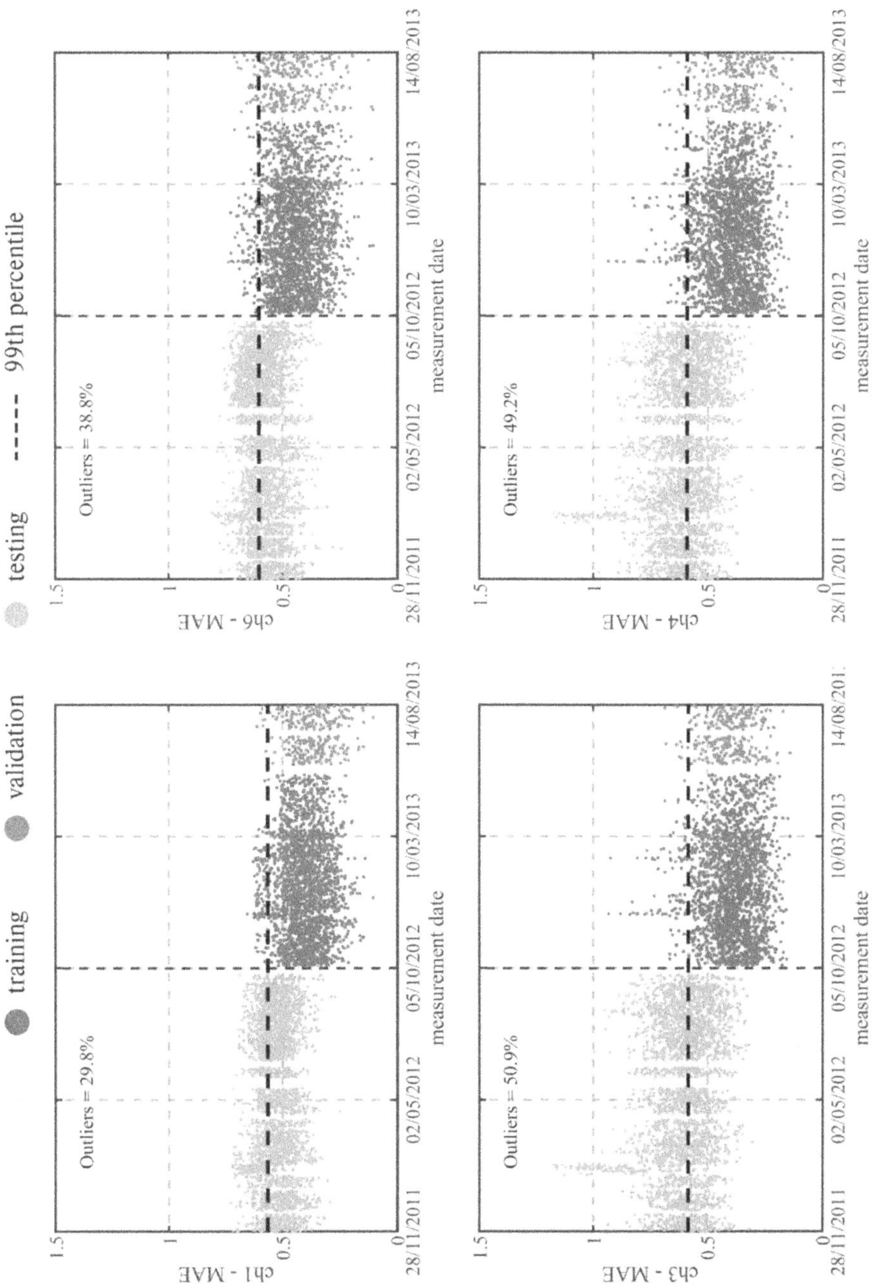

FIGURE 10.12 Mean absolute error (MAE) for the P2 training procedure.

TABLE 10.4
Average ($\overline{MAE}$) and standard deviation (σ_{MAE}) of MAE in the testing periods.

Index		Ch1	Ch2	Ch3	Ch4	Ch5	Ch6	Ch7
$\overline{MAE}$	P1	0.493	0.493	0.531	0.530	0.467	0.410	0.355
	P2	0.526	0.589	0.606	0.610	0.538	0.576	0.372
σ_{MAE}	P1	0.075	0.118	0.125	0.124	0.109	0.089	0.049
	P2	0.073	0.108	0.123	0.124	0.103	0.082	0.042

considering the two selected training periods, P1 and P2. The following should be noted:

1. In both cases, the testing set exhibits reconstruction errors that exceed the control limit, set as the 99th percentile of the training values. Moreover, up to 50% of outliers are observed in both testing periods.
2. During the testing periods, the reconstruction errors experience higher values and larger dispersion for the channels placed close to the crown of the arch (Table 10.4), so that the worse state of degradation of this region reported by Gentile and Saisi, and Cabboi et al. [16, 17] is confirmed.
3. The first two months of training (P1, Figure 10.11) are sufficient to provide a damage-sensitive feature (i.e., the MAE) that is informative of the structural degradation experienced by the bridge.
4. By extending the training period to almost one year (P2, Figure 10.12), a clearer visualization of the structural changes becomes evident, proving that EOVs (e.g., temperature) in dynamic measurements could only have been implicitly accounted for during training.

10.6 CONCLUSIONS

A new procedure for anomaly detection and localization is presented and discussed. The proposed approach is based on the availability of continuous dynamic monitoring and the definition of an appropriate SAE network. It should be remarked that, unlike traditional procedures where one SAE network is defined for each available data channel, the time series recorded by all available sensors are concurrently used to train only one network. When the structural behavior deviates from the baseline condition established during training, the SAE's ability to accurately reconstruct the signals decreases, potentially triggering an alert. In addition, the increase in the reconstruction error is conceivably more significant when the reconstructed signal refers to one sensor placed in the neighborhood of damage so that localization of critical regions is attained as well.

The reliability and accuracy of the proposed technique are demonstrated using data collected on the old ADA bridge [11–13] and the historic San Michele bridge[14–17].

The possibility of detecting the artificial damage scenarios inflicted on the ADA bridge before the infrastructure's decommissioning is first investigated: in this case, the experimental measurements were performed at constant temperature, and the scenarios are characterized by different intensities and positions of the inflicted damage, also including a partial restoration of the intact state. The results highlight that the proposed SAE network is capable of correctly classifying the differences in terms of occurrence, position, and intensity of the stiffness loss.

Since the data collected on the ADA bridge do not allow investigating the sensitivity to EOV, the second benchmark refers to the continuous monitoring of the San Michele bridge over a period of about 18 months. Within this application, the presented approach turns out (a) to be capable of detecting and localizing damage using a short training period (only two months) and (b) to implicitly capture the environmental and operational effects on the dynamic responses during the training phase.

As a final remark, the developed methodology—involving the definition of a unique SAE network from multi-channel monitoring—shows great promise in vibration-based SHM due to the quality of results and computational efficiency.

REFERENCES

[1] Farrar CR, Worden K (2007) An introduction to structural health monitoring, *Philosophical Transactions of the Royal Society A* 365(1851): 303–315.

[2] Magalhães F, Cunha Á, Caetano E (2012) Vibration based structural health monitoring of an arch bridge: From automated OMA to damage detection, *Mechanical Systems and Signal Processing* 28: 212–228.

[3] Borlenghi P, Gentile C, Pirrò M (2023) Continuous dynamic monitoring and automated modal identification of an arch bridge, in: Rizzo, P, Milazzo, A (eds) *European Workshop on Structural Health Monitoring*. EWSHM 2022. Lecture Notes in Civil Engineering, vol 254. Springer, Cham.

[4] Peeters B, De Roeck G (2001) One-year monitoring of the Z24-bridge: Environmental effects versus damage events. *Earthquake Engineering & Structural Dynamics* 30: 149–171.

[5] Goodfellow I, Bengio Y, Courville A (2016) *Deep Learning*, MIT Press, USA.

[6] Pathirage CSN, Li J, Li L, Hao H, Liu W (2018) Application of deep autoencoder model for structural condition monitoring, *Journal of Systems Engineering and Electronics* 29(4): 873–880.

[7] Finotti RP, Gentile C, Barbosa F, Cury A (2022) Structural novelty detection based on sparse autoencoders and control charts, *Structural Engineering and Mechanics* 81(5): 647–664.

[8] Wang Z, Cha YJ (2021) Unsupervised deep learning approach using a deep autoencoder with a one-class support vector machine to detect damage, *Structural Health Monitoring* 20(1): 406–425.

[9] Pirrò M (2025) *Artificial Intelligence for the Condition-Based Structural Maintenance of Green-Energy Infrastructures*. PhD Thesis, Politecnico di Milano, Milan, Italy.

[10] Pirrò M, Pereira S, Gentile C, Cunha Á (2024) Assessment of structures using dynamic monitoring and auto-encoders: Application to Baixo Sabor Dam, in: Rainieri C, Gentile C, Aenlle López M (eds) *Proceedings of the 10th International Operational Modal Analysis Conference (IOMAC 2024)*. IOMAC 2024. Lecture Notes in Civil Engineering, vol 514. Springer, Cham.

[11] Kim C, Zhang F, Chang K, McGetrick P, Goi Y (2021) Ambient and vehicle-induced vibration data of a steel truss bridge subject to artificial damage, *Journal of Bridge Engineering* 26(7).

[12] Al-Ghalib AA (2022) Structural damage detection of old ADA steel truss bridge using vibration data, *Structural Control and Health Monitoring* 29(11): e3098.

[13] Chang KC, Kim CW (2016) Modal-parameter identification and vibration-based damage detection of a damaged steel truss bridge, *Engineering Structures* 122: 156–173.

[14] Società Nazionale delle Officine di Savigliano (1889) *Il viadotto di Paderno sull'Adda (ferrovia Ponte S. Pietro-Seregno)*, Camilla e Bertolero, Torino.

[15] Gentile C, Saisi A (2011) Ambient vibration testing and condition assessment of the Paderno iron arch bridge (1889), *Construction and Building Materials* 25: 3709–3720.

[16] Gentile C, Saisi A (2014) Continuous dynamic monitoring of a centenary iron bridge for structural modification assessment, *Frontiers of Structural and Civil Engineering* 9: 26–41.

[17] Cabboi A, Gentile C, Saisi A (2014) Vibration-based SHM of a centenary bridge: A comparative study between two different automated OMA techniques, in: *Proceedings of the 9th International Conference on Structural Dynamics*. EURODYN 2014 (pp. 1461–1468). Porto.

[18] Ng A (2011) Sparse auto-encoder, *CS294A Lecture Notes* 72: 1–19.

[19] Yang L, Shami A (2020) On hyperparameter optimization of machine learning algorithms: Theory and practice, *Neurocomputing* 415.

[20] Touati R, Mignotte M, Dahmane M (2020) Anomaly feature learning for unsupervised change detection in heterogeneous images: A deep sparse residual model. *IEEE Journal of Selected Topics in Applied Earth Observations and Remote Sensing* 13: 588–600.

[21] Kingmaand DP, Ba J (2015) Adam: A method for stochastic optimization, in: *Proceedings of 3rd International Conference on Learning Representations (ICLR)*, San Diego, CA, USA.

[22] Singh D, Singh B (2020) Investigating the impact of data normalization on classification performance. *Applied Soft Computing* 97: 105524.

11 OMA-driven FE model validation

An engineering perspective on a lively footbridge

M.G. Mulas, C. Fortis, and Carmelo Gentile

11.1 INTRODUCTION

In the last decades, the availability of advanced sensors and testing procedures has promoted the development of experimental research on bridges. Associated with static tests, dynamic testing has led, on one side, to the identification of dynamic properties in proof tests, allowing the validation of finite element (FE) models and, on the other side, to the detection of uncomfortable levels of vibration in lively footbridges [1–3]. In fact, the structural conception of footbridges is often driven by aesthetic requirements so that significant slenderness is attained. Research studies have addressed the subsequent excessive vibration through performance-based serviceability assessment [4]. Particular focus has been on slender footbridges, since they can experience a modification of their modal properties when crossed by pedestrians, generating an increase in mass [5] and damping [6]. The presence of human-structure interaction affects the load conditions. New tools for pedestrian modeling and procedures of analysis appear necessary [7, 8]. In this scenario, current knowledge can advance when experimental and numerical studies, preferably based on a real structure, are carried out in parallel. The experimental identification of modal properties allows the validation of FE models. Measuring the response to human excitation drives research both in establishing and assessing serviceability conditions and in simulating the dynamic response to pedestrian loads.

Typically, benchmark studies address these aspects. The literature lists one performed in 2012 on the Podgorica footbridge in Montenegro [9]. A second dataset was collected in 2017 [10] on a 12m-long laboratory footbridge, ad hoc designed for studying human-structure interaction. Two recent studies concern the Eeklo footbridge (Belgium) [11] and the Streicker footbridge (Princeton, NJ, USA) [12]. In these studies, simple footbridges and unrestricted pedestrian traffic are considered; however, detailed data on structures, FE models, or forced vibration tests are not always available.

This chapter deals with the validation of an FE model of a footbridge located in Northern Italy, which was the subject of previous studies [13, 14]. Ambient vibration tests (AVTs), performed during proof tests, led to the identification of modal

DOI: 10.1201/9781003516941-11

properties: a critical frequency value of 1.75 Hz was associated with the first bending mode. An FE model reproduced this frequency with a practically negligible error but was unable to reproduce the frequency of the second mode, a torsional one [13]. The response to pedestrian traffic was determined through on-site tests with 52 groups of pedestrians walking at the 1st mode frequency [13]. Serviceability assessment, based on experimental data and the numerical procedure of the HiVoSS guidelines, was subsequently performed [14]. Herein, the numerical studies focus on the discrepancy on the 2nd mode frequency. The working approach for updating the FE model involves a careful study of the design data, associated with sensitivity analysis and optimization. However, the derivation of the sought model, able to simulate bending and torsion modes with the same precision, could only be achieved when engineering judgment came into play.

In the following, Sections 11.2 and 11.3 describe the case study and the results of the structural identification. The FE modeling strategy to overcome the lack of simulation of the torsional modes is presented in Sections 11.4 and 11.5. Conclusions are drawn in Section 11.6.

11.2 DESCRIPTION OF THE FOOTBRIDGE

Figure 11.1 shows a picture taken during the proof tests of the footbridge over the Lambro River (Municipality of Cerro al Lambro, Italy). The total length of 107 m is subdivided into a central span 58 m long and two side spans of 24.5 m each. The width is constant at 4.4 m.

At the two sides of the footbridge, 2-meter-high welded longitudinal I-beams provide the primary structure. The two identical beams are subdivided into 10 on-site

(a)

(b)

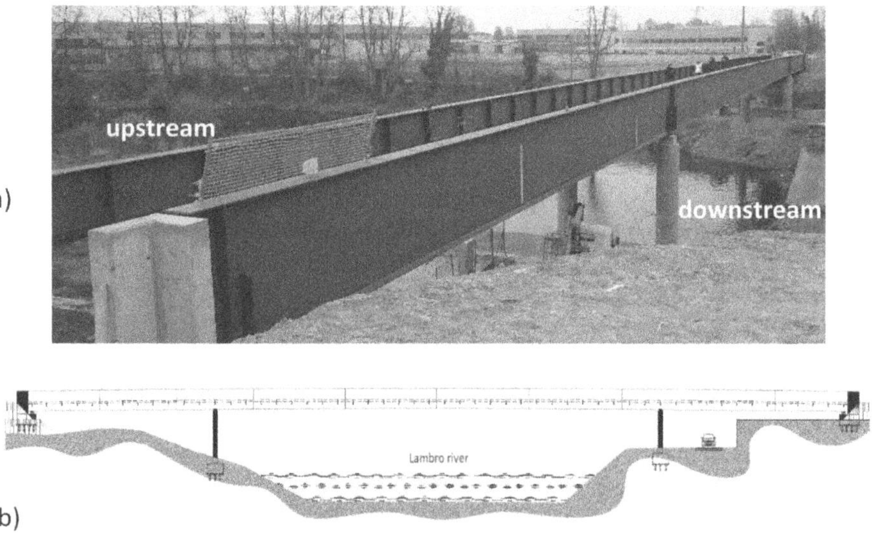

FIGURE 11.1 Lambro footbridge: (a) general view; (b) side view (adapted from Cigada et al. [13]).

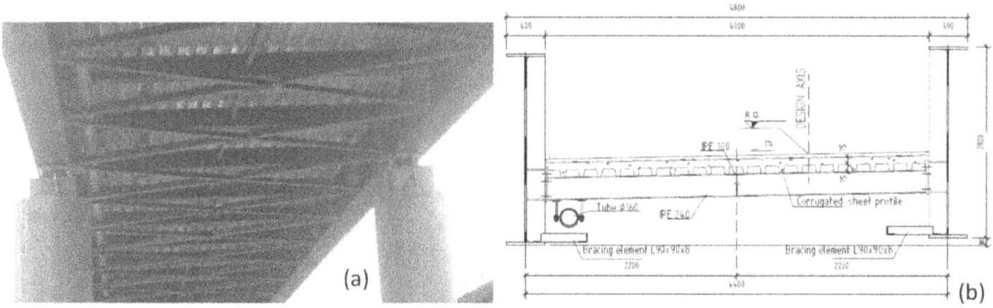

FIGURE 11.2 Lambro footbridge: (a) view from below and (b) cross section (adapted from Cigada et al. [13]).

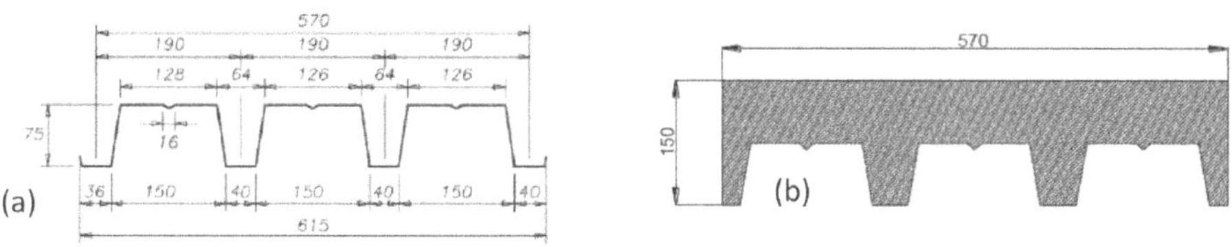

FIGURE 11.3 Cross section of the deck (units: mm): (a) corrugated sheet A75/P 570; (b) central span with cast-in-situ concrete (from Cigada et al. [13]).

welded segments, maintaining symmetry about the transverse axis at mid-span. Figure 11.2a exemplifies a view of the deck from below, including the intermediate pylons, and illustrates: (a) the cross-brace system (consisting of steel L90 × 8 profiles bolted every 2 m and on a width of 4.4 m) at the bottom flange of the longitudinal beams; (b) the transverse beams and the IPE 100 secondary beam connecting the mid-span of the transverse beams. Figure 11.2b presents a typical cross section. The transverse beams of the secondary resisting system, spaced 2 m apart, use IPE240 steel profiles in the central span and HEA240 profiles on the intermediate supports and at the side spans. These beams support a reinforced concrete (RC) slab cast on a metal corrugated sheet, overlaid with an overlying bituminous conglomerate pavement. The deck has an inclination of 2% to allow rainwater to drain into a circular PVC channel placed on one side only.

Two pairs of circular RC columns, 85 cm in diameter, provide the intermediate support (Figure 11.1b). The heights of the columns are 6.9 m on the right and 5.5 m on the left of the upstream side, respectively. The column foundations are on piles. At all supports, the bottom flange of the longitudinal beams rests on elastomeric bearings SI-H 300/52 (diameter 300 mm, height 52 mm), which in turn rest on an RC pier cap measuring 45 × 45 cm, 20 and 28 cm high on the upstream and downstream sides, respectively.

With the same A75/P750 corrugated sheet profile depicted in Figure 11.3a, two cross sections of the deck are constructed. In the central span, the sheet thickness is 0.7 mm, with a mass of 9.85 kg/m^2, and the cast-in-situ slab is 75 mm thick, giving a total deck thickness of 150 mm (Figure 11.3b). At the two ends, for a length of 4.5 m from the abutments, the sheet thickness is 1.5 mm, with a mass of 20.65 kg/m^2, and the cast-in-situ slab is 305 mm thick, giving a total thickness of 380 mm. In these zones, acting as ballast for the central span, the metal sheet is directly connected to the lower flanges of the I-profiles. A 10 mm thick waterproof sheath is interposed between the deck and a 30 mm thick concrete pavement. Their total weight is 1.2 kN/m^2.

All steel profiles are made of S335 steel, with ultimate strength f_u = 510 MPa and yield strength f_y = 335 MPa. The corrugated sheet is made of S235 steel (f_u = 360 MPa, f_y = 235 MPa). The deck is made of concrete of class C30/37, with Young's modulus E_c = 33.0 GPa, characteristic strength f_{ck} = 30.7 MPa, and design strength f_{cd} = 17.4 MPa. The class of concrete for the columns is C32/40 (E_c=33.6 GPa, f_{ck} = 33.2 MPa, f_{cd} = 18.8 MPa). Further details on the footbridge geometry can be found in [15, 16].

11.3 DYNAMIC CHARACTERISTICS FROM OPERATIONAL MODAL ANALYSIS

Ambient vibration tests (AVTs) were performed to evaluate the dynamic properties of the footbridge. Figure 11.4 shows the accelerometer layout adopted in the experimental campaign for both AVTs and forced vibration tests. During the AVTs, only vertical sensors A1-A10 were installed on the main side beams.

The modal parameters were evaluated by applying the Frequency Domain Decomposition (FDD) [17] and the data-driven Stochastic Subspace Identification (SSI-data) [18] algorithms to output-only data, implemented in the commercial software ARTeMIS [19]. As shown in Figure 11.5 and Table 11.1, the two modal

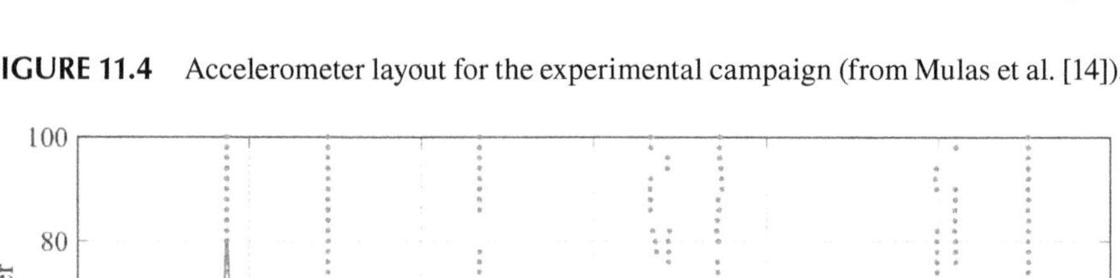

FIGURE 11.4 Accelerometer layout for the experimental campaign (from Mulas et al. [14]).

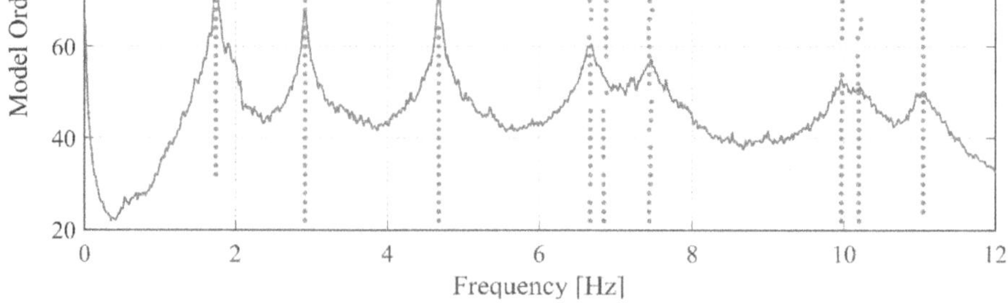

FIGURE 11.5 Stabilization diagram (SSI-data) and modal identification.

TABLE 11.1
Modal parameters (B = Bending; T = Torsion) identified through FDD and SSI.

Mode Id.	1 (B)	2 (T)	3 (B)	4 (B)	5 (T)	6 (B)	7 (T)	8 (B)	9 (T)
f_{FDD} (Hz)	1.748	2.910	4.668	6.660	6.797	7.441	9.954	10.19	11.03
f_{SSI} (Hz)	1.749	2.912	4.675	6.676	6.862	7.443	9.956	10.19	11.06
ζ_{SSI} (%)	0.135	0.758	0.249	1.001	3.084	1.389	1.373	1.774	1.172

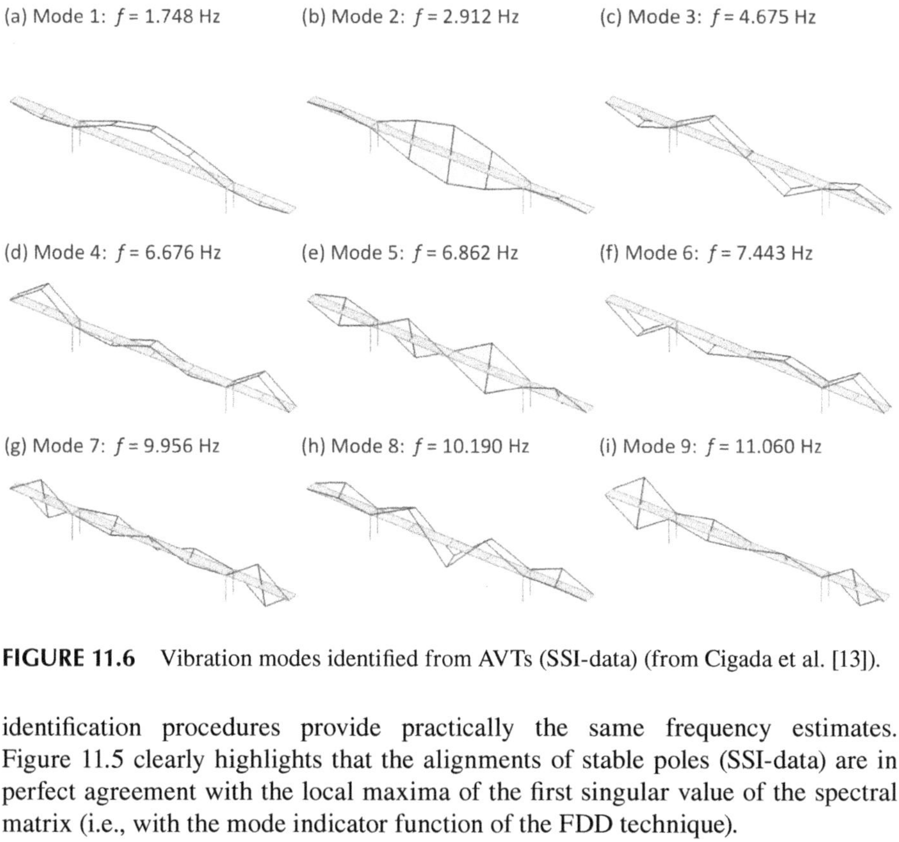

FIGURE 11.6 Vibration modes identified from AVTs (SSI-data) (from Cigada et al. [13]).

identification procedures provide practically the same frequency estimates. Figure 11.5 clearly highlights that the alignments of stable poles (SSI-data) are in perfect agreement with the local maxima of the first singular value of the spectral matrix (i.e., with the mode indicator function of the FDD technique).

Figure 11.6 shows the measurement-based mode shapes identified using the SSI-data technique. These data form the basis for the validation of an FE model of the footbridge. Further details on AVTs are found in the work of Cigada et al. [13].

11.4 FINITE ELEMENT MODEL

Previous works ([13, 15]) presented an FE model developed within the ANSYS [20] framework and based on the as-built design data extracted from the blueprints (Figure 11.7a). In this model, the x- and y-axes of the reference system are the transverse and longitudinal axes, respectively, while the z-axis is vertical upward. Two-node

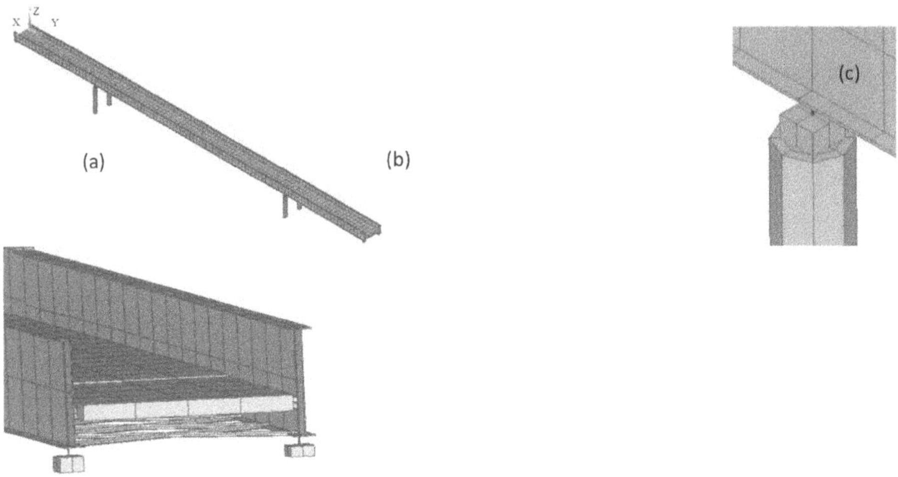

FIGURE 11.7 FE model: (a) general view, (b) detail of column with pedestal and elastomeric bearing, and (c) detail of deck modeling.

(BEAM188) Timoshenko beam elements describe longitudinal and transverse beams, pier caps, and elastomeric bearings (Figure 11.7b). Bearings are fixed at their lower end and free to rotate differently from the longitudinal beam node to which they are connected. The deck, with a mesh grid of 1 m × 1 m (Figure 11.7c), is discretized using SHELL181 elements with a rectangular cross section, with a unique value of thickness, density, and elastic modulus in both the *x*- and *y*-directions. The shape of the corrugated metal sheet (Figure 11.3a) makes the deck orthotropic. Hence, an effective shell thickness, t_e, is calibrated as that of a rectangular cross section providing the same moment of inertia about the transverse axis. Consistent mass options are adopted.

Nodes are located on three levels (Figure 11.2b). Nodes of longitudinal beams are at the level of their axes, accounting for the 2% slope of the deck; nodes of transverse beams are properly offset from them. Nodes of bracing elements are at the level of the bottom flanges of longitudinal beams. The deck is simply supported by transverse beams in the central span: their nodes share only the translational DOFs, while at the end spans, where transverse beams are included within the slab thickness, continuity is enforced.

A model updating based on an optimization process using the SSI data predicted the frequencies of the bending modes (zero error on the 1st mode) but gave a poor representation of the torsion modes (an overestimation of 7.73% on the 2nd mode). To reduce this discrepancy, further studies were conducted in three subsequent MS theses [16, 21, 22]. Zerruso [16] refined the deck mesh and inserted the rotational inertia of the longitudinal beams. The discrepancy decreased to 4.88% for the 2nd mode but increased to 1.37% for the 1st mode. Fontanilla Mercado [16], working with SAP2000©, used the ANSI approach [23] to compute the geometric properties of the composed deck. Following the design documents, an increase in the density of the transverse and longitudinal steel elements by 25% and 5%, respectively, accounted for the connections and stiffening elements in the longitudinal beams. Springs

TABLE 11.2
Natural frequencies and frequency discrepancies for models M_0, M_1, and M_2.

Mode Id.	1 (B)	2 (T)	3 (B)	4 (B)	5 (T)	6 (B)	7 (T)	8 (B)	9 (T)
f_{FDD} (Hz)	1.748	2.910	4.668	6.660	6.797	7.441	9.954	10.19	11.03
f_{M0}(Hz)	1.749	3.137	4.672	6.683	6.782	7.483	9.575	10.11	11.32
ε_f (%)	*0.06*	*7.80*	*0.09*	*0.35*	*−0.22*	*0.56*	*−3.81*	*−0.78*	*2.63*
f_{M1}(Hz)	1.741	3.084	4.651	6.662	6.701	7.462	9.450	10.08	11.12
ε_f (%)	*−0.40*	*5.98*	*−0.36*	*0.03*	*−1.41*	*0.28*	*−5.06*	*−1.08*	*0.82*
f_{M2}(Hz)	*1.630*	3.180	4.344	6.24	7.298	6.974	9.864	9.35	11.25
ε_f (%)	*−6.75*	*9.28*	*−6.94*	*−6.31*	*7.37*	*−6.27*	*−0.90*	*−8.24*	*1.99*

modeled the foundation on piles. The difference in the torsion mode decreased to 0.52%, but that in the 1st mode increased to 2.12%. Hence, in both [16, 21], the simulation of the bending response was negatively affected by the modifications.

These results pointed out the need for systematic work aimed at addressing the modeling uncertainties. Adopting the general assumptions listed earlier, Fortis [22] rebuilt, in a model named M0, the optimized ANSYS FE model [13], adopting for the deck the optimal values of the uncertain parameters: E_c = 30.57 GPa and effective thickness t_{e1} = 111 mm and t_{e2} = 347 mm for the central and side spans, respectively. From M0, two more models were derived, encompassing the features implemented in a SAP2000© FE model [16]: M1 considers an increased density of steel elements, and in M2, the equivalent deck thickness is determined using the ANSI approach.

Table 11.2 lists the natural frequencies and frequency discrepancy $\varepsilon_f = (f_{FEM} - f_{FDD})/f_{FDD}$. The performance of M1 on torsion modes improves slightly but at the expense of reduced accuracy in bending modes. For M2, the values of ε_f are beyond an acceptable range.

11.5 MODEL UPDATING

The scenario posed by the results of FE modeling is not uncommon for practitioners. The discrepancy in torsion modes, with the bending ones being well represented, raises questions about both the soundness of the engineering approach in modeling and the effectiveness of the optimization procedure. Hence, a mixed approach was used to update the model and achieve a good simulation of the torsional behavior. The basis for model updating was a thorough revision of input data concerning geometry, loads, and boundary conditions at the two abutments, implemented in a model named M3. As for the geometry, the deck width w of 4.4 m, due to node positioning, results in a value of 4.0 m in the FE model. Hence, the nominal values of weight per unit area of the corrugated steel sheet q_s, the concrete slab q_c, and the pavement q_p listed in Table 11.3 have been increased by 10%. Given the length L of the typical and abutment slab, the resultant load R on the area $L \times w$ is divided by the deck volume $L \times w \times t_{eq}$ to provide the density ρ of the shell elements. The equivalent thickness t_{eq} is the height of a rectangular cross section having the same area as the actual one.

TABLE 11.3
Loads acting on both typical and abutment slabs and related densities.

Slab	q_s [kN/m²]	q_c [kN/m²]	q_p [kN/m²]	q_{tot} [kN/m²]	$L \times w$ [m × m]	R [kN]	t_{eq} [m]	ρ [kg/m³]
Typical	0.108	2.64	1.32	4.068	98×4	1594.7	0.096	4319.6
Abutments	0.227	8.96	1.32	10.51	9×4	378.3	0.326	3285.4

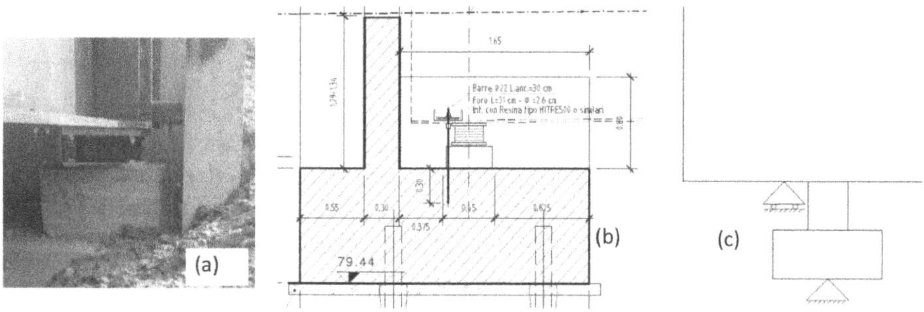

FIGURE 11.8 Boundary conditions at abutments: (a) picture, (b) technical drawing, and (c) FE modeling.

Absent in M0, the details of the boundary conditions at the bridge ends, where pier caps are placed below the elastomeric bearings and four steel rebars connect the bottom flange of the longitudinal beams to the abutment (Figure 11.8a, b), were included in M3. The pier cap is hinged on the abutment, since the two elements were cast at different times, while a roller represents the steel rebars (Figure 11.8c). The properties of the beam element describing the elastomeric bearings were derived from manufacturer data to match their vertical (1,122 × 10^3 kN/m³) and horizontal (1.90 × 10^3 kN/m³) stiffness. Finally, the effective thickness of the slab is the height of a rectangular cross section having the same inertia as the composed section, including the contribution of the steel corrugated sheet—equal to 120.6 mm and 345.2 mm for the typical and abutment slabs, respectively. Offsets of the slab nodes with respect to the longitudinal beam axis are modified accordingly. To account for the presence of connections, the density of the steel transverse elements is increased by 5%. The modal properties of M3 (Table 11.4) show a reduction in the discrepancy of torsion modes without affecting the bending modes.

Given the satisfactory effect of the input data revision, a sensitivity analysis was performed on model M3. Selected parameters include Young's modulus and effective thickness of the slab, transversal steel element density, and stiffness of elastomeric bearings, both horizontal and vertical. The results point out a strong dependency on the slab thickness. In model M4, differing from M3 in the adoption of the reduced moment of inertia of the cracked cross section, the discrepancy on the 2nd mode is reduced to 1.89%. However, two pairs of numerical modes (i.e., modes

TABLE 11.4
Modal properties of FE models with revised geometry and loads.

Mode Id.	1 (B)	2 (T)	3 (B)	4 (B)	5 (T)	6 (B)	7 (T)	8 (B)	9 (T)
f_{FDD}(Hz)	1.748	2.910	4.668	6.660	6.797	7.441	9.954	10.19	11.03
f_{M3}(Hz)	1.747	3.055	4.667	6.669	6.884	7.44	9.737	10.09	11.39
$\varepsilon_{f,M3}$ (%)	*−0.06*	*4.98*	*−0.02*	*0.14*	*1.28*	*−0.01*	*−2.18*	*−1.02*	*3.28*
f_{M4}(Hz)	1.747	2.965	4.656	**6.626**	**6.521**	7.377	**10.28**	**9.917**	11.08
$\varepsilon_{f,M4}$(%)	*−0.06*	*1.89*	*−0.26*	*−0.51*	*−4.06*	*−0.86*	*3.25*	*−2.68*	*0.43*
f_{M5}(Hz)	1.747	3.058	4.667	6.660	6.922	7.429	9.847	10.08	11.50
$\varepsilon_{f,M5}$ (%)	*−0.03*	*5.08*	*−0.03*	*0.00*	*1.83*	*−0.16*	*−1.08*	*−1.10*	*4.22*
f_{M6}(Hz)	1.747	3.015	4.662	6.65	6.73	7.412	9.485	10.01	11.24
$\varepsilon_{f,M6}$ (%)	*−0.06*	*3.61*	*−0.13*	*−0.15*	*−0.99*	*−0.39*	*−4.71*	*−1.82*	*1.93*

4 and 5, and modes 7 and 8) appear in reverse order with respect to the experimental data (Table 11.4, numbers in bold). Based on the results of the sensitivity analysis, a model optimization is performed on model M3, considering as uncertain parameters Young's modulus, the effective thickness of the slab (already considered by Cigada et al. [13]), and the density of both the steel transversal elements and the concrete slab. The latter accounts for a possible difference between the real and the design value of the slab thickness. However, the approach proved to be unsuccessful, since the modal properties of model M5 (Table 11.4), assuming the optimal values of the parameters mentioned earlier, are slightly less accurate than those of model M3. At this stage, a "quasi-final" model, M6, was obtained from M3 by engineering judgment, considering results from both the sensitivity analysis and the optimization process—namely, the average between the uncracked and cracked values of the slab moment of inertia and the optimal steel density (8272 kg/m^3), respectively. The discrepancy in the 2nd mode decreased to 3.61% for M6 (Table 11.4).

For models in Table 11.4, the presence of a large mass of bridge abutments is considered only in the boundary conditions at the ends of the longitudinal beams. Hence, the final model updating on M6 focuses on its effect. As a rough approximation, in model M7, half of their total mass (35 tons) is assigned to each of the end pier caps, producing an unexpected agreement in natural frequencies (Table 11.5). The role of the abutments and the interaction between the ground and the foundation micro-piles, modeled through a system of rotational and axial springs, is investigated by deriving three models from M6. In M7S1, a lumped mass is placed below each end pier cap and linked to the ground with springs. In M7S2, the abutment is discretized with brick elements rigidly connected to the ground. M7S3 adopts a more refined mesh for the brick elements, with springs simulating the foundation micro-piles inserted in the correct positions. The frequency discrepancy of these models in Table 11.5 shows that they produce very similar results, still affected by the same error as M6 in the 2nd mode. It was concluded that the error in the 2nd mode could reasonably be explained by the lack of modeling a mass eccentricity at the I-beam ends. Through a numerical artifice, the rotational inertia of the pier cap mass with respect to the bridge axis is added as a separate contribution to M6, producing an excellent match of modal properties in model M8.

A complete correlation analysis was carried out between experimental and numerical modal parameters of M8. The MAC index [24] in Table 11.5 shows an excellent correlation for all the modes. Figure 11.9 shows the numerical counterpart

TABLE 11.5

Modal properties of models derived from M6 with modifications at the bridge ends.

Mode Id	1 (B)	2 (T)	3 (B)	4 (B)	5 (T)	6 (B)	7 (T)	8 (B)	9 (T)
f_{FDD}(Hz)	1.748	2.910	4.668	6.660	6.797	7.441	9.954	10.19	11.03
f_{M7}(Hz)	1.747	2.903	4.664	6.655	6.81	7.416	9.523	10.01	11.25
$\varepsilon_{f,M7}$ (%)	*−0.06*	*−0.24*	*−0.09*	*−0.08*	*0.19*	*−0.34*	*−4.33*	*−1.79*	*1.95*
$\varepsilon_{f,M7S1}$ (%)	*−0.13*	*3.60*	*−0.32*	*−1.01*	*−1.04*	*−1.25*	*−5.35*	*−2.03*	*1.50*
$\varepsilon_{f,M7S2}$ (%)	*−0.07*	*3.63*	*−0.15*	*−0.29*	*−0.97*	*−0.54*	*−4.83*	*−1.84*	*1.83*
$\varepsilon_{f,M7S3}$ (%)	*−0.08*	*3.62*	*−0.18*	*−0.39*	*−0.99*	*−0.64*	*−4.97*	*−1.87*	*1.74*
f_{M8}(Hz)	1.747	2.903	4.660	6.651	6.814	7.412	9.517	10.01	11.24
$\varepsilon_{f,M8}$(%)	*−0.07*	*−0.25*	*−0.18*	*−0.14*	*0.28*	*−0.38*	*−4.39*	*−1.80*	*1.92*
MAC	*0.996*	*1.000*	*0.999*	*0.998*	*0.995*	*0.995*	*0.929*	*0.903*	*0.969*

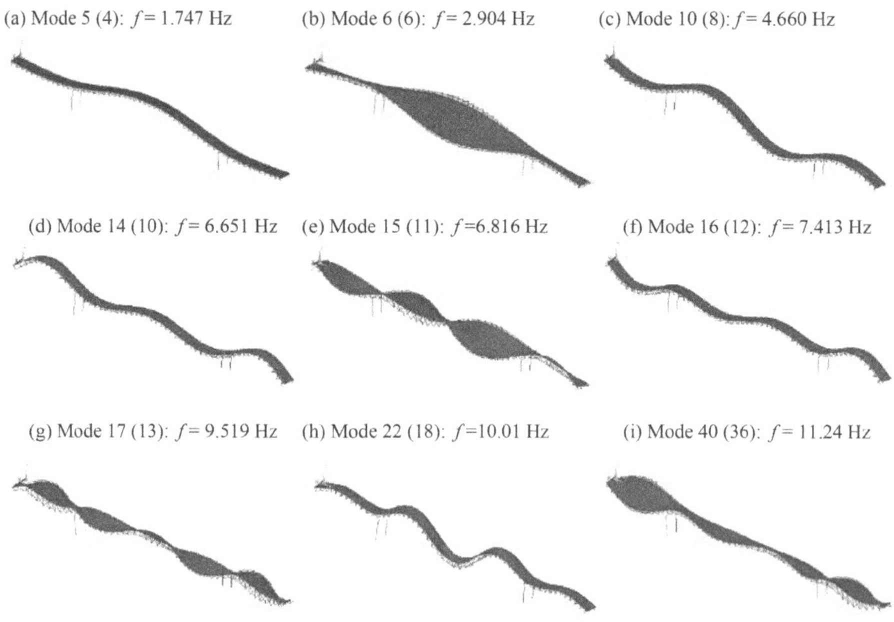

FIGURE 11.9 FE model M8: frequencies and mode shapes for numerical modes corresponding to the experimentally identified ones. Numbers in parentheses refer to those of model M0.

of Figure 11.6 for this model: the first three numerical modes are "quasi-rigid" modes—transverse, longitudinal, and rotational, respectively—involving the bearings' deformability. The numerical modes 4 (1.437 Hz) and 13 (6.508 Hz) are lateral/torsion modes, not identified experimentally but also detected in [15], although in a different order. The numerical modal shapes are practically coincident with those of model M0. M8 has been assumed as the final model, accounting in a simplified way for some undetected discrepancies with the design data in the footbridge end regions.

11.6 CONCLUSIONS

The methodology for validating the FE model of a lively steel footbridge, based on modal identification performed on a set of experimental data, is explored in this chapter. Proof tests, including AVTs, allowed the identification of nine natural modes, either in vertical bending or torsion. The fundamental mode was identified at a frequency of 1.75 Hz. The numerical FE modeling confirmed the experimentally detected presence of separate resisting mechanisms in bending and torsion. Despite the relative simplicity of the structural scheme, simulating the latter turned out to be quite difficult. The results of a previous optimization procedure [13] showed an almost null discrepancy with experimental values for the 1st bending mode and 7.80% for the 2nd and most affected torsion mode.

The aim of this work was to eliminate the error in the torsion modes without altering the precision of the bending modes. The result was tackled through a multi-step procedure, starting from a model named M0, derived from a previous optimization procedure [13]. After an unsuccessful attempt to introduce a few reasonable modifications into M0, a three-step process of model updating systematically addressed the possible reasons for the discrepancy. In the first step, a thorough revision of design data on geometry, loads, and boundary conditions led to model M3, where details at the bridge ends played a crucial role in reducing overestimation to 4.98%. The second step involved a sensitivity analysis and a subsequent optimization process, coupled with sound engineering judgment. Overestimation decreased to 3.61% without any loss of precision in the bending modes (model M6). The third step investigated the effect of modeling the masses and stiffness of the abutments. The final and successful model, M8, underestimated the experimental frequency by 0.23%.

The bridge is subdivided into a long central span and two much shorter end spans, where the latter have larger cross sections. From the engineering standpoint, the presence of a large mass at the end spans has no effect on bending modes, which are insensitive to variations in the modeling of these zones, but is determinant for the torsion mode frequency. Even though the sensitivity analysis and optimization procedure played a part in the final result, only engineering judgment could provide it, showing that numerical procedures alone cannot achieve the desired results when detached from reality. Model M8 also shows an excellent correlation between experimental and numerical modal shapes. However, lateral modes, not experimentally identified and not affecting serviceability conditions, have different frequencies in the original model M0 and in M8. This fact points out that the identification of horizontal modes, even though not relevant for serviceability conditions, could improve the process of model updating by providing a more complete set of data for FE modeling.

REFERENCES

[1] Caetano E, Cunha Á, Magalhães F, Moutinho C (2010) Studies for controlling human-induced vibration of the Pedro e Inês footbridge, Portugal. Part 1: Assessment of dynamic behaviour, *Engng Struct*, 32, 1069–1081.

[2] Lai E, Gentile C, Mulas MG (2017) Experimental and numerical serviceability assessment of a steel suspension footbridge, *J Constr Steel Res*, 132, 16–28.

[3] Ribeiro de Silva IA, Santos de Silva JG (2018) Experimental and numerical dynamic structural analysis of footbridges when subjected to pedestrians walking loads, *J Civ Struct Health Monit*, 8, 585–595.

[4] Van Nimmen K, Lombaert G, De Roeck G, Van den Broeck P (2014) Vibration serviceability of footbridges: Evaluation of current codes of practice, *Engng Struct*, 59, 448–461.

[5] Busca G, Cappellini A, Manzoni S, Tarabini M, Vanali M (2014) Quantification of changes in modal parameters due to the presence of passive people on a slender structure, *J Sound Vib*, 333, 5641–5652.

[6] Georgakis CT, Jørgensen NG (2013) Change in mass and damping on vertically vibrating footbridges due to pedestrians. In: Cunha A (eds) *Topics in Dynamics of Bridges*, Vol. 3. Proceedings of the Society for Experimental Mechanics Series. Springer, New York, NY. https://doi.org/10.1007/978-1-4614-6519-5_4.

[7] Caprani CC, Ahmadi E (2016) Formulation of human-structure interaction system models for vertical vibration, *J Sound Vib*, 377, 346–367.

[8] Mulas MG, Lai E, Lastrico G (2018) Coupled analysis of footbridge-pedestrian dynamic interaction, *Engng Struct*, 176, 127–142.

[9] Živanović S (2012) Benchmark footbridges for vibration serviceability assessment under the vertical component of pedestrian load. *ASCE J Struct Engng*, 138(10), 1193–1202.

[10] Gomez S, Marulanda J, Thomson P, Garcia J, Gomez D, Ortiz A, Dyke S, Caicedo J, Rietdyk S (2017) Benchmark problem for assessing effects of human-structure interaction in footbridges. In: *Proceedings of IMAC 35*, USA.

[11] Van Nimmen K, Van Hauwermeiren J, Van den Broeck P (2021) Eeklo footbridge: Benchmark dataset on pedestrian-induced vibrations. *ASCE J Bridge Engng*, 26(7).

[12] Venuti F, Domaneschi M, Lizana M, Glisic B (2021) Finite element models of a benchmark footbridge. *Appl Sci*, 11, 9024. https://doi.org/10.3390/app11199024.

[13] Cigada A, Gentile C, Lastrico G, Mulas MG (2020) Measuring the dynamic response of a lively footbridge to ambient and walking excitation. In: *Proceedings of XI EURODYN, 2020* (Vol. 1, pp. 1791–1804). European Association for Structural Dynamics.

[14] Mulas MG, Fortis C, Lastrico G (2023) Field testing and serviceability assessment of a lively footbridge. In: *Proceedings of XII Eurodyn*, Delft, The Netherlands.

[15] Lastrico G (2017) *Studio numerico e sperimentale della risposta dinamica di una passerella pedonale*, MS Thesis in Civil Engng, Politecnico di Milano (in Italian).

[16] Fontanilla Mercado A (2020) *Numerical assessment of the dynamic behavior of a pedestrian-cycle bridge located at Cerro al Lambro (MI)*, MS Thesis in Civil Engng, Politecnico di Milano.

[17] Brincker R, Zhang LM, Andersen P (2000) Modal identification from ambient responses using frequency domain decomposition. In: *Proceedings of the 18th International Modal Analysis Conference (IMAC), San Antonio, Texas, USA, February 7–10, 2000* (pp. 625–630).

[18] Peeters B, De Roeck G (1999) Reference-based stochastic subspace identification for output-only modal analysis, *Mech Syst Signal Pr*, 13(6), 855–878.

[19] SVS (2012) ARTeMIS Extractor 2011. http://www.svibs.com/.
[20] ANSYS® Academic Research Mechanical, Ansys Mechanical APDL 2020 R1.
[21] Zerruso L (2019) *Analisi sperimentale e simulazione numerica delle vibrazioni indotte da pedoni in transito su una passerella pedonale*, MS Thesis in Civil Engng, Politecnico di Milano (in Italian).
[22] Fortis C (2021) *Studio numerico e valutazione sperimentale delle proprietà dinamiche di una passerella pedonale*, MS Thesis in Civil Engng, Politecnico di Milano (in Italian).
[23] American National Standards Institute/Steel Deck Institute (2017) ANSI/SDI C-2017 Standard for Composite Steel Floor Deck-Slabs.
[24] Allemang RJ, Brown DL (1983) Correlation coefficient for modal vector analysis. In: *1st International Modal Analysis Conference* (pp. 247–261).

12 Development of a continuous model updating strategy for vibration-based civil SHM

I. Rosati and Carlo Rainieri

12.1 ABSTRACT

In recent years, the potential of operational modal analysis and finite element model updating for nondestructive structural evaluation has gained considerable attention within the research community. To reduce the computational demand typically associated with model updating procedures, various approximation techniques have been developed. Among these, response surface methodologies and the Douglas-Reid (DR) method have emerged as particularly prominent.

Despite their growing popularity, systematic studies investigating the parameters that influence the performance of these methods—especially in the context of structural health monitoring (SHM)—remain relatively scarce. This chapter focuses on the development of a continuous model updating strategy for vibration-based SHM of a historical structure. To this aim, the study explores how different factors affect the accuracy and reliability of the outcomes of model updating based on the DR method. The ultimate goal is the development of a continuous model updating strategy based on the DR method for the indirect monitoring of selected updating parameters in the context of SHM applications. Particular attention is paid to the robustness and precision of the method under variable environmental conditions, highlighting its promising potential for vibration-based SHM applications, even though its use in this context has been only marginally documented in the existing literature.

12.2 INTRODUCTION

Evaluating the structural condition of historical buildings often presents significant challenges due to numerous uncertainties. These include unknown or variable material properties and complex structural configurations, on one hand, and the imperative to limit the invasiveness of field investigations on the other hand [1–4]. Consequently, the reliability of structural analyses is frequently jeopardized by the scarcity of

DOI: 10.1201/9781003516941-12

accurate information and the potential presence of ongoing degradation phenomena or prior damage caused by hazardous events such as earthquakes. Additionally, signs of structural distress may be hidden by a sequence of structural interventions accumulated over time. Within this framework, methodologies enabling nondestructive evaluation of structural behavior have garnered increasing interest [5].

Among the most promising tools in this context are operational modal analysis (OMA) and finite element (FE) model updating, which have been increasingly applied over the past decades because of the opportunities they provide for assessing structural performance without physical intrusion [6–11]. By minimizing the discrepancies between experimentally identified and numerically predicted modal parameters (e.g., natural frequencies and mode shapes), it is possible to address uncertainties related to structural elements, boundary conditions, material properties, and mass distribution. The resulting calibrated model can thus be considered a reliable approximation of the true dynamic response of the structure under operational conditions [12].

FE model updating techniques are typically divided into two broad categories: direct and indirect (iterative) approaches. Direct methods involve direct modifications of the mass or stiffness matrices to match the computed response with the experimental data. While these techniques can accurately reproduce the observed structural behavior in terms of its modal parameters, they are progressively laid aside due to their tendency to yield non-physical models [13]. Indirect methods, in contrast, iteratively adjust physically meaningful parameters—such as material stiffness—to align numerical predictions with experimental observations. This approach leads to more realistic and interpretable models [13, 14].

Model updating can also be distinguished based on the updating strategy: manual or automated [15]. Manual approaches rely on engineering intuition and trial and error to adjust model parameters, typically when the initial model is already a good representation of the actual structure. Automated approaches, by contrast, utilize optimization algorithms to systematically minimize an objective function that quantifies the difference between observed and simulated modal characteristics [16]. Optimization strategies used in automated model updating are further categorized into global and local methods. Global approaches, such as genetic algorithms [17, 18] or particle swarm optimization [19], have a higher probability of locating the global optimum but often require substantial computational resources. Local methods, on the other hand, are computationally efficient, but they might converge to suboptimal solutions [20].

In iterative, optimization-based model updating, the FE model is typically reanalyzed at each iteration to evaluate the objective function. This process is exemplified in [21], where local optimization algorithms are applied by repeatedly computing modal parameters from the full model [21]. An alternative strategy involves the use of mathematical approximations—surrogate models—to simulate the response of the structure over the parameter space. These are often represented by a quadratic polynomial formulation, and they speed up optimization by reducing the number of required full-model analyses [22].

Among the various surrogate modeling strategies, the DR method is particularly attractive. It employs a quadratic function to approximate the relationship between selected updating parameters and the model response [23]. Compared to full-scale modeling or generic response surfaces, the DR method requires fewer computations to define the interpolation due to the relatively small number of coefficients and simpler calculation procedures. This efficiency has contributed to its widespread adoption in civil engineering applications, including case studies on bridges [20, 24, 25] and historical structures [26, 27].

Despite its advantages, the influence of key factors on the accuracy and reliability of the DR method has not yet been fully explored. This is especially critical in the context of vibration-based SHM, where continuous, indirect monitoring of structural parameters depends on robust and automated model updating. Although the effectiveness of the DR method in combination with optimization techniques for model updating is well recognized, a systematic investigation into the influence of factors such as the objective function, the optimization algorithm, and incomplete modal information on its performance is still lacking. Indeed, most of the available studies compare DR-based results with those obtained using different updating techniques under fixed assumptions [27], without assessing the sensitivity of model updating outcomes to changes in the aforementioned or other factors.

This chapter aims at addressing this gap by conducting an extensive evaluation of the DR method. The analyses refer to a literature case study for which benchmark results obtained through a grid search approach are available [28] and are useful for checking the effectiveness of the proposed model updating strategy with a view to its integration into continuous vibration-based SHM systems.

The latter is a topic of growing interest, although practical implementations remain limited. This is primarily due to the challenges of creating accurate structural models and the computational intensity of updating large-scale numerical models [29]. To mitigate these issues, researchers have explored the use of surrogate models to enable real-time parameter updates for damage detection [29, 30]. In contrast, the DR method—despite its low computational cost—has seen limited use in continuous monitoring applications.

Given the influence of environmental and operational conditions on the experimentally identified modal properties [31], recent studies have examined the use of the DR method for damage localization in the presence of varying environmental conditions that affect the inferred mechanical properties [32, 33]. However, these studies do not report a preliminary assessment of the reliability and accuracy of model updating outcomes under such varying conditions.

The present chapter addresses this issue by simulating time series of natural frequencies generated from an FE model of a heritage tower that incorporates the temperature dependence of material properties. These series are further perturbed with synthetic noise to replicate real-world measurement uncertainties. The DR method is then applied in a recursive framework to emulate continuous SHM, and its performance is evaluated by comparing the estimated parameters with known reference values, thus assessing both robustness to noise and overall accuracy.

The remainder of this chapter is organized as follows: the next section outlines the theoretical background of the model updating strategies applied herein, with a particular focus on the DR method. The case study is introduced afterward, followed by an in-depth discussion of systematic investigations to assess the effects of different objective functions, optimization algorithms, starting points, and missing modes on the model updating results. The methodology adopted to generate synthetic monitoring data and to assess the robustness and accuracy of continuous model updating using the DR method within an SHM context is finally presented and discussed, summarizing the main findings and their broader implications in the conclusions.

12.3 THEORETICAL BACKGROUND

Finite element (FE) model updating techniques are widely employed to reduce discrepancies between experimentally identified and numerically predicted modal properties by adjusting specific parameters that influence the mass and/or stiffness of a structure. In this chapter, particular attention is paid to the DR method [23] and its capabilities and performance. For comparative purposes, the inverse eigensensitivity (IE) method [34] is also discussed.

Originally introduced in 1982 [23], the DR method is a simplified model updating technique that aims at minimizing the discrepancy between measured and simulated modal properties by exploring combinations of selected updating parameters. Although the method is predominantly used to reduce discrepancies in natural frequencies [26, 35], applications targeting mode shapes can also be found in the literature [20, 36]. However, in the context of the present chapter, the updating parameters showed minimal sensitivity to the mode shapes; thus, only natural frequencies have been considered [28]. The extension of the method to mode shapes, while not addressed here, is extensively discussed in the literature [36].

Let f_i^{exp} ($i = 1, \ldots, N_m$) denote the experimentally derived natural frequencies from an OMA test. Rather than directly comparing these to frequencies obtained from a full FE model, a surrogate model is constructed to approximate the dependence of the numerical natural frequencies on a selected set of updating parameters θ_k (with $k = 1, \ldots, N_v$):

$$f_i^{FEM} = f_i^{FEM}\left(\theta_1, \theta_2, \theta_3, \ldots, \theta_k, \ldots, \theta_{N_v}\right) \text{ with } i = 1, \ldots, N_m \text{ and } k = 1, \ldots, N_v \quad (12.1)$$

To minimize the discrepancy between predicted and observed values, the DR method introduces a quadratic approximation of the model response within predefined bounds of the updating parameters:

$$f_i^{SM} = \gamma_i + \sum_{k=1}^{N} (\alpha_{i,k}\theta_k + \beta_{i,k}\theta_k^2) \text{ with } i = 1, \ldots, N_m \text{ and } k = 1, \ldots, N_v \quad (12.2)$$

Here, γ_i, $\alpha_{i,k}$, and $\beta_{i,k}$ are the unknown coefficients to be determined through a limited number of FE simulations. These simulations are performed by varying each updating parameter independently—alternatively at the lower θ_k^L, nominal θ_k^B, and upper θ_k^U values—while keeping the others fixed at their nominal values. The corresponding numerical frequencies are used to establish a set of equations that allows for the estimation of the DR coefficients, which can be expressed in the matric form as follows:

$$\left\{f^{FEM}\right\} = [C]\{K\} \quad (12.3)$$

where $\{K\}$ is the vector of DR coefficients and [C] is a matrix holding the values of the updating variables. From Eq. (12.3), the DR coefficients can be computed as follows:

$$\{K\} = [C]^{-1}\{f^{FEM}\} \tag{12.4}$$

The obtained DR model can replace the full FE model in the computation of the objective function, significantly reducing computational costs in model updating routines.

12.3.1 The inverse eigensensitivity method

The IE method [34] offers a linearized alternative to model updating by approximating the relationship between natural frequencies and updating parameters using a first-order Taylor expansion:

$$\{f_{exp}\} = \{f(\{\theta_0\})\} + [S](\{\theta\} - \{\theta_0\}) \tag{12.5}$$

Here, $\{f_{exp}\}$ is the vector of the experimental natural frequencies, $f(\{\theta_0\})$ the vector of the natural frequencies resulting from the FE model at the initial parameter guess $\{\theta_0\}$, and [S] the sensitivity matrix. The generic element of [S], $\frac{\partial f_i}{\partial \theta_k}$, describes how each natural frequency responds to changes in the selected updating parameter.

The IE method iteratively updates the parameters according to the following expression:

$$\{\theta_{p+1}\} = \{\theta_p\} + [H](\{f_{exp}\} - \{f(\{\theta_p\})\}) \tag{12.6}$$

where the gain matrix [H] is obtained as the Moore–Penrose pseudoinverse of the sensitivity matrix:

$$[H] = [S]^T([S][S]^T)^{-1} \tag{12.7}$$

This iterative approach can rapidly converge, although its performance is highly dependent on the accuracy of the initial parameter estimates and the linearity of the system response in the region of interest.

12.4 THE INVESTIGATED BUILDING

The goal of the study is to assess the performance of the DR method for model updating in view of its potential integration within a continuous vibration-based SHM system. To this aim, analyses are carried out with reference to a previously investigated heritage building, so that independently obtained reference results are available for comparison.

The selected case concerns the Tower of the Nations, a heritage reinforced concrete (r.c.) structure located within the area of the Mostra D'Oltremare urban park in Naples, Italy [37]. From an architectural point of view, the Tower is characterized by two blind sides and two fully see-through sides (Figure 12.1a). The building also features elevator shafts and stairwells located in its central part. Except for the first, second, third, and tenth floors, the remaining levels consist of alternating half-floor layouts, resulting in a unique structural configuration.

Over the years, extensive surveys and structural investigations have been carried out to support both restoration and seismic retrofitting design. An accurate numerical model of the Tower was previously obtained by refining it through a grid search approach, as reported by Rainieri et al. [28].

Based on this previous study, two structural parameters were selected for model updating: the elastic modulus of concrete (E_c) and the shear modulus of the tuff masonry infill walls (G_t). These walls are embedded within the outer r.c. frames on the blind sides of the Tower. The elastic modulus of concrete was considered to vary between 13,000 MPa and 30,000 MPa, in agreement with the variability observed in core compression tests [28]. The shear modulus of tuff masonry ranged between 300 MPa and 420 MPa, in agreement with Italian seismic code provisions for existing buildings. As a result, the Young's modulus of the tuff masonry (E_t) varied between 900 MPa and 1,260 MPa.

Numerical modal analyses were performed using the commercial software SAP2000® [38]. The FE model (Figure 12.1b) adopted beam elements to represent the structural frames (beams, columns, and braces) and shell elements to model floors, r.c. walls, and masonry infills.

Reference experimental modal parameters were obtained from an OMA test. The ambient vibration response of the Tower was recorded by ten force balance

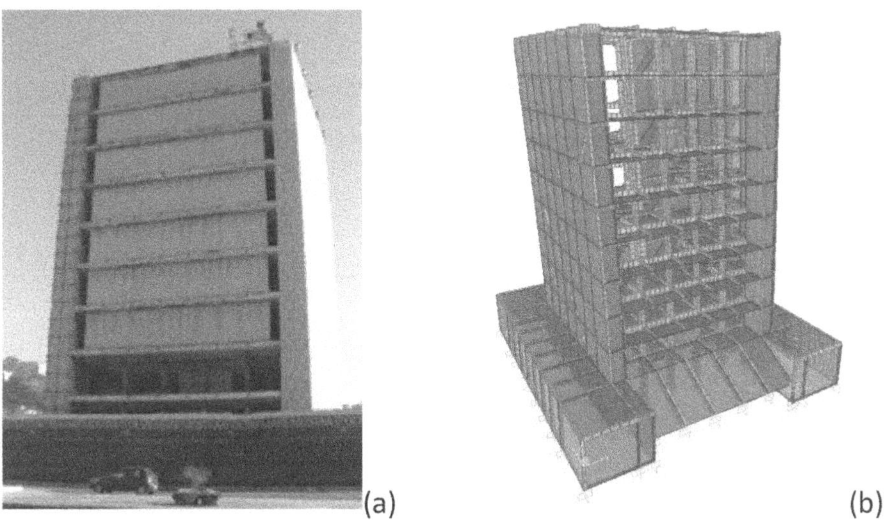

FIGURE 12.1 View of the Tower of the Nations (a) and its FE model (b).

accelerometers with a ±0.25 g full-scale range, 10 V/g sensitivity, and a dynamic range of 140 dB. These were connected to a high-performance data acquisition system characterized by 24-bit resolution, a dynamic range exceeding 114 dB, and a built-in analog anti-aliasing filter. The sensors were deployed in two orthogonal directions at the fourth, fifth, and top floors of the building, according to a layout designed to ensure the observability of both translational and torsional vibration modes [28].

The data were sampled at 100 Hz; 30-minute time histories were collected and processed using well-established OMA techniques—namely the frequency domain decomposition (FDD) [39] and the stochastic subspace identification (SSI) [40, 41]—to extract the fundamental modal parameters. For the sake of completeness, the experimental results from Rainieri et al. [28] are also reported in Table 12.1, along with the results of modal analyses carried out by means of an FE model of the Tower refined according to a grid search approach in combination with the following objective functions:

$$H_f' = \frac{100}{N_m} \sum_{i=1}^{N_m} \left\| \left(\frac{f_i^{num} - f_i^{exp}}{f_i^{exp}} \right) \right\| \tag{12.8}$$

$$H_f'' = \frac{100}{N_m} \sum_{i=1}^{N_m} \left\| \left(\frac{f_i^{num} - f_i^{exp}}{f_i^{exp}} \right) \right\| PMR_i \tag{12.9}$$

are reported in Table 12.2. The latter serves as an independent benchmark for the comparative assessment conducted in the present study.

TABLE 12.1
The Tower of the Nations: reference OMA results [28].

Mode #	Shape description	Natural frequency [Hz]
I	Bending (open side)	0.80
II	Bending (blind side)	1.33
III	Torsion	1.66
IV	Bending (open side)	2.96
V	Bending (blind side)	4.23
VI	Torsion	4.90

TABLE 12.2
The Tower of the Nations: modal analysis results from the refined models in Rainieri et al. [28].

		Natural frequency [Hz]					
Objective function	Value of variables in the refined model	I	II	III	IV	V	VI
H_f'	E_c = 24000 MPa, G_t = 360 MPa	0.73	1.36	1.63	2.97	4.02	4.88
H_f''	E_c = 28500 MPa, G_t = 300 MPa	0.79	1.37	1.64	3.21	4.04	4.91

In Eqs. (12.8) and (12.9), f_i^{num} and f_i^{exp} denote the numerical and experimental natural frequencies of the i-th mode, respectively; N_m is the total number of modes considered, and PMR_i is the participating mass ratio associated with the *i*-th mode.

The first objective function, H'_f, provides the average of the modulus of the normalized relative errors, while the second, H''_f, introduces a weighting factor in the form of the modal PMR to emphasize the modes contributing the most to the overall seismic response.

12.5 FACTORS AFFECTING MODEL UPDATING RESULTS

This section discusses the main outcomes of comparative analyses aimed at exploring how various factors—such as the objective function, optimization algorithm, initial parameter estimates, and number of reference modes—affect the performance of the DR model updating method, with a focus on its applicability to continuous vibration-based SHM systems. As mentioned in the previous section, findings from a previous study [28] are taken as references to check the consistency of the obtained results.

For the application of the DR method, it is first necessary to define nominal values (indicated by the subscript $_{nom}$) and admissible bounds for the updating parameters. In agreement with [28], the upper and lower bounds for the elastic moduli of concrete and tuff masonry were set as discussed in Section 12.4, while the nominal values were set at the midpoint of the respective ranges; that is, $E_{c,nom}$ = 22,500 MPa and $E_{t,nom}$ = 1,080 MPa.

The analysis began by examining how the choice of the objective function affects the model updating results. In this context, the following objective functions were evaluated:

$$H_{f,I} = \frac{1}{2}\sum_{i=1}^{N_m}\left(\frac{f_i^{SM} - f_i^{exp}}{f_i^{exp}}\right)^2 \tag{12.10}$$

$$H_{f,II} = \frac{100}{M}\sum_{i=1}^{N_m}\left|\left(\frac{f_i^{SM} - f_i^{exp}}{f_i^{exp}}\right)\right| \tag{12.11}$$

$$H_{f,III} = \frac{100}{M}\sum_{i=1}^{N_m}\left|\left(\frac{f_i^{SM} - f_i^{exp}}{f_i^{exp}}\right)\right| \cdot PMR_i \tag{12.12}$$

where the superscript ^{SM} refers to the parameters estimated by the Douglas-Reid surrogate model. Mode shapes are not considered due to their limited sensitivity to the updating variables in the present case [28]. However, although mode shapes were not incorporated in this application, in general, including additional data (such as mode shapes) or introducing regularization techniques such as the Tikhonov regularization term or its variants [30] could help mitigate any potential ill-conditioning problems. Among the objective functions, Eq. (12.10) is the most frequently adopted for model updating, as reported in several studies [12, 20, 26, 35, 36]. Nevertheless, some researchers have proposed modified objective functions, deviating from the

TABLE 12.3
Model updating results for different objective functions and for the IE method.

Δf [%] E_{upd} [MPa]	$H_{f,I}$	$H_{f,II}$	$H_{f,III}$	IE
Δf_1	–4.60	–8.59	–2.11	–8.09
Δf_2	2.80	2.14	2.12	5.27
Δf_3	0.61	–0.04	–0.05	3.05
Δf_4	4.13	–0.06	6.69	0.57
Δf_5	–4.46	–5.03	–5.18	–2.06
Δf_6	0.22	–0.37	–0.56	2.73
Δf_{TOT}	16.82	16.23	16.71	21.77
$E_{c,upd}$	26,101	23,835	27,600	24,031
$E_{t,upd}$	1,007	1,091	908	1,176

traditional formulations [12, 36]. To ensure consistency with the earlier study on the Tower, the objective functions in Eqs. (12.11) and (12.12) were also considered. In particular, Eq. (12.12) can be viewed as a variant of Eq. (12.11), where the participating mass ratios (PMR) act as weighting coefficients. The results obtained from applying the DR method in combination with the aforementioned objective functions are presented in Table 12.3. These results were obtained by minimizing the objective function using the Interior Point optimization algorithm, with the number of reference modes N_m equal to six. Specifically, Table 12.3 shows the percentage scatter between the reference experimental natural frequency estimate of the i-th mode f_i^{exp} and the corresponding optimized numerical value f_i^{upd} obtained from the updated model, as described in Eq. (12.13).

$$\Delta f_i = \frac{f_i^{upd} - f_i^{exp}}{f_i^{exp}} \quad \text{with}\, i = 1,\ldots,N_m \tag{12.13}$$

Additionally, the cumulative frequency error is computed as per Eq. (12.14). Finally, Table 12.3 also reports the refined values of the updating variables $E_{c,upd}$ and $E_{t,upd}$.

$$\Delta f_{TOT} = \sum_{i=1}^{M} |\Delta f_i| \tag{12.14}$$

Despite the inherent approximations associated with the DR method and the finite increment size for the updating variables in the grid search method, the two approaches produced consistent optimized values for the updating variables. This consistency can be observed by comparing the results presented in Tables 12.2 and 12.3 for the objective formulations H_f' and $H_{f,II}$, and H_f'' and $H_{f,III}$, respectively.

Table 12.3 further highlights the influence of the selected objective function on the model updating outcomes. By carefully analyzing the modal frequency scatters, it is evident that different objective functions result in distinct distributions of errors across the modes. While all three objective functions yield a similar cumulative frequency error, $H_{f,I}$ produces a more uniform distribution of the frequency scatter across the modes, with errors consistently below 5%. In contrast, the results derived from $H_{f,II}$ and $H_{f,III}$ are characterized by relatively larger errors in the first and fourth modes, respectively. Regarding the optimized values of the updating parameters, both $H_{f,I}$ and $H_{f,III}$ led to consistent results. However, discrepancies were noted between the optimized parameters obtained from $H_{f,I}$ and $H_{f,II}$. This difference can likely be attributed to the non-uniqueness of the solution associated with the least absolute deviation method [42]. The issue is mitigated in $H_{f,III}$ by the inclusion of the modal PMRs as weighting coefficients, which provide additional stability to the solution.

Another significant observation can be drawn by analyzing the results presented in Table 12.3: while the solutions obtained using the selected objective functions differ from one another, they indicate that the optimal sets of parameters represent various balances between the contributions of reinforced concrete and tuff masonry to the overall stiffness of the structure. Specifically, an increase in the optimized elastic modulus of concrete corresponds to a decrease in the optimized value of the elastic modulus of tuff masonry and vice versa. This outcome is not surprising, as it aligns with findings from previous studies in the literature [43] and is further supported by the similar cumulative natural frequency errors associated with the different objective functions. Recognizing this characteristic behavior, the model refinement process was also carried out using the IE method, which serves as an independent updating approach in this context. The IE method utilizes the results of the full FE model and applies an iterative approximation procedure to find the solution. The optimized parameters and modal frequency errors, as defined by Eq. (12.13), align with those obtained from the DR method using the $H_{f,II}$ objective function (see Table 12.3). This consistency likely arises because both methods focus on minimizing the difference between experimental and numerical natural frequencies. However, a slightly larger cumulative frequency error, as shown in Eq. (12.14), is observed in the results obtained from the IE method compared to those from the DR method. This discrepancy can be attributed to the inherent approximations of the IE method, including the use of finite differences rather than actual derivatives, as well as the errors introduced by the threshold set to stop the iterations. Nevertheless, the IE results confirm that the choice of objective function plays a dominant role in determining the model updating outcomes, while the procedure used to reach the solution primarily affects the precision of the results.

Regarding the influence of the optimization procedure, the focus was placed on gradient-based algorithms, as these have been widely used in model updating since its early applications. Gradient-based algorithms are generally fast, but it is well recognized that they may converge to local minima rather than global ones [20, 21]. To address this issue, the results obtained by applying different gradient-based optimization algorithms were compared and analyzed to provide practical insights into the integration of DR model updating into SHM systems.

Table 12.4 shows the results obtained by setting $E_{c,nom} = 22{,}500\,MPa$ and $E_{t,nom} = 1{,}080\,MPa$ as the initial values for the updating variables. These results are generally

TABLE 12.4
Sensitivity of model updating results to the optimization algorithm.

Objective function	Updating results	Interior Point	SQP	Active set	Trust Region
$H_{f,I}$	$E_{c,upd}$ [MPa]	26,101	26,102	26,101	26,101
	$E_{t,upd}$ [MPa]	1,007	1,007	1,007	1,007
$H_{f,II}$	$E_{c,upd}$ [MPa]	23,835	23,834	23,833	23,915
	$E_{t,upd}$ [MPa]	1,091	1,091	1,092	1,086
$H_{f,III}$	$E_{c,upd}$ [MPa]	27,600	27,784	27,784	24,267
	$E_{t,upd}$ [MPa]	908	900	900	1,048
	$H^{upd}_{f,III}$	0.77	0.75	0.75	1.18

consistent across all objective functions, except for the case where the Trust Region algorithm was used to minimize the $H_{f,III}$ objective function. In this case, a suboptimal solution was obtained, remarked by the higher value of the H_{upd} corresponding to the optimized parameters. This outcome confirms that gradient-based algorithms can be trapped in local minima. Therefore, the use of different optimization methods for the same inverse problem, as well as selecting different initial points, as discussed in the next section, should be considered viable strategies to ensure that the optimized variables correspond to the global minimum. Given a gradient-based optimization algorithm, applying it multiple times with different initial points is a potential strategy to verify whether the obtained solution corresponds to a local or global minimum. To investigate this, the sensitivity of the solutions to the initial point for each objective function was explored. Five distinct initial points were considered (corresponding to five pairs of $E_{c,nom}$ and $E_{t,nom}$):

1. ($E_{c,nom}$ = 22,500, $E_{t,nom}$ = 1,080)
2. ($E_{c,nom}$ = 18,000, $E_{t,nom}$ = 960)
3. ($E_{c,nom}$ = 27,000, $E_{t,nom}$ = 1,200)
4. ($E_{c,nom}$ = 18,000, $E_{t,nom}$ = 1,200)
5. ($E_{c,nom}$ = 27,000, $E_{t,nom}$ = 960)

The first corresponds to the previously defined nominal values of the updating variables, while the remaining four were chosen near the boundaries of the search space to create different convergence paths toward the minimum. The Interior Point algorithm was employed to determine the optimal solutions. The results of these variations are presented in Table 12.5.

The objective function $H_{f,I}$ provided the most stable solution, with only minor variations attributed to the finite increment size and the allowed tolerance. In contrast, $H_{f,II}$ exhibited the highest sensitivity to the initial point. Similarly, $H_{f,III}$ showed some variability in the optimized values of the updating parameters. Based on these findings, $H_{f,I}$ appears to provide the most robust results, making it particularly suitable for continuous SHM applications (as further discussed in Section 12.6).

TABLE 12.5
Sensitivity of model updating results to the initial point.

Updating results	1	2	3	4	5
$E_{c,upd}$ [MPa]	26,101	25,992	26,176	26,031	26,166
$E_{t,upd}$ [MPa]	1,007	1,005	1,010	999	1,010
$H_{f,I}^{upd}$	0.0033	0.0033	0.0033	0.0032	0.0033
$E_{c,upd}$ [MPa]	23,835	23,758	23,895	23,877	26,551
$E_{t,upd}$ [MPa]	1,091	1,087	1,114	1,082	959
$H_{f,II}^{upd}$	2.67	2.68	2.72	2.60	2.74
$E_{c,upd}$ [MPa]	27,600	26,807	27,656	26,859	26,969
$E_{t,upd}$ [MPa]	908	938	908	938	939
$H_{f,III}^{upd}$	0.77	0.85	0.78	0.84	0.86

This robustness is especially important when an unfortunate selection of the initial point occurs, which may otherwise lead to convergence to a local minimum rather than the global optimum. Given the observed sensitivity to the initial point, it is essential to employ a robust strategy to prevent the solution from being trapped in a local minimum when applying gradient-based algorithms [20, 21]. To demonstrate the stability of the solution provided by the $H_{f,I}$ objective function, the optimization procedure was repeated using the Interior Point method, with the starting point randomly varied within the search space. A total of 300 different initial points, uniformly distributed across the design space of the two variables, were selected using the Latin hypercube sampling method. The results confirmed that the previously obtained optimal values of the updating parameters (reported in Table 12.5) were already quite satisfactory. However, by repeating the optimization with varying initial points, the solution was refined, leading to optimized values of $E_{c,upd}$ = 25,932 MPa and $E_{t,upd}$ = 982 MPa, corresponding to a minimum value of the objective function $H_{f,I}^{upd} = 0.00317$. While this approach ensured greater robustness, it also significantly increased the computational time. This trade-off must be carefully considered when integrating model updating into continuous SHM systems, as the computational time must be compatible with the system's performance requirements, especially in terms of response time.

When model updating is going to be integrated into continuous vibration-based SHM systems, it is possible that incomplete modal data are obtained from the application of an automated OMA procedure [44]. The impact of the number of experimentally identified modes on the model updating results was, therefore, investigated.

Although the number of measured degrees of freedom can also affect the results in several applications, it was not considered here due to the very limited sensitivity of mode shapes to the updating variables in the present case. The results of model updating, progressively decreasing the number of identified modes (from six to four) and minimizing the objective functions by the Interior Point method starting from $E_{c,nom}$ = 22,500 MPa and $E_{t,nom}$ = 1,080 MPa, are summarized in Table 12.6.

TABLE 12.6
Sensitivity of model updating results to incomplete modal information.

Number of modes	Updating variable	$H_{f,I}$	$H_{f,II}$	$H_{f,III}$
6	$E_{c,upd}$ [MPa]	26,101	23,835	27,600
	$E_{t,upd}$ [MPa]	1,007	1,091	908
5	$E_{c,upd}$ [MPa]	26,095	23,835	27,068
	$E_{t,upd}$ [MPa]	1,013	1,091	932
4	$E_{c,upd}$ [MPa]	26,204	23,915	27,211
	$E_{t,upd}$ [MPa]	921	1,026	926

The findings indicate that if the number of identified modes falls below five, the updated variables are noticeably affected, particularly for the $H_{f,I}$ and $H_{f,II}$ objective functions. To prevent inconsistent results that could lead to false alarms in practical SHM applications, it is crucial to thoroughly inspect and verify the estimates derived from incomplete modal information to ensure their validity before relying on them for decision-making.

12.6 CONTINUOUS MODEL UPDATING FOR SHM APPLICATIONS

The integration of model updating into continuous vibration-based SHM systems is an area of growing research interest, primarily due to the potential it offers for damage detection. However, the sensitivity of modal parameters to environmental factors also introduces variability in the results of model updating [32, 33]. This variability remarks the need of assessing the reliability and accuracy of continuous model updating outcomes under changing environmental conditions, especially within the context of vibration-based SHM.

It is important to note that the continuous model updating approaches proposed here are deterministic in nature. Although probabilistic methods (e.g., Bayesian methods [45, 46]) represent another attractive option to account for the variability in experimental modal properties, their outcomes are typically limited to posterior probability distributions of the uncertain parameters. As such, they do not provide information about the evolution of these parameters over time, which is a key consideration for continuous SHM systems. Thus, two possible approaches for continuous model updating are compared here in terms of accuracy and computational effort. Accuracy is evaluated by comparing the results of continuous model updating—specifically the optimized values of the updating variables—with the input values of the FE model used to simulate the natural frequency time series. These simulations account for the influence of temperature and measurement noise on the system.

To simulate the influence of temperature on the system, expressions for the elastic moduli of concrete and tuff masonry as functions of temperature were defined. First, a measured temperature time series (Figure 12.2) was considered, and a series of numerical modal analyses was performed to generate the fundamental natural frequencies corresponding to the elastic moduli at each temperature value in the time series.

The effect of experimental uncertainties was modeled by adding random noise to the simulated natural frequency time histories. The perturbation equation used for this purpose is as follows [31, 47]:

$$f_i^{pert}(t) = f_i(t) + \chi \cdot f_i^{rms} \cdot \Gamma(0,1) \tag{12.15}$$

where

- f_i and f_i^{pert} represent the unperturbed and perturbed natural frequencies of the i-th mode at time t, respectively;
- f_i^{rms} is the root mean square (RMS) value of the natural frequency of the i-th mode over the entire observation period;
- $\Gamma(0,1)$ represents a zero-mean, unit-variance Gaussian white noise.
- χ is a scaling factor that accounts for the magnitude of the added noise. The noise magnitude was set to achieve a signal-to-noise ratio of approximately 57 dB across all natural frequency time series.

Realistic modeling of the variability of Young's modulus with temperature is complex, as there are limited studies in the literature on this topic. An extensive literature review revealed that the Young's modulus of concrete varies almost linearly with temperature over a wide range (–20°C to 60°C) [48]. To account for this influence, the following relationship was used to describe the temperature dependence of the elastic modulus of concrete:

$$E_c(T) = -125T + 29130 \tag{12.16}$$

where $E_c(T)$ is the elastic modulus of concrete expressed in MPa as a function of temperature T (in °C).

In contrast, a similar relationship for tuff masonry was not found in the literature. As no references were available to improve the formulation, a linear variation of the elastic modulus with temperature was assumed for simplicity. The relationship used to model the temperature dependence of the Young's modulus of tuff masonry was

$$E_t(T) = 9T + 920 \tag{12.17}$$

where $E_t(T)$ is the elastic modulus of tuff expressed in MPa as a function of temperature T (in °C).

Eq. (12.17) provides a reasonable trend for the evolution of the elastic modulus of tuff masonry with temperature. The increase in stiffness with rising temperature has been observed in several monitoring applications [49, 50], making this linear relationship physically plausible.

Once the temperature-dependent relationships for concrete and tuff masonry were defined, simulated monitoring results in terms of natural frequency time series were obtained by performing a series of numerical modal analyses. Each analysis was based on a pair of values for E_c and E_t corresponding to a specific temperature value.

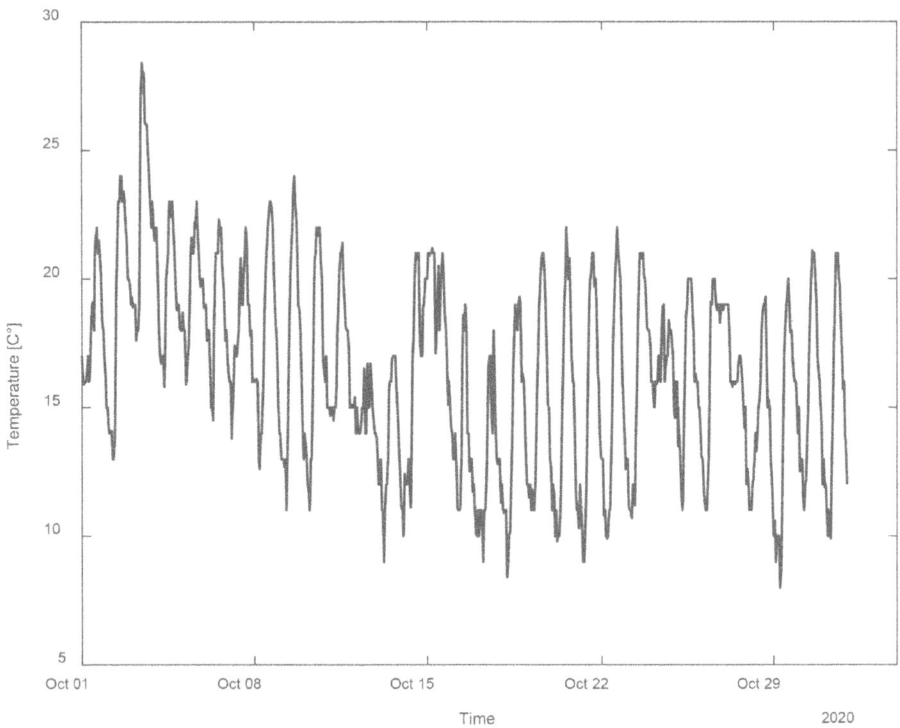

FIGURE 12.2 Temperature time series.

For the simulations, the temperature time history recorded in Naples in October 2020 was used (Figure 12.2). The simulated natural frequency time series, with added random noise, are shown in Figure 12.3.

The simulated monitoring data (744 samples, corresponding to one frequency estimate per hour) were then used for continuous model updating using the DR method in order to assess its robustness to noise in the data and its ability to accurately estimate structural parameters under varying environmental conditions. Notably, no compensation for the environmental influence on natural frequencies was applied during this phase of the analysis.

Two continuous model updating approaches were tested:

1. In the first approach, the quadratic expression of Eq. (12.2) is computed only once at the beginning of the monitoring period.
2. In the second approach, the quadratic expression of Eq. (12.2) is updated at each iteration using the nominal values of the updating variables obtained from the model updating process in the previous iteration.

The boundary values of the updating variables were the same for both approaches. Both methods were evaluated in terms of their accuracy and computational efficiency.

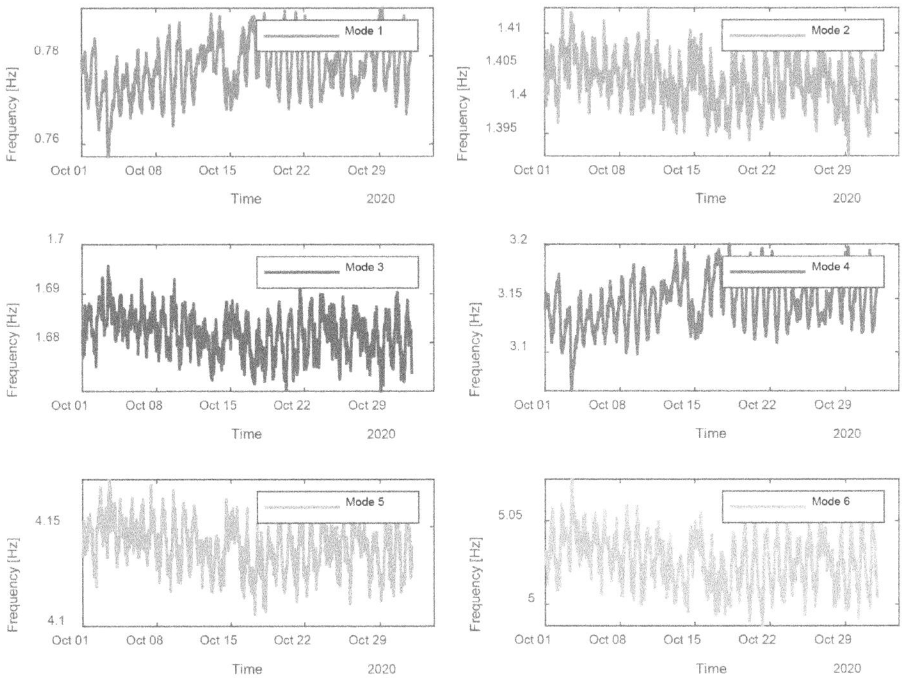

FIGURE 12.3 Simulated natural frequency time series with added random noise.

Both continuous model updating approaches demonstrated very similar levels of accuracy, with maximum deviations of around 0.8% for the Young's modulus of concrete and 1.5% for the Young's modulus of tuff masonry in both cases.

However, there was a significant difference in the computational time required by the two approaches. The method that recursively updated the nominal values completed the analysis in 65,466 seconds, with respect to the 208 seconds spent by the approach based on fixed nominal values. The computational time was evaluated by running the codes associated with the two approaches on a regular laptop equipped with an Intel Core i7 CPU, 8 GB of RAM, and the Windows 10 operating system.

By comparing the computational time between the two methods, it becomes evident that the response time of the algorithms is primarily affected by the analyses needed to continuously update the DR model at each cycle, rather than by the sampling interval of the monitoring data. Since the significantly longer computational time (about 300 times greater) does not yield a corresponding improvement in accuracy, the approach with fixed nominal values can be recommended for practical SHM applications.

The accuracy of this method is visually demonstrated in Figure 12.4, where the estimation errors of the updating variables are plotted. Additionally, Figure 12.5 shows that the continuous model updating results are not biased, as all points are roughly evenly distributed along the bisector. However, some variability is observed

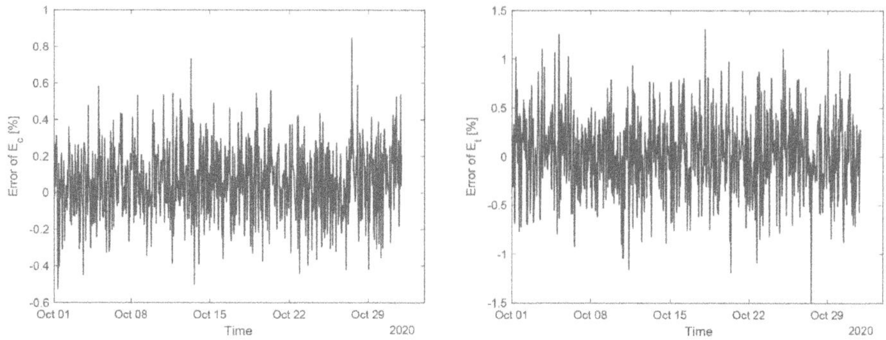

FIGURE 12.4 Estimation errors for the updating variables versus time.

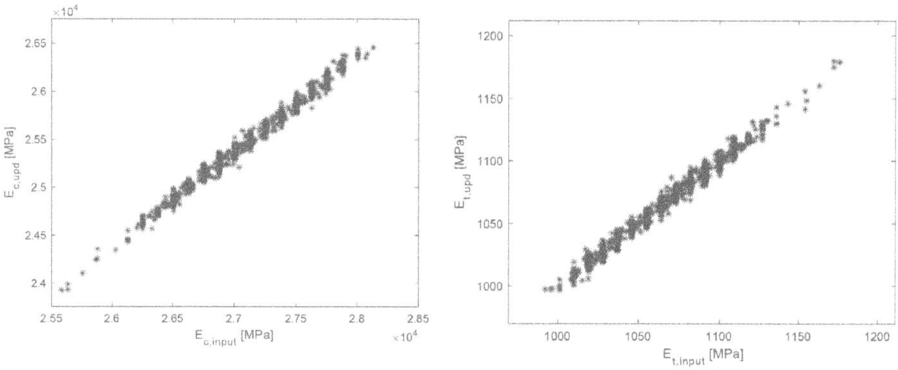

FIGURE 12.5 Comparison of the optimized updating variables and the corresponding input values.

in the results, which can be addressed to the random errors affecting the natural frequencies.

12.7 CONCLUSIONS

This chapter investigated the performance of the DR method in the context of model updating, offering a comprehensive assessment of the influence of different factors on the accuracy and consistency of model updating outcomes. Moreover, the study explored the suitability of DR for integration into continuous vibration-based SHM systems. To this aim, two alternative strategies for continuous model updating were evaluated in terms of their effectiveness under changing environmental variations impacting the natural frequencies.

The key conclusions drawn from this investigation can be outlined as follows:

- The $H_{f,I}$ objective function demonstrated significant robustness, making it a suitable choice for real-world applications.

- Gradient-based optimization algorithms, thanks to their low computational demand, can significantly reduce processing time in model updating tasks, thereby supporting efficient integration within SHM systems. Nonetheless, to ensure reliable and repeatable outcomes, appropriate strategies must be adopted to guarantee convergence toward the global minimum.
- Incomplete modal datasets may affect the accuracy of model updating results. In the context of automated, continuous SHM processes, it is essential to assess the impact of missing modes in advance and implement corrective strategies to ensure the reliability of monitoring outputs.
- The DR method proved to be effective for continuous model updating even under environmental variability and measurement noise, with minimal computational requirements. This makes it especially advantageous for SHM purposes. Time-dependent tracking of updating parameters is feasible, provided that their values remain within the bounds defined for the construction of the DR model—bounds that are generally wide enough to encompass the expected variations.
- Using fixed nominal values for updating parameters throughout the monitoring process maintains satisfactory accuracy while greatly reducing computational demands, making it a practical solution for ongoing SHM applications.

12.8 ACKNOWLEDGMENTS

The present study is part of the research activities developed by the first and third authors within the framework of the PNRR Program, CN00000023 National Center for Sustainable Mobility, SPOKE 7 "CCAM, Connected Networks and Smart Infrastructure"—WP4 (CUP B43C22000440001).

REFERENCES

[1] UNESCO. Operational guidelines for the implementation of the world heritage convention. 2019. Available online at: http://whc.unesco.org/en/guidelines/.

[2] ICOMOS. Principles for the analysis, conservation and structural restoration of architectural heritage. 2003. Available online at: https://www.icomos.org/images/DOCUMENTS/Charters/structures_e.pdf.

[3] Ministero per i Beni e le Attività Culturali. Linee Guida per la valutazione e riduzione del rischio sismico del patrimonio culturale allineate alle nuove Norme Tecniche per le Costruzioni, D.M. 14 gennaio 2008, Circolare 26/2010; 2010 (in Italian).

[4] Lerna M, Sabbà MF, Diaferio M, Carnimeo L, Ivorra S, Foti D. Seismic risk assessment of a medieval tower: The case study of Craco. In: Papadrakakis M, Fragiadakis M, Papadimitriou C, editors. Proceedings of the XI International Conference on Structural Dynamics EURODYN 2020, Athens, Greece; 2020, pp. 4911–4921.

[5] Rainieri C, Marra A, Rainieri GM, Gargaro D, Pepe M, Fabbrocino G. Integrated non-destructive assessment of relevant structural elements of an Italian heritage site: The Carthusian monastery of Trisulti. Journal of Physics: Conference Series 2015; 628: Article number 012018.

[6] Conde B, Ramos LF, Oliveira DV, Riveiro B, Solla M. Structural assessment of masonry arch bridges by combination of non-destructive testing techniques and three-dimensional numerical modelling: Application to Vilanova bridge. Engineering Structures 2017; 148: 621–638.
[7] Marasco S, Noori AZ, Domaneschi M, Cimellaro GP. Seismic vulnerability assessment indices for buildings: Proposals, comparisons, methodologies at collapse limit states. International Journal of Disaster Risk Reduction 2021; 63: Article number 102466.
[8] Domaneschi M, Noori AZ, Pietropinto MV, Cimellaro GP. Seismic vulnerability assessment of existing school buildings. Computers and Structures 2021; 248: Article number 106522.
[9] Milani G, Clementi F. Advanced seismic assessment of four masonry Bell Tower in Italy after Operational Modal Analysis (OMA) identification. International Journal of Architectural Heritage 2021; 15(1): 157–186.
[10] Bru D, Ivorra S, Baeza FJ, Reynau R, Foti D. OMA dynamic identification of a masonry chimney with severe cracking condition. In: Proceedings of the 6th International Operational Modal Analysis Conference, Gijon, Spain; 2015.
[11] Diaferio M, Foti D, Giannoccaro NI, Sabbà MF. Dynamic identification on an irregular structure. Applied Sciences 2022; 12: Article number 3445.
[12] Aguilar R, Marques R, Sovero K, Martel C, Trujillano F, Boroschek R. Investigations on the structural behaviour of archaeological heritage in Peru: From survey to seismic assessment. Engineering Structures 2015; 95: 94–111.
[13] Elyamani A, Roca P. A review on the study of historical structures using integrated investigation activities for seismic safety assessment. Part 2: Model updating and seismic analysis. Scientific Culture 2018; 4(1): 29–51.
[14] Abdullah NAZ, Sani MSM, Rahman MM, Zaman I. A review on model updating in structural dynamics. IOP Conference Series: Materials Science and Engineering 2015; 100: Article number 012015.
[15] Atamturktur S, Hemez F, Unal C. Calibration under uncertainty for finite element models of masonry monuments. Technical Report LA-14414, Los Alamos National Laboratory, Los Alamos, NM, USA; 2010.
[16] Atamturktur S, Laman JA. Finite element model correlation and calibration of historic masonry monuments: Review. The Structural Design of Tall and Special Buildings 2012; 21(2): 96–113.
[17] Standoli G, Salachoris GP, Masciotta MG, Clementi F. Modal-based FE model updating via genetic algorithms: Exploiting artificial intelligence to build realistic numerical models of historical structures. Construction and Building Materials 2021; 303: Article number 124393.
[18] Bianconi F, Panagiotis Salachoris G, Clementi F, Lenci S. A Genetic algorithm procedure for the automatic updating of FEM based on ambient vibration tests. Sensors 2020; 20: Article number 3315.
[19] Shabbir F, Omenzetter P. Particle swarm optimization with sequential niche technique for dynamic finite element model updating. Computer-Aided Civil and Infrastructure Engineering, 2014; 30(5): 359–375.
[20] Bautista-De Castro Á, Sánchez-Aparicio LJ, Ramos LF, Sena-Cruz J, González-Aguilera D. Integrating geomatic approaches, operational modal analysis, advanced numerical and updating methods to evaluate the current safety conditions of the historical Bôco bridge. Construction and Building Materials 2018; 158: 961–984.
[21] Ferrari R, Froio D, Rizzi E, Gentile C, Chatzi EN. Model updating of a historic concrete bridge by sensitivity- and global optimization-based Latin hypercube sampling. Engineering Structures 2019; 179: 139–160.

[22] Vincenzi L, Savoia M. Coupling response surface and differential evolution for parameter identification problems. Computer-Aided Civil and Infrastructure Engineering 2015; 30(5): 376–393.
[23] Douglas BM, Reid WH. Dynamic tests and system identification of bridges. ASCE Journal of the Structural Division 1982; 108(10): 2295–2312.
[24] Gentile C. Modal and structural identification of a R.C. arch bridge. Structural Engineering and Mechanics 2006; 22(1): 53–70.
[25] Zordan T, Briseghella B, Liu T. Finite element model updating of a tied-arch bridge using Douglas-Reid method and Rosenbrock optimization algorithm. Journal of Traffic and Transportation Engineering (English Edition) 2014; 1(4): 280–292.
[26] Peña F, Lourenço PB, Mendes N, Oliveira DV. Numerical models for the seismic assessment of an old masonry tower. Engineering Structures 2010; 32(5): 1466–1478.
[27] Gentile C, Saisi A. Ambient vibration testing of historic masonry towers for structural identification and damage assessment. Construction and Building Materials 2007; 21(6): 1311–1321.
[28] Rainieri C, Fabbrocino G, Verderame GM. Non-destructive characterization and dynamic identification of a modern heritage building for serviceability seismic analyses. NDT&E International 2013; 60: 17–31.
[29] García-Macías E, Ierimonti L, Venanzi I, Ubertini F. An innovative methodology for online surrogate-based model updating of historic buildings using monitoring data. International Journal of Architectural Heritage 2021; 15(1): 92–112.
[30] García-Macías E, Venanzi I, Ubertini F. Metamodel-based pattern recognition approach for real-time identification of earthquake-induced damage in historic masonry structures. Automation in Construction 2020; 120: Article number 103389.
[31] Rainieri C, Magalhaes F, Gargaro D, Fabbrocino G, Cunha A. Predicting the variability of natural frequencies and its causes by second-order blind identification. Structural Health Monitoring 2019; 18(2): 486–507.
[32] Cabboi A, Gentile C, Saisi A. From continuous vibration monitoring to FEM-based damage assessment: Application on a stone-masonry tower. Construction and Building Materials 2017; 156: 252–265.
[33] Cabboi A, Gentile C, Saisi A. Automated modal and structural identification of a stone-masonry bell-tower for SHM. In: Proceedings of the 6th World Conference on Structural Control and Monitoring, Barcelona, Spain; 2014.
[34] Friswell MI, Mottershead JE. Finite element model updating in structural dynamics. Dordrecht, The Netherlands: Kluwer Academic Publishers; 1995.
[35] Gentile C, Saisi A, Cabboi A. Structural identification of a masonry tower based on operational modal analysis. International Journal of Architectural Heritage 2014; 9(2): 98–110.
[36] Baggio C, Sabbatini V, Santini S, Sebastiani C. Comparison of different finite element model updates based on experimental onsite testing: The case study of San Giovanni in Macerata. Journal of Civil Structural Health Monitoring 2021; 11: 767–790.
[37] VV.AA. La Mostra d'Oltremare nella Napoli occidentale: ricerche storiche e restauro del moderno. In: Aveta A, Castagnaro A, Mangone F, editors. Naples, Italy: FedOAPress; 2021.
[38] Computers and Structures Inc. CSI analysis reference manual for SAP2000®, ETABS®, SAFE® and CSiBridge®, Berkeley, CA, USA; 2017.
[39] Brincker R, Zhang L, Andersen P. Modal identification of output-only systems using frequency domain decomposition. Smart Materials and Structures 2001; 10: 441–445.
[40] Peeters B. System Identification and Damage Detection in Civil Engineering [Ph.D. thesis] Katholieke Universiteit Leuven, Leuven, Belgium; 2000.

[41] Van Overshee P, De Moor B. Subspace identification for linear systems: Theory—implementation—applications. Dordrecht, The Netherlands: Kluwer Academic Publishers; 1996.

[42] Yong L. Novel global harmony search algorithm for least absolute deviation. Journal of Applied Mathematics 2014; 2014: Article number 632975.

[43] Boscato G, Russo S, Ceravolo R, Zanotti Fragonara L. Global sensitivity-based model updating for heritage structures. Computer-Aided Civil and Infrastructure Engineering 2015; 30: 620–635.

[44] Rainieri C, Fabbrocino G. Development and validation of an automated operational modal analysis algorithm for vibration-based monitoring and tensile load estimation. Mechanical Systems and Signal Processing 2015; 60–61: 512–534.

[45] Monchetti S, Viscardi C, Betti M, Bartoli G. Bayesian-based model updating using natural frequency data for historic masonry tower. Probabilistic Engineering Mechanics 2022; 70: Article number 103337.

[46] Kovačević VC, Monchetti S, Betti M, Borri C. Metamodels in computational mechanics for Bayesian FEM updating of ancient high-rise masonry structures. In: Carcaterra A, Paolone A, Graziani G, editors. Proceedings of XXIV AIMETA Conference 2019, Lecture Notes in Mechanical Engineering, Springer; Cham; 2020.

[47] Jin S-S, Cho S, Jung H-J. Adaptive reference updating for vibration-based structural health monitoring under varying environmental conditions. Computers & Structures 2015; 158: 211–224.

[48] Jiao Y, Liu H, Wang X, Zhang Y, Luo G, Gong Y. Temperature effect on mechanical properties and damage identification of concrete structure. Advances in Materials Science and Engineering 2014; 2014: Article number 191360.

[49] Saisi A, Gentile C, Guidobaldi M. Post-earthquake continuous dynamic monitoring of the Gabbia Tower in Mantua, Italy. Construction and Building Materials 2015; 81: 101–112.

[50] Ramos LF, Marques L, Lourenco PB, De Roeck G, Campos-Costa A, Roque J. Monitoring historical masonry structures with operational modal analysis: Two case studies. Mechanical Systems and Signal Processing 2010; 24(5): 1291–1305.

13 Measuring structural vibration behavior via deep learning and wavelet transform in laminated composite plates with nonlinearity in dimensions and boundary conditions

Wael A. Altabey, Mohammad Noori, and Ahmed Silik

13.1 INTRODUCTION

Recent progress in modal analysis and vibration diagnostics has been significantly influenced by the integration of advanced methods, including deep learning and innovative signal processing techniques. These developments have enabled more efficient monitoring of structural integrity and enhanced strategies for predictive maintenance. They also provide improved means for identifying and forecasting vibrational behavior in both mechanical and civil engineering structures, as well as for analyzing noise and vibration data.

This chapter presents a novel approach for evaluating the vibrational response of complex structures. The method integrates the wavelet transform (WT) with deep learning algorithms. Specifically, it is applied to a complex structure: a skew laminated composite plate (SLCP) featuring variable thickness and intermediate elastic support. Such structures are commonly found in mechanical and civil engineering applications.

The methodology begins by extracting vibration characteristics and features from the plate using WT. These extracted features are then used to train a custom-designed deep neural network based on convolutional neural network (CNN) architecture. This network aims to capture and interpret the vibrational behavior of the plate, emphasizing critical aspects related to the mode shapes of SLCPs.

DOI: 10.1201/9781003516941-13

By incorporating deep learning, the proposed technique significantly reduces the computational load and iterative steps typically required when using the WT method alone. The findings indicate that this combined approach is particularly effective for analyzing complex structures under varying environmental conditions, showing the potential of deep learning as a powerful tool in structural vibration analysis.

13.2 LITERATURE REVIEW

The use of composite materials in various engineering sectors—such as aerospace, civil, and marine industries—is driven by their distinctive properties, including a high strength-to-weight ratio and resistance to corrosion, especially in harsh environmental conditions. Despite these advantages, structures made from these materials are susceptible to defects like matrix cracking, fiber fracture, and delamination. These internal flaws, which are often not visible on the surface, can lead to critical structural failures [1–4].

Other significant failure modes in fiber-reinforced polymers (FRPs) include those caused by thermal effects, mechanical stresses (bending or tension), impact loads, and improper installation. Such failures are often complex and difficult to detect, particularly when multiple stress factors are involved [5–7].

To address these challenges, structural health monitoring (SHM) systems play a crucial role in ensuring safety and identifying potential damage in composite plates. SHM systems typically encompass three main functions: identifying the structural system, extracting relevant features for damage detection and prediction, and evaluating reliability and risk [8–19].

Traditionally, defect detection in composite structures has relied on vibration analysis through conventional modal analysis techniques. Among these, mode shapes serve as vital indicators in SHM, as variations in vibration responses corresponding to different damage types and locations can be identified by analyzing changes in vibrational parameters. In recent years, substantial research has focused on innovative methods for determining the natural frequencies of composite plates, reflecting the growing interest in this area [20–34].

To analyze mode shapes under various boundary conditions (BCs) and with intermediate elastic supports, numerical or experimental approaches are often employed. Several studies have investigated the dynamic behavior of multi-span plates using a range of analytical techniques. Notably, Altabey [35, 36] applied the first shear deformation theory (FSDT) as a semi-analytical tool to study the vibrational characteristics of basalt FRP laminated rectangular plates with variable thickness and intermediate elastic supports. That study demonstrated enhanced result accuracy and reduced computational demands by integrating artificial neural networks (ANNs) with the response surface (RS) methodology.

In the current chapter, a novel CNN is proposed for predicting the vibrational behavior of SLCPs that feature variable thickness and intermediate elastic support. The analysis considers four different BC scenarios—SSSS, CCCC, SSFF, and CCFF—and varying values of the elastic restraint coefficient (K_T). The rectangular SLCP is composed of basalt fiber-reinforced polymer (BFRP) layers arranged in five symmetric plies with stacking angles of [45°/–45°/45°/–45°/45°] and includes

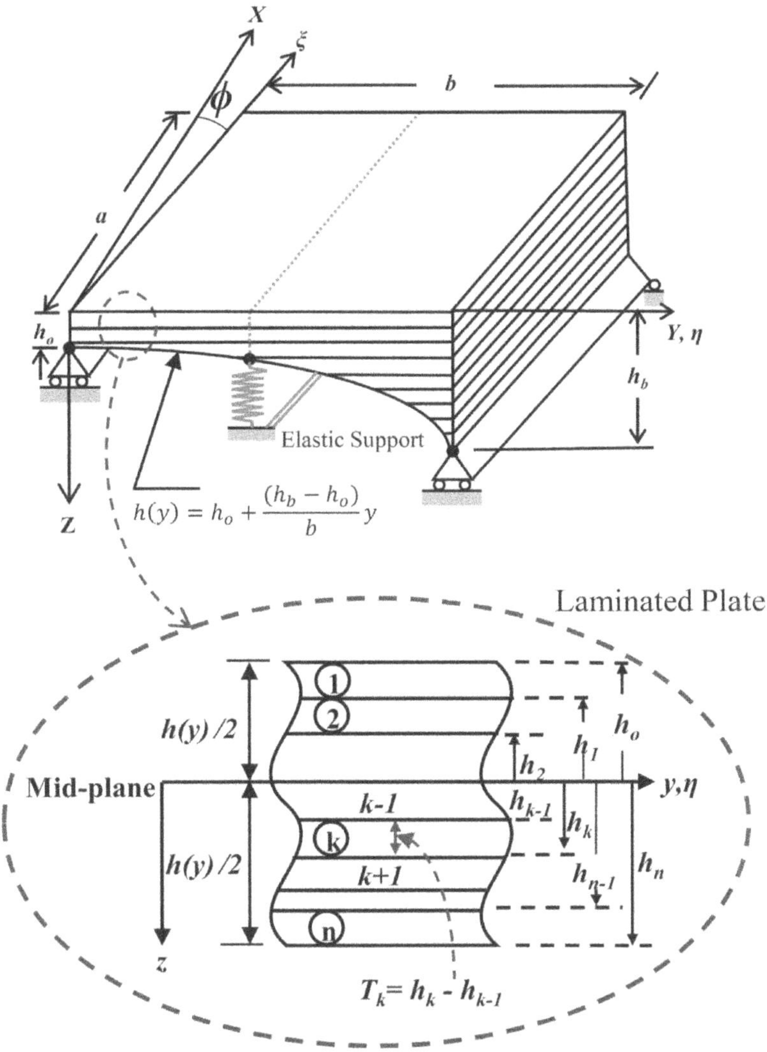

FIGURE 13.1 The geometry of the rectangular SLCP with variable thickness and intermediate elastic support.

a variable thickness distribution $h(y)$. The intermediate elastic support is located along the plate's midline, as depicted in Figure 13.1.

Initially, the WT is applied to extract key vibrational features of the plate. These features are then input into a specially trained CNN model, which is used to predict the vibrational response of the SLCP based on different values of the elastic restraint coefficient (K_T). The CNN's predictions closely match the outcomes derived from the WT approach, demonstrating strong alignment between the two methods. These findings suggest that the proposed deep learning model offers a robust and efficient alternative for vibration analysis of complex composite structures.

13.3 MODEL OVERVIEW

This section investigates the dynamic behavior of a laminated composite plate using a detailed numerical model. The material selected for the plate is a unidirectional fiber-reinforced composite, and its elastic and shear properties are essential for accurately simulating its mechanical response. Table 13.1 summarizes the key material properties used in the model, including the longitudinal and transverse elastic moduli (E_{11}, E_{22}), Poisson's ratios (ν_{12}, ν_{21}), shear moduli (G_{12}, G_{21}), and mass density (ρ). These parameters provide the foundation for simulating the vibrational characteristics of the composite structure.

The normalized partial differential equation describing the vibration behavior of the plate system shown in Figure 13.1 is derived under the assumptions of classical plate theory. This equation is expressed in terms of the plate deflection $w_o(x,y,t)$, using the nondimensional variables ξ and η, which are related to the skew coordinate system (u,ν,ϕ). The coordinate transformations are defined as $u = xsec(\phi)$, $\nu = y - xtan(\phi)$, and the nondimensional variables are $\xi = \frac{u}{a}$, $\eta = \frac{v}{b}$. After the necessary derivations, the governing equation can be formulated as follows:

$$\begin{aligned}
&\psi_1 h^3(\eta)\frac{1}{a^4}W_{\xi\xi\xi\xi} - 4\beta(\sin\phi)\ \psi_3 h^3(\eta)W_{\xi\xi\xi\eta} + \nu\beta^2\psi_2\ \left(\cos^2\phi\right)\ \frac{\partial^2 h^3(\eta)}{\partial\eta^2}W_{\xi\xi} \\
&+\ 2\beta^2\psi_2\ \left(\cos^2\phi\right)\ \frac{\partial h^3(\eta)}{\partial\eta}\ W_{\xi\xi\eta} + 2\beta^2\psi_2 h^3(\eta)\ \left(3\sin^2\phi + \cos^2\phi\right)\ W_{\xi\xi\eta\eta} \\
&+2\beta^4\ \left(\cos^2\phi\right)\ \frac{\partial h^3(\eta)}{\partial\eta}\ W_{\eta\eta\eta} -\ 2\beta^3\psi_4\nu\ \left|\sin\phi\ \cos^2\phi\right|\ \frac{\partial^2 h^3(\eta)}{\partial\eta^2}\ W_{\xi\eta} \\
&-\ 4\beta^3\psi_4\ \left(\sin\phi\right)\ h^3(\eta)\ W_{\xi\eta\eta\eta} +\ \beta^4\ \left(\nu\tan^2\phi + 1\right)\ \cos^4\phi\ \frac{\partial^2 h^3(\eta)}{\partial\eta^2}\ W_{\eta\eta} \\
&+\beta^4 h^3(\eta)\ W_{\eta\eta\eta\eta} - 4\beta^3\psi_4\left(\sin\phi\cos^2\phi\right)\frac{\partial h^3(\eta)}{\partial\eta}W_{\xi\eta\eta} = -\Omega^2 h(\eta)h_o^2\left(\cos^4\phi\right)\ W_{tt}
\end{aligned} \tag{1}$$

$$\text{where } \psi_1 = \frac{D_{11}}{D_{22}},\ \psi_2 = \frac{\left(D_{12} + 2D_{66}\right)}{D_{22}},\ \psi_3 = \frac{D_{16}}{D_{22}},\ \psi_4 = \frac{D_{26}}{D_{22}}$$

In this study, the BCs along both the x and y directions are modeled using various combinations of classical BCs. As the primary focus of this chapter is the treatment of intermediate elastic support conditions, these are discussed in greater detail. For the numerical analysis, four classical BC configurations are considered: SSSS, CCCC, SSFF, and CCFF, as illustrated in Figure 13.2. The parameters associated with these BCs are listed in Table 13.2.

TABLE 13.1
Model property.

E_{11} (GPa)	E_{22} (GPa)	υ_{12}	υ_{21}	G_{12} (GPa)	G_{21} (GPa)	rho kg/m^3
96.74	22.55	0.3	0.6	10.64	8.73	2700

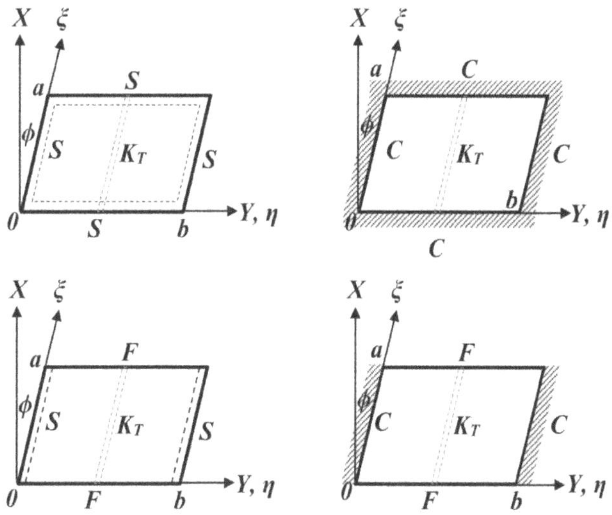

FIGURE 13.2 Representation of different support conditions for the analysis.

TABLE 13.2
The boundary conditions parameters.

No.	Type	The boundary conditions
1.	Simply supported edges	$w_o\big\|_{\xi=0,1}=0,\ \frac{1}{a^2}\frac{\partial^2 w_o}{\partial \xi^2}\Big\|_{\xi=0,1}=0$ $w_o\big\|_{\eta=0,1}=0,\ \frac{1}{b^2}\frac{\partial^2 w_o}{\partial \eta^2}\Big\|_{\eta=0,1}=0$
2.	Clamped supported edges	$w_o\big\|_{\xi=0,1}=0,\ \frac{1}{a}\frac{\partial w_o}{\partial \xi}\Big\|_{\xi=0,1}=0$ $w_o\big\|_{\eta=0,1}=0,\ \frac{1}{b}\frac{\partial w_o}{\partial \eta}\Big\|_{y=0,1}=0$
3.	Free edges	$\frac{1}{a^3}\frac{\partial^3 w_o}{\partial \xi^3}\Big\|_{\xi=0,1}=0,\ \frac{1}{a^2}\frac{\partial^2 w_o}{\partial \xi^2}\Big\|_{\xi=0,1}=0$ $\frac{1}{b^3}\frac{\partial^3 w_o}{\partial \eta^3}\Big\|_{\eta=0,1}=0,\ \frac{1}{b^2}\frac{\partial^2 w_o}{\partial \eta^2}\Big\|_{\eta=0,1}=0$
4.	Intermediate elastic support at $y=b/2$	$K_T w_o=-2\psi_3\frac{\partial^3 w_o}{\partial x^3}-\frac{\partial^3 w_o}{\partial y^3}-\psi_5\frac{\partial^2 w_o}{\partial x^2 \partial y}-4\psi_4\frac{\partial^3 w_o}{\partial x \partial y^2}$ $\frac{\partial w_o}{\partial y}\Big\|_{\eta=1^-/2}=\frac{\partial w_o}{\partial y}\Big\|_{\eta=1^+/2}$

where $\psi_1 = \frac{D_{11}}{D_{22}}$, $\psi_2 = \frac{(D_{12} + 2D_{66})}{D_{22}}$, $\psi_3 = \frac{D_{16}}{D_{22}}$, $\psi_4 = \frac{D_{26}}{D_{22}}$, $K_T = \frac{T_{b/2}b^3}{D_{22}}$, $\psi_5 = \frac{(D_{12} + 4D_{66})}{D_{22}}$.

13.4 DETERMINATION OF VIBRATION BEHAVIOR USING WT

In this section, the mode shapes of the SLCP are extracted using the WT method with an adjustable frequency parameter. To enhance the accuracy of the extraction, the frequency parameter is optimized based on the minimization of WT entropy.

13.4.1 Continuous Wavelet Transform

The continuous wavelet transform (CWT) is a convolution-based technique that involves the transformation of a data sequence using a family of wavelets generated by continuous scaling and translation of a base function, known as the mother wavelet $\psi(t)$ [37]. The translation operation provides time localization of the wavelet across the input signal *x (t)*, effectively smoothing the signal over its duration. The scaling operation compresses or stretches the wavelet, allowing the analysis to be conducted at various resolutions.

A stretched wavelet is used to detect slow or gradual variations in the signal, while a compressed wavelet is effective in capturing rapid or abrupt changes. This multi-resolution capability enables detailed time-frequency analysis of the signal. However, achieving higher resolution comes with increased computational cost and memory usage, as the process involves computing the wavelet coefficients for each scale and time step by multiplying the input signal with appropriately scaled and shifted versions of the mother wavelet.

In mathematical terms, the wavelet function $\psi(t)$ is scaled by a dilation parameter s and shifted in time by a translation parameter τ. These operations allow the wavelet to move along the time axis and adjust in scale to extract signal features at different resolutions [38–45].

$$\psi_{s,\tau}(t) = \frac{1}{\sqrt{s}}\psi\left(\frac{t-\tau}{s}\right) \quad \tau \in \Re, \quad S > 0 \tag{13.1}$$

Let $x(t)$ represent the system's shape function response, where t denotes time. The CWT of a function $x(t) \in L^2(\Re)$, where $L^2(\Re)$ denotes the space of square-integrable functions on the real line, is denoted by $W_{s,\tau}$, and is defined as follows:

$$W_{s,\tau} = \langle x(t), \psi_{s,\tau}(t) \rangle = \frac{1}{\sqrt{s}} \int_{\infty}^{-\infty} x(t).\psi^*\left(\frac{t-\tau}{s}\right) dt \tag{13.2}$$

where the wavelet scale s and the period τ are used to adjust the frequency and time location. $W_{s,\tau}$ shows how closely $\psi_{s,\tau}(t)$ is correlated with $x(t)$. By inverse CWT, the signal $x(t)$ can be regenerated

Here, the wavelet *scale* s and *translation* (or period) τ are used to adjust the frequency content and time localization of the wavelet, respectively. The CWT coefficient $W_{s,\tau}$ represents the correlation between the scaled and translated wavelet $\psi_{s,\tau}(t)$ and the original signal $x(t)$. A higher magnitude of $W_{s,\tau}$ indicates a stronger similarity between the wavelet and the signal at the corresponding scale and time [46].

The original signal x(t) can be reconstructed from its wavelet transform using the inverse CWT [37], which is given by

$$x(t)=\frac{1}{C_{\psi}}\int_{-\infty}^{+\infty}\int_{-\infty}^{+\infty}W_{s,\tau}\ \psi_{s,\tau}\ \frac{dsd\tau}{S^2} \tag{13.3}$$

As illustrated in Figure 13.1, the free-response equation for a single beam system along the ξ-direction can be expressed in the following form:

$$x(t)=X_i(\xi)Y_i(\eta)e^{j\omega\tau} \tag{13.4}$$

The WT of Eq. (12.4) is as follows:

$$\left|W_{s_0,\tau}\right|=\frac{\sqrt{s_0}}{2}X_i(\xi)Y_i(\eta).\psi^*(S_0\omega)e^{j\omega\tau} \tag{13.5}$$

The logarithm of Eq. (13.5) gives

$$\ln\left|W_{s_0,\tau}\right|=j\omega\tau+\ln\left|\frac{\sqrt{s_0}}{2}X_i(\xi)Y_i(\eta)\left|\psi^*(S_0\omega)\right|\right| \tag{13.6}$$

By analyzing the slope of the straight line obtained from the logarithm of the WT modulus, the natural frequency of the system can be determined. It is expressed as

$$Arg\left(W_{s_0,\tau}\right)=\omega\tau\Rightarrow\frac{d}{d\tau}Arg\left(W_{s_0,\tau}\right)=\omega \tag{13.7}$$

The plot of $\frac{d}{d\tau}Arg\left(W_{s_0,\tau}\right)$ remains constant over time and corresponds to the natural frequency ω. The nondimensional frequency parameter (NDFP), denoted as Ω, is defined as

$$\Omega=\left(\frac{m_o h(\eta)\omega^2a^4}{h_o D_{22}}\right)^{1/2} \tag{13.8}$$

The first six mode shapes corresponding to the natural frequencies of the intact plate are shown in Figure 13.3.

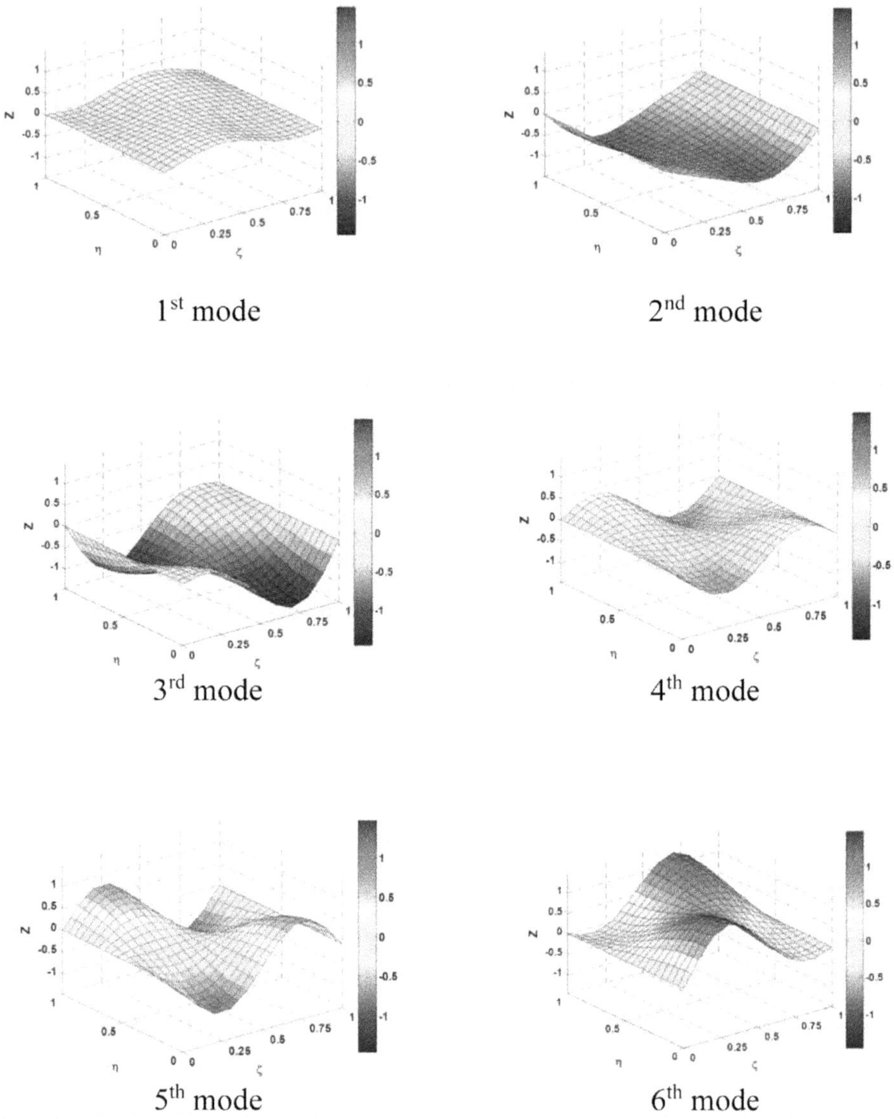

FIGURE 13.3 The first six mode shapes of an intact plate.

13.5 DEEP LEARNING ALGORITHMS

In recent years, deep learning, characterized by neural networks with multiple hidden layers, has been increasingly applied to solving and identifying both ordinary and partial differential equations [47, 48]. As a subset of artificial intelligence (AI), deep learning algorithms are particularly effective in addressing complex, nonlinear problems [49].

These networks consist of interconnected computational nodes (neurons), each of which performs a specific operation on the input it receives. The result is then passed on to subsequent nodes through output connections. In many cases, the output of one node serves as the input for one or more other nodes, forming a deeply layered architecture capable of learning hierarchical patterns from data.

In this section, we utilize the results obtained in Section 13.3, specifically the vibration behavior of the SLCP extracted using the WT at selected values of the elastic restraint coefficient (K_T), to generate training data. The trained deep learning model is then used to predict the vibrational behavior of the SLCP for different values of K_T that were not included in the training dataset.

13.5.1 The convolutional neural network

The architecture of the CNN is illustrated in Figure 13.4. As shown, the CNN structure typically consists of multiple layers, including convolutional layers, pooling layers, activation functions, and fully connected layers. The input to the network comprises the vibration parameter features of the plate. These features are processed through the convolutional and pooling layers, which are responsible for extracting significant local patterns and reducing spatial dimensions while preserving key information. Finally, the processed data is passed through the fully connected layers to generate the predicted vibration behavior of the plate.

As depicted in Figure 13.5, the fully connected layers are positioned between the input (feature set) and the output (target prediction) layers. These layers play a crucial role in learning high-level representations of the input data during the training process, enabling accurate prediction of the vibration response based on the extracted features.

Using weights w and biases b, the CNN is trained to identify and extract more meaningful nonlinear features from the input data. In the proposed CNN model, this process is defined as

$$x_j^l = f\left(\sum_i x_i^{l-1} w_{ij}^l + b_j^l\right) \tag{13.9}$$

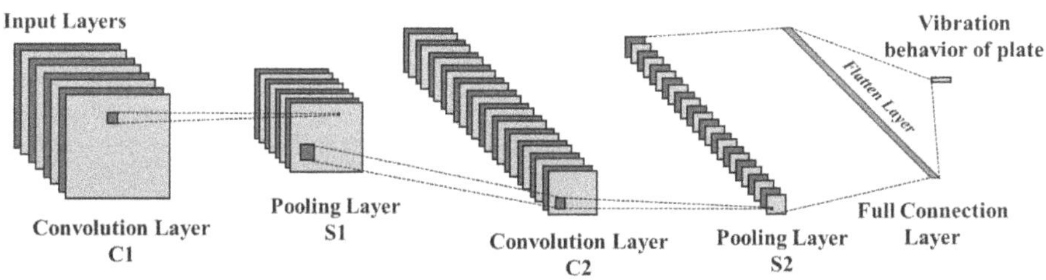

FIGURE 13.4 The architecture of a typical CNN.

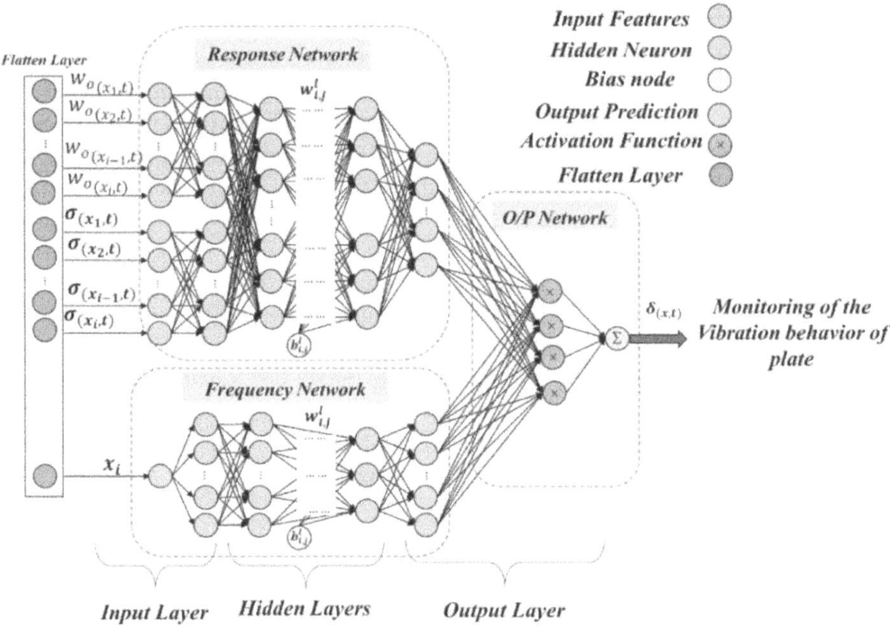

FIGURE 13.5 The architecture of the proposed CNN for the NDFP (Ω) prediction.

where x_j^l is output feature map in layer; x_i^{l-1} is the i-th output feature map from the previous layer l-1; w_{ij}^l denotes the weight connecting the i-th feature map in layer l-1 to the j-th feature map in layer l; b_j^l is the bias associated with the j-th feature map in layer l; and $f(\cdot)$ is a nonlinear activation function applied element-wise.

The pooling layer is usually arranged between sequential convolution layers. It is used to reduce the feature maps' locative size. This is also called undersampling, by which network overfitting can be controlled. The operations that can be used for undersampling include maximum pooling and average pooling. It can be expressed as the average pooling feature of the pooling layer in Eq. (13.10), assuming the pooling size is c, j-th is the region, and l-th is the number of pooling layers.

The pooling layer is typically placed between successive convolutional layers and serves to reduce the spatial dimensions of the feature maps. This process, also referred to as undersampling, helps control overfitting by decreasing the number of parameters and computational load. Common pooling operations include max pooling and average pooling, both of which summarize local regions of the feature maps. The average pooling operation in a pooling layer can be expressed as shown in Eq. (13.10), assuming a pooling window of size c, where j denotes the index of the pooled region and i represents the index of the pooling layer.

$$x_j^l = f\left(B_j^l \, mean\left(x_j^{l-1}\right) + b_j^l\right) \tag{13.10}$$

Here, B_j^l represents a multiplicative constant, and mean($\cdot$) denotes the average operation. The convolution and pooling layers work together to detect local patterns,

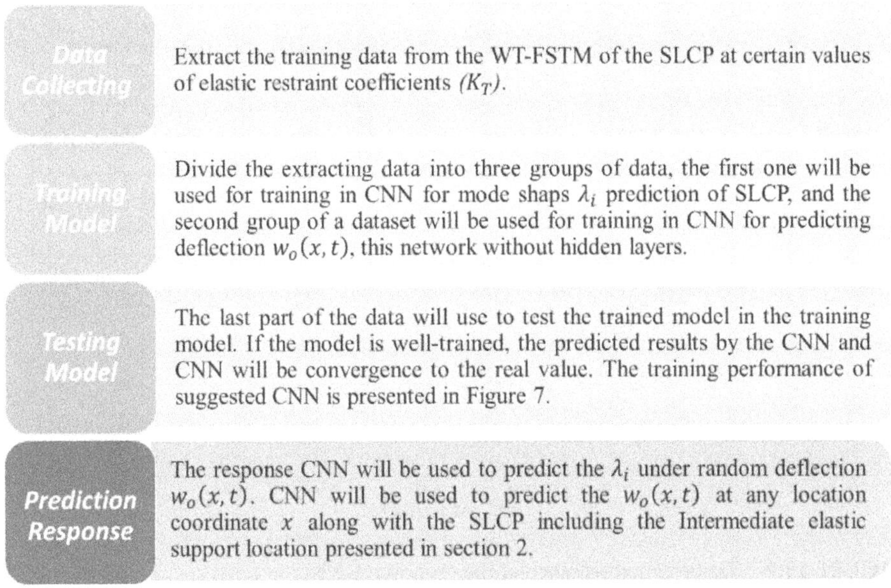

FIGURE 13.6 The steps of CNN training to predict NDFP (Ω).

merge similar features, and eliminate irrelevant or redundant information from the input data. The steps involved in predicting the NDFP, denoted as Ω, are shown in Figure 13.6.

It is essential for the neural network designer to evaluate whether the performance of the proposed CNN is adequate. This is commonly assessed using the mean squared error (MSE), which is defined by the following formula:

$$MSE = \sum\left(\left(\Omega\right)_{NN} - \Omega\right)^2 / n \tag{13.11}$$

Therefore, although there is only one global minimum for the performance index based on the input feature vectors, the loss function may also contain multiple local minima at finite input values. These local minima cannot be ignored when training a CNN. A local minimum can still be considered acceptable if it is sufficiently close to the global minimum and yields a low MSE.

To address the risk of the network becoming trapped in a shallow local minimum, the designer must choose an appropriate optimization method. One effective strategy is the use of momentum, which enables the network to consider not only the current gradient but also recent trends in the error surface. This helps the model maintain its direction across flat or shallow regions of the loss function and escape poor local minima. Without momentum, the network may converge prematurely and fail to reach optimal performance. Figure 13.7 illustrates the performance curves during training for three different learning data groups, highlighting the impact of optimization on convergence behavior.

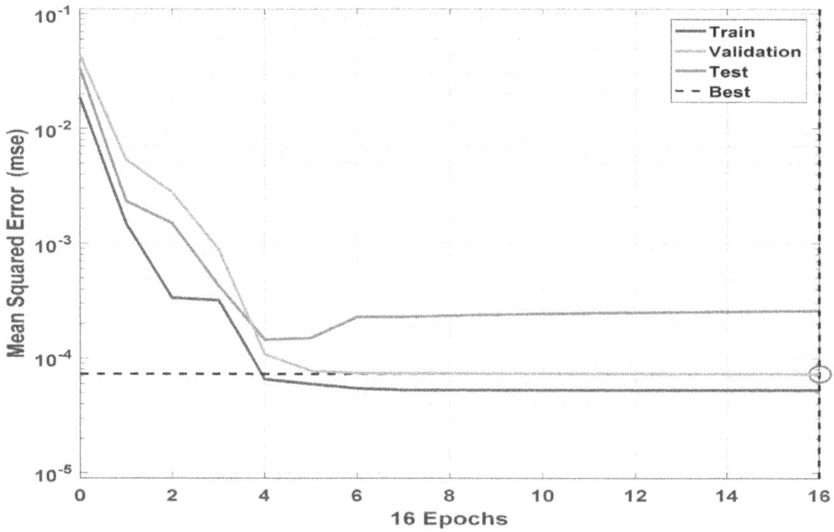

FIGURE 13.7 Training performance of the proposed CNN.

13.6 RESULTS AND DISCUSSION

In this section, we first review relevant findings from the existing literature. Then, the vibration behavior of the SLCP with variable thickness, analyzed using the WT approach as outlined in Section 13.2, is presented for selected values of the elastic restraint coefficient (K_T), representing intermediate elastic support conditions.

These WT-based results are subsequently used to generate training data for the proposed CNN. The trained CNN is then employed to investigate the influence of varying intermediate elastic support stiffness (i.e., different values of K_T) on the natural frequencies of the SLCP. This dual-stage approach enables both validation of the WT method and efficient prediction of vibrational characteristics under a broader range of support conditions.

13.6.1 Convergence study and accuracy

It is essential to compare the results of the present work with those available in the literature to validate the accuracy and reliability of the proposed technique. In this subsection, the WT method is applied to a clamped-clamped (CCCC) skew plate with variable thickness. The analysis is conducted for a taper ratio $\beta = 0.5$, thickness variation parameter $\Delta = (0,\ 0.2,\ 0.4,\ 0.5)$, and skew angles $\phi = (30°, 45°, 60°)$.

The convergence behavior of the WT results is illustrated in Figure 13.8, where they are compared with results obtained using the FSDT method as reported in [35]. This comparison serves to validate the accuracy and consistency of the WT-based approach used in this study.

As shown in the figure, the results obtained using the proposed WT method demonstrate excellent agreement with those reported by other accurate methods in the literature [23–26, 35], particularly after convergence is achieved. This validates the

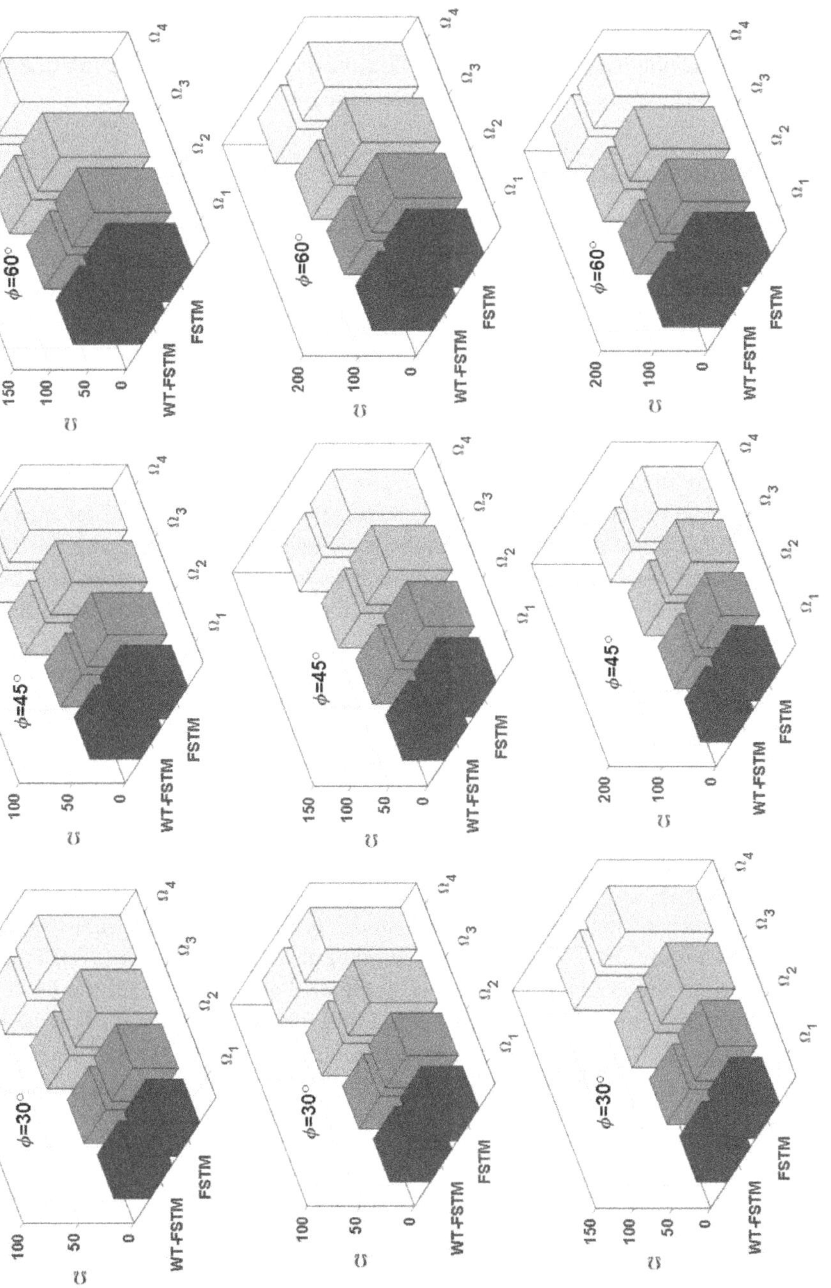

FIGURE 13.8 Comparison of the first four natural frequencies of CCCC skew plates $\beta = 0.5$.

reliability of the WT-based approach for analyzing the vibration behavior of skew laminated composite plates.

Additionally, the effects of key geometric parameters, namely the skew angle (ϕ), taper ratio (Δ), and aspect ratio (β), on the NDFP (Ω) are clearly observed. Specifically, (Ω) increases with higher values of ϕ, β, and Ω, indicating that these parameters significantly influence the dynamic response of the plate.

13.6.2 Proposed method results

In the present study, numerical computations based on the WT method were employed to extract the vibration behavior of the SLCP. However, due to the computational complexity of the WT approach, which involves a large number of calculations and iterations, it may not be the most efficient method for rapid and accurate vibration analysis. To address this, a new CNN model was developed to learn from the WT-derived results. The CNN was trained to identify and extract vibration behavior features that capture essential information about the mode shapes of the SLCP.

The proposed approach was trained using data corresponding to only two values of the elastic restraint coefficient K_T: one at $K_T = 50$ and the other at $K_T = 750$. These values were used to compute the NDFP (Ω), serving as the training set for the CNN.

The NDFP Ω was calculated for various skew angles φ while keeping the aspect ratio $\beta = 0.5$ and taper ratio $\Delta = 0.5$. This analysis was conducted across four classical boundary conditions—SSSS, CCCC, SSFF, and CCFF—to study the influence of skew angle on the natural frequencies.

Figures 13.9 and 13.10 present the comparison between the WT-computed NDFP values and those predicted by the CNN for $K_T = 50$ and $K_T = 750$, respectively, under the four specified boundary conditions. The results demonstrate that the CNN model provides highly satisfactory prediction accuracy, validating its effectiveness for vibration behavior prediction in complex structural systems.

13.6.3 CNN performance

The performance of the proposed CNN model is summarized in Table 13.3 and illustrated in Figure 13.7. The MSE and prediction accuracy for the NDFP Ω are calculated using Eq. (13.11). As shown in Table 13.3 and Figure 13.7, the MSE and accuracy for the training dataset are *7.2 E-5* and 99.7%, respectively. For the validation dataset, the MSE is *6.2 E-5*, and the accuracy reaches 99.8%. These results confirm that the proposed CNN achieves high predictive performance, demonstrating its effectiveness and reliability in modeling the vibration behavior of the SLCP.

13.6.4 CNN predicting results

In this subsection, the primary objective of designing the CNN to predict the vibration behavior of the SLCP under varying elastic restraint coefficients K_T is successfully achieved. A total of seven different K_T values were selected for analysis across four classical boundary condition configurations: SSSS, CCCC, SSFF, and CCFF. The CNN-predicted results for the first six natural frequencies of the SLCP, with an aspect ratio $\beta = 0.5$ and a taper ratio $\Delta = 0.5$, are presented in Figure 13.11.

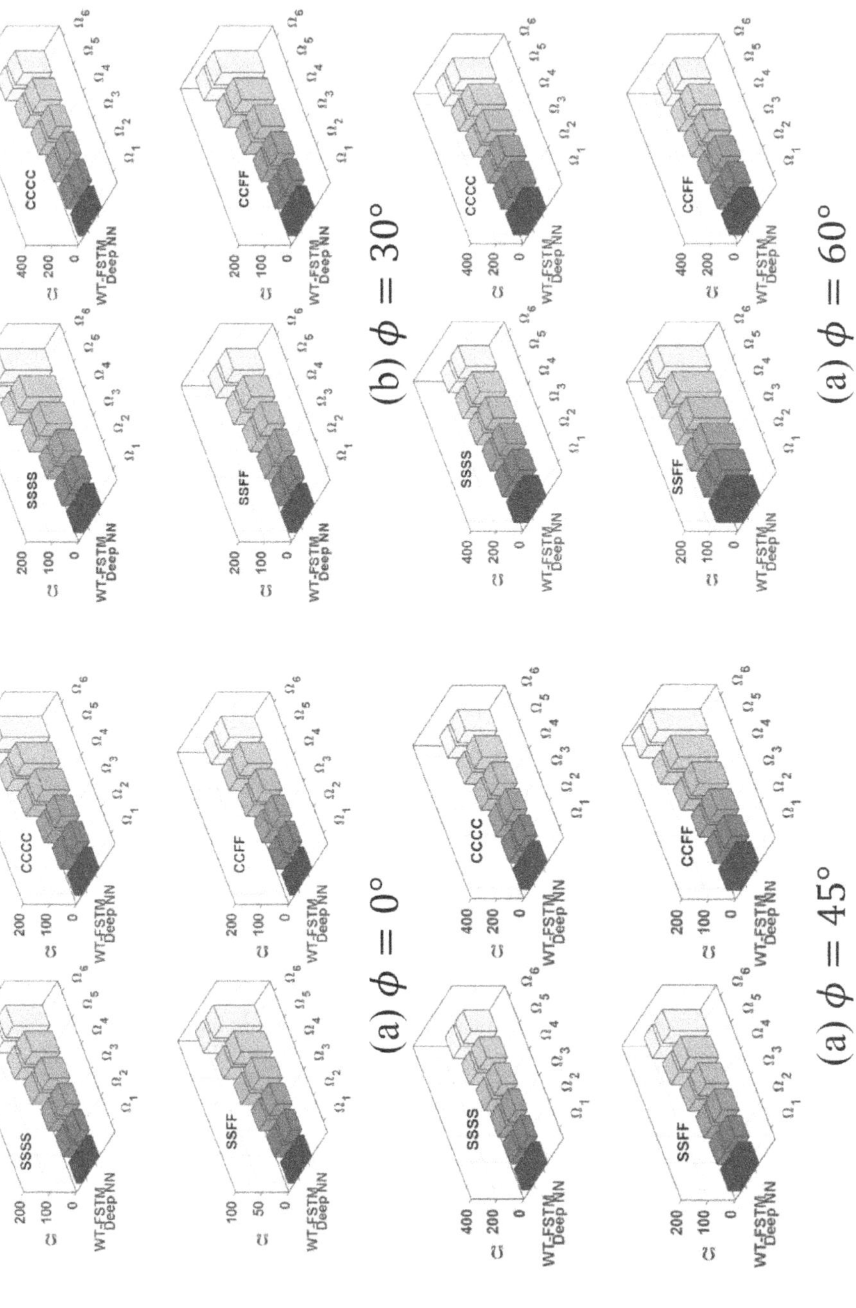

FIGURE 13.9 Comparison between the WT data and CNN-predicted data for $K_T = 50$.

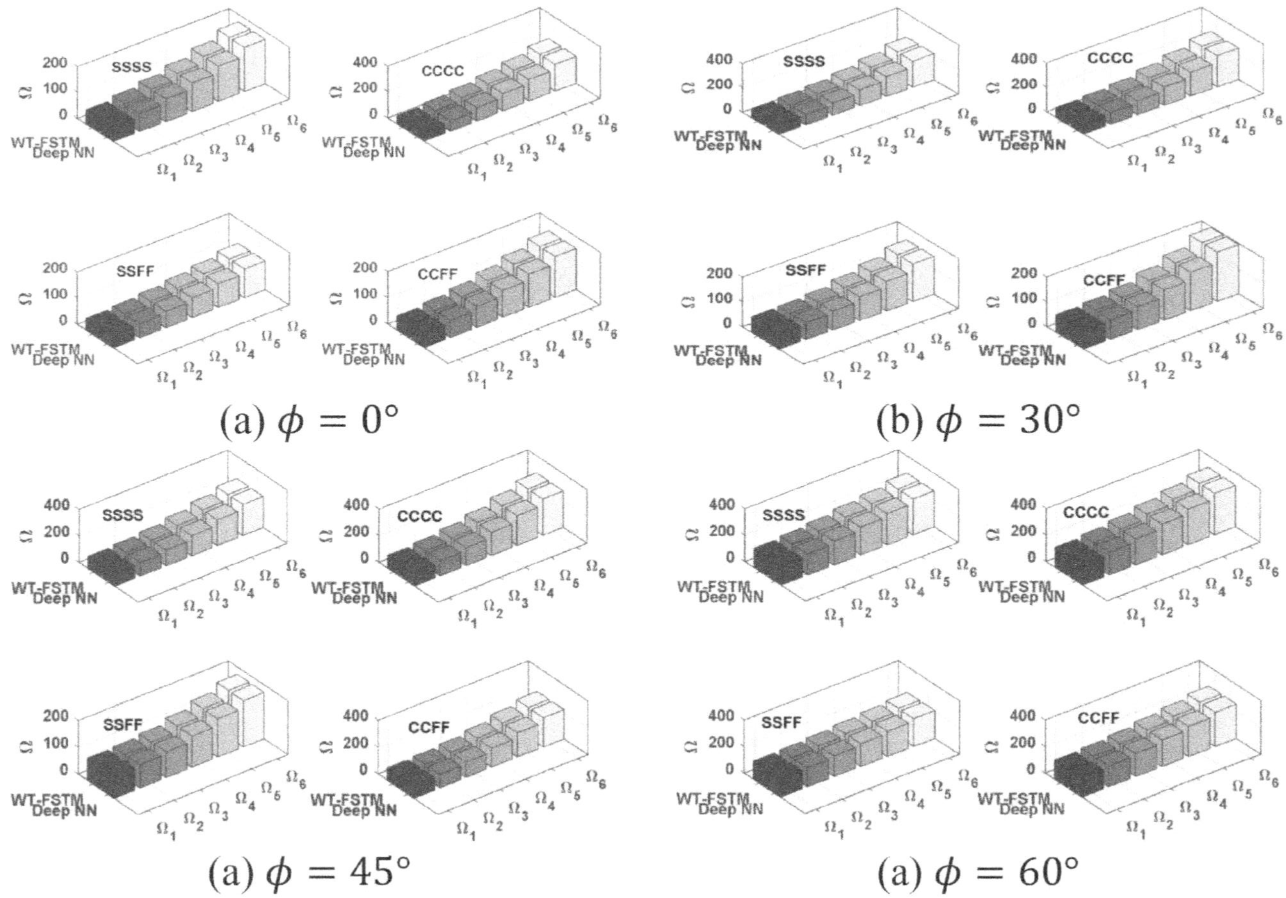

FIGURE 13.10 Comparison between the WT data and CNN-predicted data for $K = 750$.

TABLE 13.3
Mean square error (MSE) and accuracy values.

Data	MSE	Accuracy
Training	7.2 E-5	99.7%
Validating	6.2 E-5	99.8%

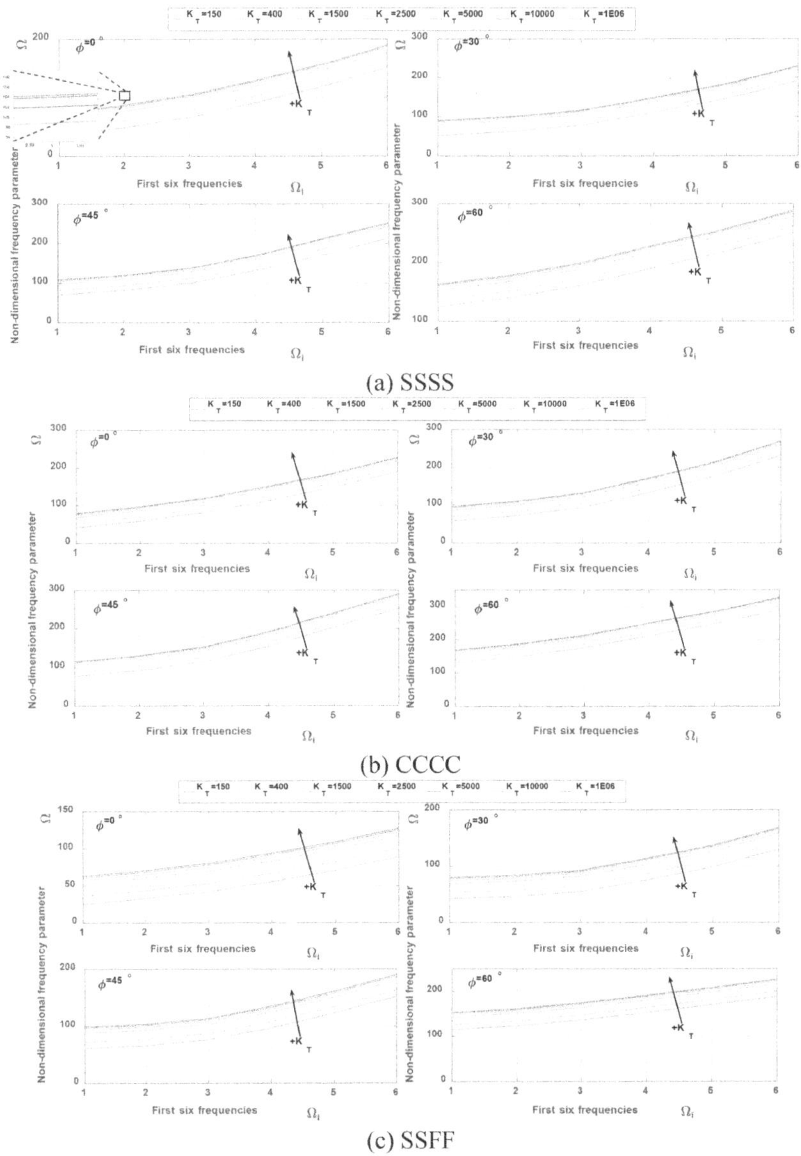

FIGURE 13.11 The CNN-predicted results of NDFP (Ω).

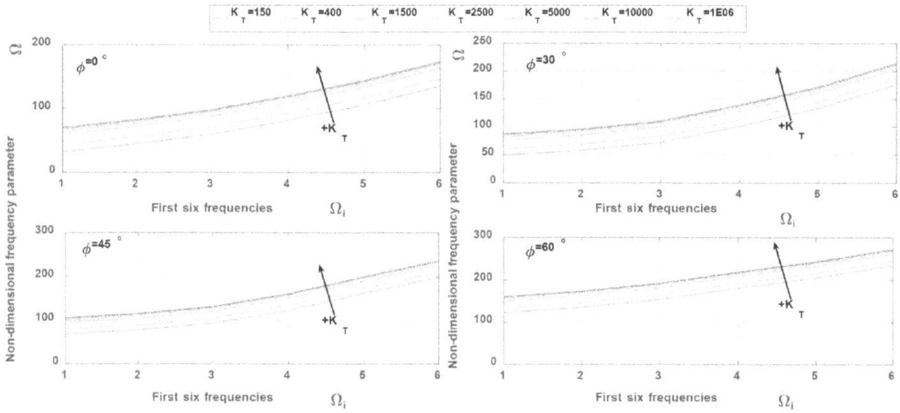

(d) CCFF

FIGURE 13.11 (Continued)

Moreover, the influence of intermediate elastic support on the vibration behavior of SLCPs with variable thickness is illustrated in Figure 13.11. As shown in the figure, for all values of skew angle ϕ and all boundary condition types, the first six natural frequencies increase as the elastic restraint coefficient K_T increases. Notably, the frequencies exhibit a rapid increase at lower values of K_T, while the effect of the intermediate elastic support becomes negligible at higher values of K_T.

Additionally, across all skew angles ϕ, the fully clamped (CCCC) plates exhibit the highest natural frequencies, whereas the semi-simply supported (SSFF) plates show the lowest. The other two boundary conditions, simply supported on all edges (SSSS) and clamped on two edges with free edges (CCFF), fall between these two extremes. Furthermore, the skew angle ϕ is shown to have a significant impact on the NDFP Ω. Specifically, Ω increases with increasing skew angle, indicating that the angular geometry of the plate plays an important role in its vibrational characteristics.

13.7 CONCLUSIONS

A combination of the WT method and a deep learning algorithm was used to extract the vibration behavior of an SLCP with variable thickness and intermediate elastic support. The plate is made of BFRP laminate. First, to evaluate the accuracy and reliability of the proposed technique, a convergence study was performed by comparing the results of the present work with those available in the literature, thereby validating its accuracy and credibility. However, due to the computational intensity of the proposed method, which requires numerous calculations and a large number of iterations, it may not be ideal for the rapid and efficient extraction of vibration behavior. To address this, a new CNN model was developed to learn and test the vibration behavior features, capturing key information about the mode shapes in SLCPs. The effects of β, Δ, φ, and K_T on the predicted normalized dominant frequency parameter (Ω) were investigated under four different support conditions: SSSS, CCCC, SSFF, and CCFF.

Based on the WT and CNN-predicted results, we conclude that the CNN-predicted results of the NDFP (Ω) are in very good agreement with the WT results, with the accuracy of the training and validation data being 99.7% and 99.8%, respectively.

13.8 APPENDIX A NOTATIONS

SLCP/SLCPs	skew laminated composite plate/plates
WT/WTs	wavelet transform/transforms
BCs	boundary conditions
CWT	continuous wavelet transform
MSE	mean square error
CNN	convolutional neural network
SSSS	simply supported at all edges
CCCC	clamped supported at all edges
SSFF	simply supported at two edges and free at the other
CCFF	clamped supported at two edges and free at the other
a, b	length and width of the plate
$h(y)$	plate thickness as a function of y (see Appendix A)
h_o, h_b	plate thickness at $y = 0$ and $y = b$, respectively
h_{ok}	distance from the middle plane of the plate according to h_o to the bottom of the h_{oth} layer.
m_o	$= \rho h_o$, mass density per unit area of the plate
p	dynamic load function
$w_o(x, y, z)$	deflection at any point (x, y)
D_{ij}	plate flexural rigidity
$\overline{Q_{ij}^k}$	plane stress transformed reduced stiffness coefficients of the lamina
E	Young'ss modulus
E_{11}	longitudinal Young's moduli parallel to the fiber orientation
E_{22}	transverse Young's moduli perpendicular to the fiber orientation
G_{12}, G_{21}	plane shear modulus of elasticity
Q	number of terms used in the series solution
$W(x, y, z)$	time-dependent deflection at any point (x, y)
M_n	the normal bending moment at the boundary
X_m	classical beam functions in the x-direction
Y_n	a series solution in the y-direction
K_T	elastic restraint coefficient
T	translational stiffness per unit length
$W(\xi, \eta, t)$	shape function
$Y_i(\eta)$	unknown function to be determined
$X_i(\xi)$	chosen a priori, the basic function in the ξ-direction
$\upsilon_{12}, \upsilon_{21}$	Poisson's ratio
ρ	density per unit area of the plate
ω	natural frequency (*rad/sec*)

(*Continued*)

(Continued)

ξ, η	$= u / a$ and v / b, respectively, nondimensional variables
ϕ	skew angle
β	$= a / b$, plate aspect ratio
Δ	$= (h_b - h_o) / h_o, (h_o)$, plate taper ratio
$(\Omega)_{NN}$	predicted nondimensional frequency parameter (PNDFP)
n	number of CNN data
$\partial w_o / \partial n$	the slope at the boundary
rho	material density

REFERENCES

[1] Zhao Y., Noori M., Altabey W. A., Ghiasi R., Wu Z. A Fatigue Damage Model for FRP Composite Laminate Systems Based on Stiffness Reduction. Structural Durability and Health Monitoring, 2019, Vol. 13, Issue 1, P. 85–103, http://dx.doi.org/10.32604/sdhm.2019.04695.

[2] Zhao Y., Noori M., Altabey W. A., Naiwei L. Reliability Evaluation of a Laminate Composite Plate under Distributed Pressure Using a Hybrid Response Surface Method. International Journal of Reliability, Quality and Safety Engineering, 2017, Vol. 24, Issue 3, 1750013, http://dx.doi.org/10.1142/S0218539317500139.

[3] Altabey W. A., Noori M. Fatigue Life Prediction for Carbon Fibre/Epoxy Laminate Composites under Spectrum Loading Using Two Different Neural Network Architectures. International Journal of Sustainable Materials and Structural Systems (IJSMSS), 2018, Vol. 3, Issue 1, http://dx.doi.org/10.1504/IJSMSS.2017.10013394.

[4] Wang T., Noori M., Altabey W. A. Identification of Cracks in an Euler-Bernoulli Beam Using Bayesian Inference and Closed-form Solution of Vibration Modes. Proceedings of the Institution of Mechanical Engineers, Part L: Journal of Materials: Design and Applications, Online Published: November 2, 2020, https://doi.org/10.1177/1464420720969719.

[5] Altabey W. A. An Exact Solution for Mechanical Behavior of BFRP Nano-Thin Films Embedded in NEMS. Advances in Nano Research, 2017, Vol. 5, Issue 4, P. 337–357. https://doi.org/10.12989/anr.2017.5.4.337.

[6] Altabey W. A. A Study on Thermo-Mechanical Behavior of MCD through Bulge Test Analysis. Advances in Computational Design, 2017, Vol. 2, Issue 2, P. 107–119, https://doi.org/10.12989/acd.2017.2.2.107.

[7] Ghannadi P., Kourehli S. S., Noori M., Altabey W. A. Efficiency of grey wolf optimization algorithm for damage detection of skeletal structures via expanded mode shapes. Advances in Structural Engineering, 2020, Vol. 23, Issue 13, P. 2850–2865, https://doi.org/10.1177/1369433220921000.

[8] Ghiasi R., Ghasemi M. R., Noori M., Altabey W. A. A Non-Parametric Approach toward Structural Health Monitoring for Processing Big Data Collected from the Sensor Network, Structural Health Monitoring 2019: Enabling Intelligent Life-Cycle Health Management for Industry Internet of Things (IIOT). Proceedings of the 12th International Workshop on Structural Health Monitoring (IWSHM 2019), September 10–12, 2019, Stanford, CA, USA. http://dx.doi.org/10.12783/shm2019/32395.

[9] Altabey W. A. Detecting and Predicting the Crude Oil Type Inside Composite Pipes Using ECS and ANN. Structural Monitoring and Maintenance, 2016, Vol. 3, Issue 4, P. 377–393.

[10] Altabey W. A. FE and ANN Model of ECS to Simulate the Pipelines Suffer from Internal Corrosion. Structural Monitoring and Maintenance, 2016, Vol. 3, Issue 3, P. 297–314, http://dx.doi.org/10.12989/smm.2016.3.3.297.
[11] Altabey W. A., Noori M. Detection of Fatigue Crack in Basalt FRP Laminate Composite Pipe Using Electrical Potential Change Method. Journal of Physics: Conference Series, 2017, Vol. 842, 012079. http://dx.doi.org/10.1088/1742-6596/842/1/012079.
[12] Zhao Y., Noori M., Altabey W. A., Zhishen W. Fatigue Damage Identification for Composite Pipeline Systems Using Electrical Capacitance Sensors, Smart Materials and Structures, 2018, Vol. 27, Issue 8, 085023, http://dx.doi.org/10.1088/1361-665x/aacc99.
[13] Altabey W. A. Delamination Evaluation on Basalt FRP Composite Pipe by Electrical Potential Change. Advances in Aircraft and Spacecraft Science, 2017, Vol. 4, Issue 5, P. 515–28. http://dx.doi.org/10.12989/aas.2017.4.5.515.
[14] Altabey W. A. EPC method for delamination assessment of basalt FRP pipe: Electrodes number effect. Structural Monitoring and Maintenance, 2017, Vol. 4, Issu 1, P. 69–84. http://dx.doi.org/10.12989/smm.2017.4.1.069.
[15] Altabey W. A., Noori M., Alarjani, A., Zhao, Y. Nano-Delamination Monitoring of BFRP Nano-Pipes of Electrical Potential Change with ANNs. Advances in Nano Research, 2020, Vol. 9, Issue 1, P. 1–13, http://dx.doi.org/10.12989/anr.2020.9.1.001.
[16] Altabey W. A., Noori M., Alarjani A. Zhao Y. Tensile Creep Monitoring of Basalt Fiber-Reinforced Polymer Plates Via Electrical Potential Change and Artificial Neural Network. Scientia Iranica, Transactions on Mechanical Engineering (B), 2020, Vol. 27 Issue 4, P. 1995–2008, http://dx.doi.org/10.24200/SCI.2020.52754.2874.
[17] Altabey W. A. The Thermal Effect on Electrical Capacitance Sensor for Two-Phase Flow Monitoring, Structural Monitoring and Maintenance, 2016, Vol. 3, Issue 4, P. 335–347, http://dx.doi.org/10.12989/smm.2016.3.4.335.
[18] Altabey W. A. An Exact Solution for Acoustic Simulation Based Transmission Loss Optimization of Double-Chamber Silencer. Sound & Vibration (SV), 2020, Vol. 54, Issue 4, P. 215–224, http://dx.doi.org/10.32604/sv.2020.011516.
[19] Ghiasi R., Noori M., Altabey W. A., Silik A., Wang T., Zhishen W. Uncertainty Handling in Structural Damage Detection Via Non-Probabilistic Meta-Models and Interval Mathematics, a Data-Analytics Approach. Applied Sciences, In Press, https://www.mdpi.com/journal/applsci.
[20] Laura PA. A, Grossi RO., Soni SR. Free Vibrations of a Rectangular Plate of Variable Thickness Elastically Restrained against Rotation along Three Edges and Free on the Forth Edge. Sound & Vibration, 1979, Vol. 62, P. 493–503.
[21] Mukhopadhyay M. Free Vibration of Rectangular Plates with Edges Having Different Degrees of Rotational Restraint. Sound & Vibration, 1979, Vol. 67, Issue 4, P. 459–468.
[22] Malhotra S. K., Ganesan N., Veluswami M. A. Vibration of Orthotropic Square Plate Having Variable Thickness (Parabolic Shape). Sound & Vibration, 1987, Vol. 119, P. 184–188.
[23] Dokainish M. A., Kumar K. Vibration of Orthotropic Parallaogramic Plates with Variable Thickness. AIAA Journal, 1973, Vol. 11, P. 1618–1621.
[24] Bardell N. S. The Free Vibration of Skew Plates Using the Hierarchical Finite Element Method. Computers & Structures, 1992, Vol. 45, P. 841–874.
[25] Singh B., Saxcna V. Transverse Vibration of Skew Plate with Variable Thickness. Sound & Vibration, 1997, Vol. 206, P. 1–13.
[26] Mizusawa T., Kajita T., Naruoka M. Vibration of Skew Plates by Using B-Spline Functions. Sound & Vibration, 1979, Vol. 62, P. 301–308.

[27] Chopra I., Durvasula S. Natural Frequencies and Modes of Tapered Skew Plates, International Journal of Mechanical Sciences, 1971, Vol. 13, P. 935–944.
[28] Liu W. H., Chang I. B. Vibration of Non-Uniform Skewed Cantilever Plats by the Method of Finite Element Transfer Matrix. Sound & Vibration, 1990, Vol. 136, P. 157–163.
[29] Garg A. K., Khare R. K., Kant T. Free Vibration of Skew Fiber Reinforced Composite and Sandwich Laminates Using a Shear Deformable Finite Element Model. Journal of Sandwich Structures & Materials, 2006, Vol. 8, P. 33–52.
[30] Li Z., Noori M., Altabey W. A. An Experimental Study on the Seismic Performance of Adobe Walls, In Press, Structural Durability and Health Monitoring, 2020, http://www.tspsubmission.com/index.php/sdhm.
[31] Zhao Y., Noori M., Altabey W. A. Reaching Law Based Sliding Mode Control for a Frame Structure under Seismic Load. In Press, Earthquake Engineering and Engineering Vibration, 2020, https://www.springer.com/journal/11803.
[32] Narjabadifam P., Hoseinpour R., Noori M., Altabey W. Practical Seismic Resilience Evaluation and Crisis Management Planning through GIS-Based Vulnerability Assessment of Buildings. Earthquake Engineering and Engineering Vibration, 2021, Vol. 20, Issue 1.
[33] Setoodeh A. R., Karami G. A. A Solution for the Vibration and Buckling of Composite Laminates with Elastically Restrained Edges. Computers & Structures, 2003, Vol. 60, P. 245–253.
[34] Cheung Y. K., Zhou D. Vibration Analysis of Symmetrically Laminated Rectangular Plates with Intermediate Line Supports. Computer & Structures, 2001, Vol. 79, P. 33–41.
[35] Al-Tabey W.A. Vibration Analysis of Laminated Composite Variable Thickness Plate Using Finite Strip Transition Matrix Technique and MATLAB Verifications, Chapter (21). In: MATLAB Applications for the Practical Engineer. InTech, World's Largest Science, Technology & Medicine Open Access Book Publisher, 2014, P. 283–620, http://dx.doi.org/10.5772/57384.
[36] Altabey W.A. Prediction of Natural Frequency of Basalt Fiber Reinforced Polymer (FRP) Laminated Variable Thickness Plates with Intermediate Elastic Support Using Artificial Neural Networks (ANNs) Method. Journal of Vibroengineering, 2017, Vol. 19, Issue 5, P. 3668–3678, https://doi.org/10.21595/jve.2017.18209.
[37] Silik, Ahmed, Mohammad Noori, Wael A. Altabey, Ramin Ghiasi, Zhishen Wu. Comparative Analysis of Wavelet Transform for Time-Frequency Analysis and Transient Localization in Structural Health Monitoring. Structural Durability & Health Monitoring, 2021, Vol. 15, Issue 1:
[38] Noori M., Haifegn W., Altabey W. A., Ahmad I. H. S. A Modified Wavelet Energy Rate Based Damage Identification Method for Steel Bridges, Scientia Iranica. International Journal of Science and Technology, Transactions on Mechanical Engineering (B), 2018, Vol. 25, Issue 6, P. 3210–3230, https://doi.org/10.24200/sci.2018.20736.
[39] Zhao Y., Noori M., Altabey W. A., Seyed B. B. Mode Shape Based Damage Identification for a Reinforced Concrete Beam Using Wavelet Coefficient Differences and Multi-Resolution Analysis. Structural Control & Health Monitoring, 2017, Vol. 25, Issue 3, P. 1–41. https://doi.org/10.1002/stc.2041.
[40] Zhao Y., Noori M., Altabey W. A. Damage Detection for a Beam under Transient Excitation via Three Different Algorithms. Structural Engineering and Mechanics, 2017, Vol. 63, Issue 6, P. 803–817. https://doi.org/10.12989/sem.2017.64.6.803.
[41] Zhao Y., Noori M., Altabey W. A., Awad T. A Comparison of Three Different Methods for the Identification of Hysterically Degrading Structures Using BWBN Model. Frontiers in Built Environment, 2019, Vol. 4, P. 80. https://doi.org/10.3389/fbuil.2018.00080.

[42] Silik A., Noori M., Altabey W. A., Ghiasi R. Comparative Analysis of Wavelet Transform for Time-Frequency Analysis and Transient Localization in Structural Health Monitoring. In Press, Structural Durability & Health Monitoring, 2020, http://www.tspsubmission.com/index.php/sdhm.

[43] Silik A., Noori M., Altabey W. A. Ghiasi R. Choosing Optimum Levels of Wavelet Multi-Resolution Analysis for Time-Varying Signals in Structural Health Monitoring. In Press, Structural Control & Health Monitoring, 2020, https://onlinelibrary.wiley.com/journal/15452263.

[44] Ashory M. R., Khatibi M. M., Jafari M., Malekjafarian A. Determination of Mode Shapes Using Wavelet Transform of Free Vibration Data. Archive of Applied Mechanics, 2013, Vol. 83, P. 907–921.

[45] Miranda F. J. Wavelet Analysis of Lightning Return Stroke. Journal of Atmospheric and Solar-Terrestrial Physics, 2008, Vol. 70, Issue 11–12, P. 1401–1407.

[46] Silik A., Noori M., Ghiasi R., Wang T., Kuok S.-C., Farhan N. S. D., Dang J., Wu Z., Altabey W. A. Dynamic Wavelet Neural Network Model for Damage Features Extraction and Patterns Recognition. Journal of Civil Structural Health Monitoring, 2023, Vol. 13, Issue 4, P. 925–945.

[47] Wang T., Altabey W. A., Noori M., Ghiasi R. A Deep Learning Based Approach for Response Prediction of Beam-Like Structures. In Press, Structural Durability & Health Monitoring, 2020, Vol. 14, Issue 4, P. 315–338, https://doi.org/10.32604/sdhm.2020.011083.

[48] Kost A., Altabey W. A., Noori M., Awad T. Applying Neural Networks for Tire Pressure Monitoring Systems. Structural Durability & Health Monitoring, 2019, Vol. 13, Issue 3, P. 247–266, http://dx.doi.org/10.32604/sdhm.2019.07025.

[49] Zhao Y., Noori M., Altabey W. A., Ghiasi R., Wu Z. Deep Learning-Based Damage, Load and Support Identification for a Composite Pipeline by Extracting Modal Macro Strains from Dynamic Excitations. Applied Sciences, 2018, Vol. 8, Issue 12, 2564, https://doi.org/10.3390/app8122564.

14 Vibration-based assessment of the effect of prehydration on the early-age stiffness of cement pastes

P. Cassese, L. Cieri, and Carlo Rainieri

14.1 INTRODUCTION

Ordinary Portland Cement (OPC) is commonly used in the formulation of mortar and concrete. OPC is designed to react chemically with water to set and harden. The exposure of cement to the surrounding atmosphere during manufacture, handling, or storage causes the absorption of moisture from the air, which might activate premature hydration (prehydration) with subsequent development of irreversible chemical interactions of the cement with water [1]. Such an unintentional reaction induces (i) the physical adsorption of water on the cement particles, on one hand, and (ii) the activation of chemical reactions involving water and the anhydrous OPC minerals to form solid hydration products such as calcium silicate hydrate, gypsum, and ettringite, on the other hand. These phenomena start on the exposed cement surfaces and diffuse into the internal portions after prolonged exposure [2, 3]. Environmental temperature and relative humidity (RH), as well as time of exposure and particle fineness, have been recognized as key variables affecting the degree of cement prehydration [4]. The main prehydration effect is the reduced reactivity of cement with the liquid water of the mix, as proved by the reduced rheological properties, increased setting time, and delayed strength development [5]. The prehydration phenomenon is generally associated with the handling and storage of OPC in the absence of appropriate countermeasures. However, it can already occur during manufacturing activities [6]. The rather detrimental influence of accidental prehydration of cement on its mechanical response has been well known since the beginning of the last century [7]. For this reason, common recommendations for the storage of Portland cement and its dry mixtures suggest using airtight boxes or dehumidified chambers. Some methods have also been proposed for extending the cement shelf life, which is commonly limited to no more than six months [8]. So far, research studies dealing with the assessment of the influence of prehydration on cementitious materials are rather limited, with only a few chemical-physical experimental approaches adopted and

DOI: 10.1201/9781003516941-14

specific materials considered. Furthermore, not fully convergent conclusions have been drawn [3, 4, 9–12]. Among the investigated materials, no specific attention has been paid to pozzolanic-based dry admixtures, although the use of such a binder, also in combination with hydrated lime, provides a solution characterized by high strength, ductility, and durability performance with a lower amount of ordinary cement [13, 14]. Specifically, pozzolan has been largely used in regions like Southern Italy because of the large supply availability. Moreover, its considerable performance in terms of chemical protection and mechanical compatibility with masonry block is well-known [15, 16], and it has motivated the use, in recent years, of pozzolanic and natural lime mortars to obtain inorganic matrices for the application of fiber-reinforced strengthening systems, with promising results in terms of structural and sustainability performance [17, 18].

The effect of cement prehydration is commonly recognizable from the early-age mechanical response, which is relevant to many structural applications such as the retrofit of existing structures based on the use of cementitious materials. Therefore, a reliable and actual assessment of the evolution of Young's modulus during the early-age phase after casting due to hydration reactions can be critical for the safe and accurate definition of the work sequence (e.g., setting time), as well as for predicting the long-term stiffness [19].

The experimental characterization of the early-age mechanical behavior of cementitious materials by means of nondestructive tests (NDTs) has been an object of study for many decades, and many methodologies have been developed based on different technological approaches. However, for the application of most of them, the setting of the material is necessary, and generally, several days of curing are required [20]. The hydration state, for example, is correlated with the temperature of the cement paste measured on different days of curing within the so-called maturity method; this method is largely used for on-site investigations, but its efficiency is strictly related to the reliability of the adopted maturity functions. The mechanical properties of cement-based materials during the hydration phase can also be derived by measuring their dielectric parameters in the case of electrical methods or the reflection coefficients if ultrasonic tests are carried out; both techniques offer almost qualitative indications of the stiffness evolution after the setting time. Recently, an innovative methodology based on the measurement of Radon exhalation since the casting of cement paste has been proposed, although more results are needed for a robust evaluation of its efficiency in monitoring the hydration process [21].

A mechanical NDT procedure based on vibration measurements for continuous monitoring of the Young's modulus (E) of concrete, cement paste, and mortar since casting was proposed in 2010 and 2011 by Azenha et al. [22, 23]. The methodology, known as the E-modulus measurement through ambient response method (EMM-ARM), is widely recognized as a reference among NDTs because it has been extensively used for various materials, and its reliability compared to destructive and other NDT methods has been proven through specific studies [24].

The EMM-ARM is applied herein to study the effects of prehydration on the early-age mechanical response of cementitious materials, with specific attention to the case of pozzolanic cement.

The chapter provides a detailed description of the experimental program, focusing on (i) the test methodology background through a detailed literature review on EMM-ARM, (ii) the characterization of specimens, setup, and data processing, and (iii) the obtained response curves. A discussion of the results is finally reported, and the main conclusions are drawn.

14.2 EXPERIMENTAL PROGRAM

An experimental program has been carried out at the Laboratory of the Neapolitan Branch of the Construction Technologies Institute of the National Research Council of Italy to investigate the effect of prehydration level on the mechanical response of cement-based materials through EMM-ARM tests [23]. Section 14.2.1 provides a concise report of a comprehensive literature review on EMM-ARM experimental experiences available in the literature. The most relevant information is considered and discussed in chronological order based on the reviewed studies. Detailed descriptions of the specimens and setup, materials, measurement system, and data processing procedure are reported in Sections 14.2.2, 14.2.3, and 14.2.4. The experimental results are presented and critically discussed in Section 14.2.5.

14.2.1 The EMM-ARM Methodology: Background and Experimental Validation

EMM-ARM is one of the NDT methods aimed at the characterization of the mechanical response of cementitious materials (concrete, mortar, cement paste, and others) [22, 23]. Unlike the majority of NDTs, this methodology can be applied almost since the casting of the material. Indeed, EMM-ARM provides a vibration-based indirect measure of the Young's modulus of the material under study: The latter is poured into a specific hosting beam, properly fixed at the ends to configure a known dynamic structural model. The dynamic response to ambient vibration of the composite beam, i.e., the combined system of the hosting beam and the investigated material, is continuously monitored, and the fundamental frequency is determined starting from the very early age after mixing, when the solid state has not yet been reached.

Some details about the theory, setup, and measurement procedure are given in Section 14.2.3, whereas an extensive description can be found in the work of Azenha et al. [22, 23]. In the following, an extensive literature review of the experimental studies that adopted EMM-ARM is presented in order to demonstrate the validation of the method and to identify case studies not yet investigated.

In its first proposal, dated 2010, the EMM-ARM methodology was devoted to the characterization of the stiffness (i.e., Young's modulus or E-modulus) development of concrete starting from the early age, that is, before the setting time [22]. To validate their results, the authors provided a comparison of the vibration-based E-modulus values with those derived from standard compression tests performed on various days after the setting time, with satisfactory approximation. Cement type CEM I 42.5 R was used as the binder, and the specimens were cured at a temperature (T) and RH of 20°C and 50%, respectively. One year later, the authors proposed a new version of the methodology [23], with a distinct experimental layout aimed at assessing the E-modulus of cement pastes and mortars, with special attention to evaluating the effect of chemical additives. Two different types of cement were studied, namely CEM I 52.5 and CEM IV/A 32.5R (FA), whereas no information about the curing environment was provided.

In 2012, EMM-ARM was applied by the same authors on cement pastes with several mineral admixtures and corresponding water-to-cement (w/c) ratios to estimate the variation in Young's modulus [26]. Cement type CEM I 42.5R was used as the binder for all specimens, while limestone, fly ash, silica fume, and metakaolin were considered as admixtures. The tests were performed under controlled curing conditions (T = 20°C and RH = 50%). The results showed a considerable influence of mix composition and w/c ratio on the development of the E-modulus, as expected. In the same year, an attempt was made to adapt the original experimental layout for concrete to in-situ conditions [27]. Some critical issues were identified, and further investigation was necessary.

In 2014, the applicability of EMM-ARM was also extended to sand-cement material derived from soil extraction [28]. CEM I 42.5 type was used, and the study was specifically devoted to assessing the influence of the geometry of the samples as well as the extraction techniques on the reliability of the obtained results. During the same year, a significant contribution to the validation of EMM-ARM for cement paste was proposed through a systematic study in which the results of many nondestructive and destructive test methodologies on the same materials were compared with those derived from EMM-ARM [29]. Two different binders were considered in the analyzed mixes, CEM I 42.5R and CEM II/B-L 32.5N, respectively; these were used for manufacturing the cement pastes. The tests were carried out under two different temperature values (i.e., 20°C and 40°C).

The experimental in-situ experience was continued and extended in 2016 [30]. Different soil admixtures and mortars, obtained by using both CEM I 42.5R and CEM II/B-L 32.5N types of cement, were studied to establish the combined interference of mold geometry and mix composition. Meanwhile, the validation of EMM-ARM as an NDT methodology for concrete was decisively demonstrated through a round-robin study carried out among three experimental laboratories [24]. Cement type CEM I 52.5N PMES CP2 was introduced into the concrete mix. The results of EMM-ARM showed satisfactory correspondence with those of seven different experimental techniques, both destructive and nondestructive, capable of monitoring the Young's modulus of cement-based materials from the early ages.

In 2017, a further refinement of the EMM-ARM setup for in-situ applications on concrete was proposed. Three different types of cement were considered (namely, CEM I 52.5 N PMES CP2, CEM II 42.5R, and CEM II/A-L 42.5R) under a controlled curing temperature of 20 ± 2°C. Some changes to the geometry and constituent material of the original EMM-ARM mold, a new support system, and the use of a small excitation applied to the beam to decrease the method's sensitivity to environmental noise were introduced [31].

The refined EMM-ARM layout was used for carrying out two experimental campaigns [32, 33] in which the authors assessed the influence on the early-age E-modulus of concrete replacing part of the cement in the mix with sustainable biomass ash and ordinary aggregates with recycled ones. CEM I 52.5R-SR 3 was used as the binder in the mix, and the effect of curing temperature was also investigated, considering two alternative values of 20°C and 40°C, respectively. The results showed good agreement with other conventional NDTs.

Recently, EMM-ARM was applied in combination with conventional cyclic compression tests [34, 35] and the innovative H-nuclear magnetic resonance [36], further proving its reliability. In Garijo et al. [34], the method was applied to hydraulic lime mortar, and the change in E-modulus evolution as a function of the water-to-lime ratio was assessed. Mixed lime-cement mortars were studied in the work of Ramesh et al. [35], in which cement type CEM I 42.5 R was partially replaced with lime. Following the same line, the influence of curing temperature on the development of E-modulus was studied in the work of Ramesh et al. [37] on lime-cement mortars with CEM I 42.5 R cement, varying the lime amounts. A considerable reduction in setting time was observed as the lime content increased.

In 2024, EMM-ARM was used to study the effect of adding expansive additives to cement paste on the microstructure and E-modulus evolution at an early age by Gupta et al. [38]. In particular, a slightly revised version of the original methodology was implemented in order to account for the effect of the external restraint on the material. OPC was used with free lime-based expansive additives and several combinations of w/c ratios and amounts of expansive additives. The results of EMM-ARM demonstrated a crucial influence of the restraint condition on the mechanical response of the material. Meanwhile, in order to reduce the costs and complexity of the setup, a new version of the EMM-ARM was developed by Araghi et al. [39], identified as EMM-IRM. The main change consisted of the introduction of an impulse excitation system that allowed the use of cheaper accelerometers with lower sensitivity, while also enhancing signal clarity and reducing interference from background noise. The system was validated through experimental comparisons with conventional EMM-ARM, cyclic compression, and traditional impulse excitation tests. Cement paste with OPC type CP V-ARI was used as the test material. The results showed good estimates from the EMM-IRM, with a higher signal-to-noise ratio.

The last developments on EMM-ARM by Azenha et al. are related to the implementation of a new test system, particularly in terms of data acquisition components and post-processing software, with considerably reduced costs and elaboration complexity [40, 41]. The proposed new system was validated through a comparison with the original EMM-ARM. A specific experimental program was carried out on cement paste with CEM II/B-L 32.5 cement and different w/c ratios. Promising results were obtained, and all the resources necessary for its full reproduction are available as open source [42, 43].

The extensive literature review reported earlier proves that AMM-ERM, although rather innovative and recent, can be considered a well-established and recognized NDT methodology capable of continuously monitoring the Young's modulus of cement-based materials, beginning from the pouring phase. However, so far, such a methodology has not yet been used to investigate the influence of prehydration of OPC and pozzolanic cement on the early-age E-modulus evolution due to storage outside airtight boxes or dehumidified chambers. Herein, this last issue is studied through an experimental campaign, as described in the following sections.

14.2.2 Preparation of specimens and test setup

According to the original EMM-ARM for the characterization of cement pastes [23], a specific test setup must be implemented to investigate the cement's early-age behavior, as shown in Figure 14.1. Immediately after the mixing phase is completed, the fluid mixture is poured into a plastic pouch with a cap (Figure 14.1a). The cap is tightly connected to a clear plastic hose. The hose is stuck into a clear acrylic tube, plugged on the opposite end through a polystyrene leaf (weighing less than 0.1 g) previously sealed with cyanoacrylate glue. By keeping the tube inclined at approximately 15° with respect to the vertical, the mixture is gently poured starting from the capped end (Figure 14.1b). The pressure is manually modulated by pushing the pouch. As the tube is gradually filled, the hose is lifted, and simultaneously, the mixture already poured is vibrated using a vibrator kept in contact with the tube.

At the end of the pouring phase, the free end of the tube is closed with a purge system made of a steel washer glued to a hex nut. The system is attached to the top of the tube, which has been previously cleaned and ground, using cyanoacrylate glue. Once the tube is placed vertically, a steel bolt is slowly screwed in to remove the remaining air bubbles until a drop of mixture becomes visible outside the tube. In Figure 14.2, the specimen (Figure 14.2a) and a detail of the purge system (Figure 14.2b) are shown.

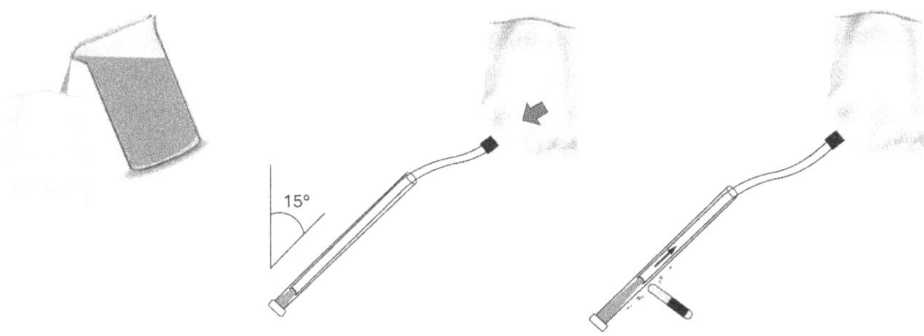

(a) (b) (c)

FIGURE 14.1 Preparation of the specimen: pouring of cement mixture.

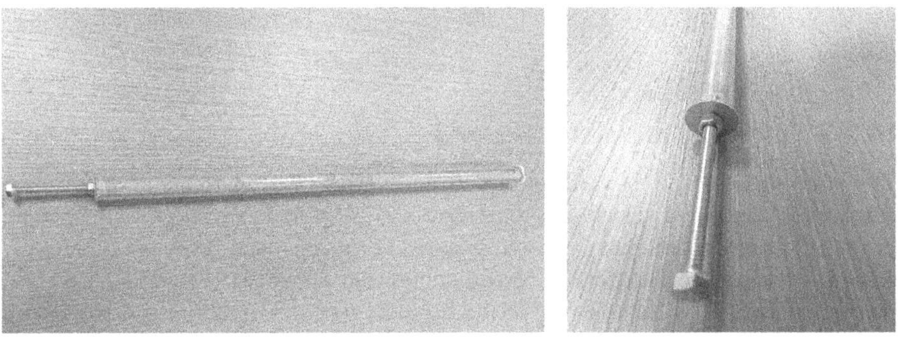

(a) (b)

FIGURE 14.2 Preparation of the specimen: purge system installation.

FIGURE 14.3 Specimen installed into the restraint system before the start of the test.

The acrylic tube represents a critical part of the setup since (i) it acts as a mold for the cement mixture in its fluid state, and (ii) it participates in the vibration-based measurements during the setting and hardening phases. After preparation, the specimen, consisting of the external tube and internal cement paste, is placed horizontally and installed with the end characterized by the presence of the purge system into a clamping device. The latter consists of a double steel plate, the largest one rigidly connected to the floor of the laboratory by four dowels and characterized by four vertical threaded rods. Once the specimen is carefully positioned, the upper plate is placed on top of the lower one and constrained to it by tightening the nuts with a torque wrench. The tightening operation is very sensitive, since there is a possibility of inducing brittle damage with the resulting loss of waterproofness. The specimen, in its final test configuration, behaves like a composite polymeric-cementitious cantilever beam. The acrylic tubes employed in the experimental program are characterized by length (L), diameter (Φ), and thickness (t) equal to 500 mm, 20 mm, and 2 mm, respectively. The clamping device has plane dimensions equal to 100 mm (in the direction of the beam) and 120 mm (in the opposite direction). Therefore, the final free span of the specimen is equal to 400 mm. Figure 14.3 shows a specimen before the start of the test.

14.2.3 Materials and test conditions

The main focus of the experimental program was to investigate the effect of accidental prehydration of cement due to uncontrolled storage, i.e., without using airtight boxes or dehumidified chambers, on its early-age mechanical response. Therefore, cement paste was prepared by using cement powder under two different conditions: (i) immediately after the opening of the original sack (t_0 condition) and (ii) after a certain period following the initial opening (t_i condition). During the period between t_0 and the test time t_i, the cement powder was stored in the original sack inside a chamber characterized by controlled temperature and RH conditions of 20 ± 0.5°C and 35 ± 5%, respectively. In this way, the storage influence was directly associated with its duration ($t_i - t_0$).

The study considered two different typologies of cement according to the European standard [44]:

1. CEM IV/B (P) 32.5 R
2. CEM II/A LL 42.5 R

The first is a pozzolanic cement and the second one a limestone Portland cement, both largely available and used in Southern Italy. The cement paste was obtained by adding water and a commercial shrinkage-reducing agent (Mapei Mapecure SRA). The mix proportions for a specimen were as follows:

1. Cement = 225 g;
2. Water = 112.5 g (w/c = 0.5);
3. Curing additive = 0.75 g (0.33% of the weight of cement, according to the manufacturer's instructions).

Overall, six specimens were tested. Four tests were carried out on pozzolanic cement pastes at 0, 30, 240, and 360 days from the initial opening. Two tests were carried out on Portland cement paste at 0 and 240 days. Notice that fewer tests were performed for Portland cement, since, as already discussed in Section 14.1, pozzolanic cement has been scarcely investigated in the literature regarding prehydration effects and early-age behavior.

Two additional tests were carried out on pozzolanic cement pastes for a preliminary study of the influence of curing temperature on the early-age mechanical response. These two tests were performed under controlled temperature conditions by applying two distinct values, of 15°C and 20°C, respectively.

14.2.4 Test procedure and measurement system

The procedure adopted in the present study is in line with the original EMM-ARM [23]. Accordingly, the Young's modulus of the cement paste contained in the acrylic tube is derived through the resolution of an inverse problem. Starting from the equation of a free vibrating cantilever with a constant cross section and by introducing the boundary conditions, Eq. (9.1) can be obtained [23].

$$a^3\left[\cosh(aL)\cos(aL)+1\right]+\frac{\omega^2 m_1}{EI}\left[\cos(aL)\sinh(aL)-\cosh(aL)\sin(aL)\right]=0 \quad (14.1)$$

In Eq. (14.1), E is the Young's modulus of the composite cement-acrylic material, I is the second moment of inertia of the composite cross section, $\omega = 2\pi f$ is the circular natural frequency of the fundamental mode, L represents the free-span length (equal to 400 mm), and m_1 is a concentrated mass at the free end of the cantilever.

The coefficient a in Eq. (14.2) depends on the circular natural frequency ω and the constants EI and $\bar{m}$, namely the bending stiffness and the uniformly distributed mass of the composite cross section, respectively.

$$a = \sqrt[4]{\frac{\omega^2 \bar{m}}{EI}} \tag{14.2}$$

The constant terms EI, $\bar{m}$, L, m_1 can be assumed to be time-independent; conversely, the frequency is a function of time and can be identified by operational modal analysis [45]. To this aim, the specimen is equipped with a piezoelectric accelerometer characterized by high sensitivity (1 V/g) and very low weight (7.5 g), attached to the free end of the tube by using cyanoacrylate glue. The accelerometer measured the ambient vibration response of the cantilever at a sampling frequency of 100 Hz. The acceleration time series were measured by a data acquisition system characterized by 24-bit resolution, an analog anti-aliasing filter, and control by the S2-DDA commercial software v.2.0.0 [46]. In order to reduce noise and enhance the vibration of the cantilever, an electric fan was employed. The fan was selected with reduced power to avoid altering the broadband nature of the excitation. During the test, the acquired acceleration time series were divided into sequential blocks with a duration of 15 minutes. For each block, the normalized power spectral density (NPSD) was computed according to the Welch procedure, considering 50% overlap and a Hanning window. The fundamental natural frequency was identified by implementing an Automatic Peak Detection (APD) algorithm; accordingly, sequential blocks of data were fitted with a quadratic polynomial as a function of a certain threshold. The use of the APD algorithm represents a slight modification to the original EMM-ARM [23] procedure, aimed at improving the reliability of the identified vibration frequency.

The automated peak-picking dynamic identification procedure described earlier was implemented in software developed in the LabVIEW programming environment [47].

The identified natural frequency is introduced in Eq. (14.1) to obtain the Young's modulus by iterative resolution. The coefficients EI, $\bar{m}$, L, m_1 of Eq. (14.1) can be simply determined: in particular, the length L and the weight of the specimen are systematically measured. From the weight of the specimen, its mass per unit length ($\bar{m}$) is derived. The concentrated mass m_1 is equal to the mass of the applied accelerometer. The latter is negligible compared to the mass of the specimen. Therefore, the initial value of the bending stiffness for the iterative procedure can be computed as per Eq. (14.3).

$$(EI)_0 \cong \frac{\bar{m} L^4 \omega^2}{1.8751^4} \tag{14.3}$$

At the end of the iterative procedure, the final value EI is obtained. Then, the elastic modulus of the cement paste E_c can be derived according to Eq. (14.4), where ϕ_i and ϕ_e are the internal and external diameter values of the acrylic tube, whereas E_a is its elastic modulus.

$$E_c = \frac{64EI}{\pi \phi_i^4} - E_a \frac{\phi_e^4 - \phi_i^4}{\phi_i^4} \tag{14.4}$$

E_a is experimentally assessed by carrying out a preliminary test on the tube filled only with water, following the testing procedure described earlier, and immediately before starting the preparation of the corresponding specimen.

The analytical processing procedure is based on the classic elastic beam theory. Thus, loss of adhesion and premature cracking of specimens must be monitored. In this regard, at the end of the test, the specimen is removed from the restraint and cut into several portions, which are accurately inspected.

14.3 PRELIMINARY TESTS AND RESULTS

A series of preliminary tests were performed to refine the setup and the data processing procedure. Figure 14.4 shows the comparison between the frequency response curves obtained for a pozzolanic cement paste according to the original EMM-ARM procedure for the peak detection [23] and the procedure reported in Section 14.2.4. In general, a good match can be observed, with an almost negligible scatter for the black curve (modified EMM-ARM) compared to the other one. This confirms the increased reliability of the proposed method.

To further check the experimental procedure, a pilot testing activity was implemented. In particular, an interesting case study was investigated. Two specimens containing Portland limestone cement (CEM II/A LL 42.5 R) and pozzolanic cement (CEM IV/B (P) 32.5 R), respectively, were tested in parallel following the procedure of Section 14.2.4. The main objective of the test was to reproduce the expected theoretical results: Portland cement is characterized by a reduced hardening time compared to pozzolanic cement due to its higher clinker content, that is, a higher Young's modulus reached after a given time. The results obtained are presented in Figure 14.5 in terms of E-modulus versus time. Notice that in Figure 14.5: (i) the

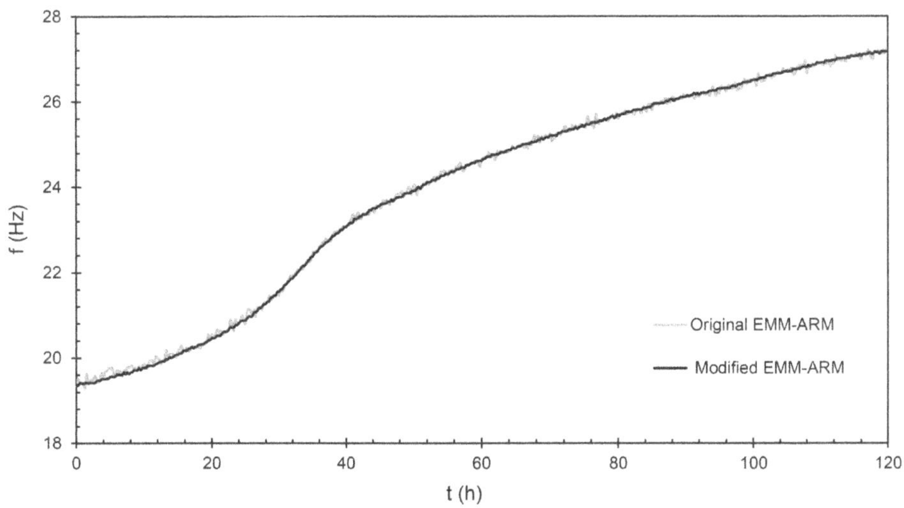

FIGURE 14.4 Frequency estimation according to the original and modified EMM-ARM data processing procedures.

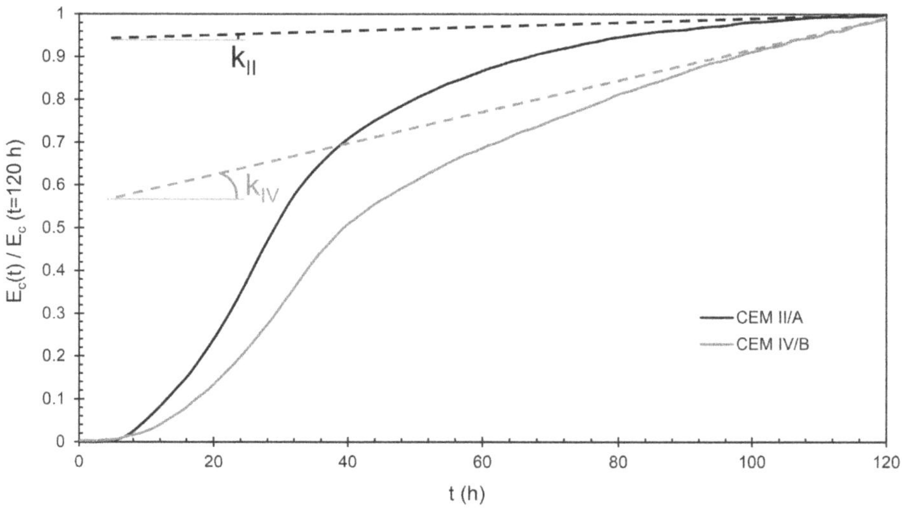

FIGURE 14.5 Normalized Young's modulus evolution for Portland and pozzolanic cement paste during the first 120 h since pouring.

results up to 120 h are reported for the sake of convenience; (ii) on the *y*-axis, the normalized value is reported, i.e., the ratio of E_c at time t to the maximum recorded value $E_{c,max}\left(t = 120h\right)$ in the considered time interval for the considered specimen, to allow a clearer comparison between the two curves, given the marked stiffness difference; (iii) a 23-term moving average was applied to minimize the noise in the curves for the sake of clarity of presentation.

Figure 14.5 confirms the theoretical expectations. The normalized elastic modulus grows faster for the Portland cement paste. For example, after 24 h, it is equal to 0.2 for CEM IV-type and 0.35 for CEM II-type. At the end of the observation, namely after 120 h, the Portland cement paste achieved a large part of its final stiffness, as confirmed by the almost negligible slope (see K_{II} in Figure 14.5) compared to the pozzolanic cement (see K_{IV} in Figure 14.5).

14.4 RESPONSE CURVES AND DISCUSSION OF RESULTS

Each specimen was associated with a standardized identification code. The alphanumeric string type cMMdNNN was used for specimens subjected to prehydration, whereas the cMMtPP string was adopted for specimens used to investigate the temperature effect. The meaning of the symbols is reported as follows:

1. "c": cement paste;
2. "MM": cement class (i.e., II or IV);
3. "d" and "t": experimental variables (i.e., prehydration and temperature, respectively);
4. "NNN": storage days up to test start;
5. "PP": value of curing temperature

Figures 14.6 and 14.7 show the E_c-t response curves of the specimens containing prehydrated Portland and pozzolanic cement paste, respectively, produced after different storage ages. In Figure 14.8, the E_c-t response curves of pozzolanic cement paste subjected to different curing temperatures are shown. For the sake of convenience, to achieve a better comparison among the results, the t-axis of all graphs is limited to 120 h.

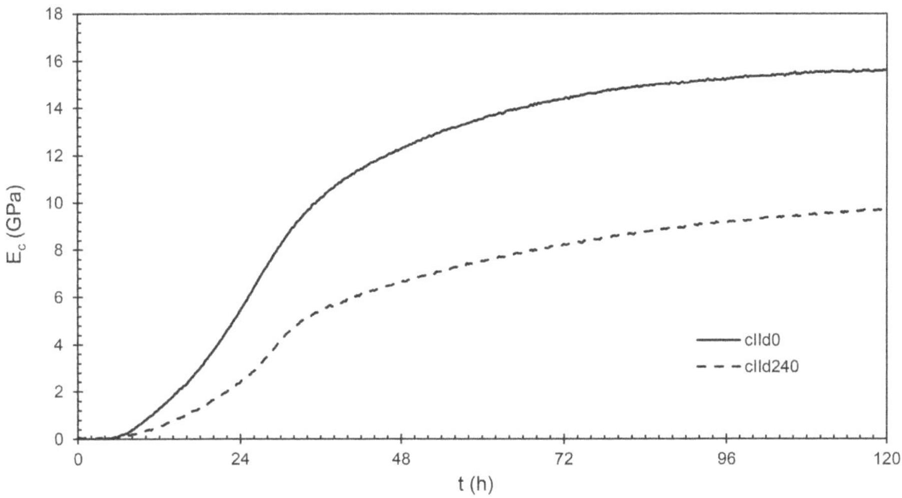

FIGURE 14.6 Effect of prehydration duration on the elastic modulus evolution at early ages of Portland cement paste.

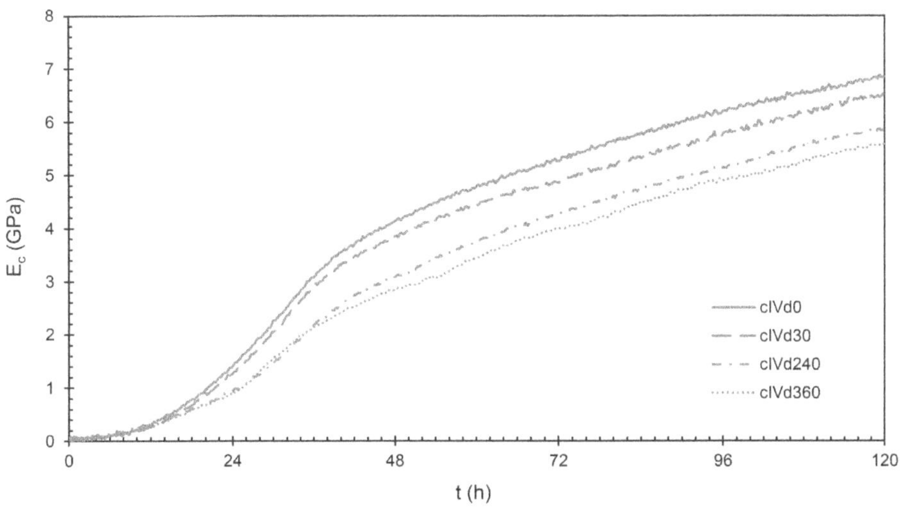

FIGURE 14.7 Effect of prehydration duration on the elastic modulus evolution at early ages of pozzolanic cement paste.

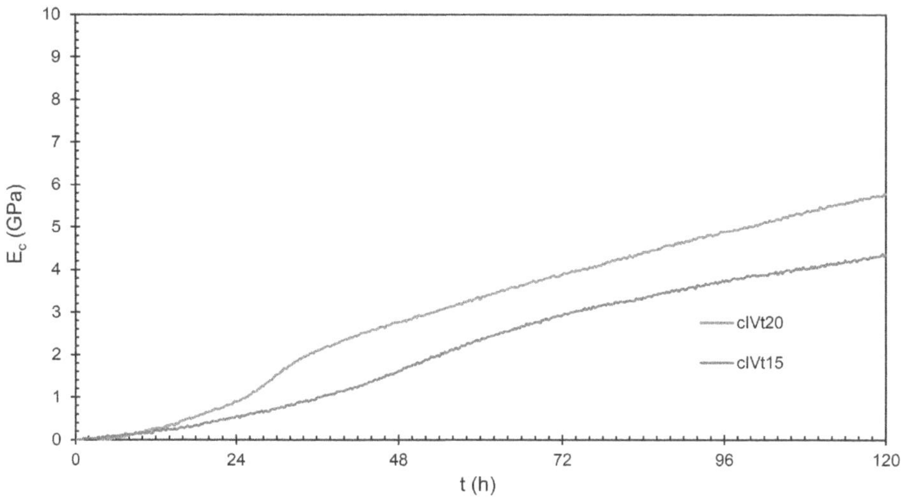

FIGURE 14.8 Effect of curing temperature on the elastic modulus evolution at early ages of pozzolanic cement paste.

By inspecting Figure 14.6, the effect of prehydration can be clearly appreciated for the Portland cement paste. Two specimens containing CEM II cement paste were tested according to the provisions of Section 14.2.4. The first test (see curve cIId0) was performed immediately after the opening of the cement bag, whereas the second test (see curve cIId240) was started after 240 days (8 months) of storage. A horizontal initial part characterizes both curves, in line with the theoretical expectations: the activation of the hydration process begins immediately after the mixing phase, and after the so-called dormant period, the first chemical reactions occur until the setting time (where the E modulus starts growing) is reached, with the formation of C-S-H gel and calcium sulfoferrite [48].

The setting time was equal to 4 h for the specimen cIId0 and 5 h for the specimen cIId240. After the setting point, the hardening phase starts, and an increase in the elastic modulus E_c with time can be observed in both tests. Nonetheless, a substantial divergence in the response characterizes the tests, with a considerably limited increase of E_c for the specimen cIId240. The difference between E_c (cIId0) and E_c (cIId240) is equal to zero before the setting time and afterward increases with time until 72 h; then, it remains almost constant at about 6 GPa. At 24 h, the E_c values were 5.44 GPa and 2.38 GPa for the specimens cIId0 and cIId240, respectively. After two days, the corresponding values were 12.28 GPa and 6.62 GPa. At the end of the observation period (120 h), the E_c values were 15.6 GPa and 9.7 GPa. Therefore, after five days, the modulus E_c of the specimen cIId240 was equal to 62% of that of the specimen cIId0. Moreover, at 120 h, both curves were characterized by an almost horizontal slope, so the elastic modulus was almost approaching its final value.

In Figure 14.7, four curves are provided, corresponding to the cement storage ages at the beginning of the tests equal to 0, 30, 240, and 360 days, associated with the specimens cIVd0, cIVd30, cIVd240, and cIVd360, respectively. Notice that, for the pozzolanic cement paste specimens, a larger number of tests were performed compared to CEM II/B since, as argued in Section 14.1, the research studies available in the literature focused on assessing the effect of prehydration of cement are largely devoted to the latter.

Typical features of cement maturity curves (namely dormant period, setting point, and hardening) can also be identified in Figure 14.7 for all curves. The setting points can be collocated within 1 and 2 hours from the end of the mixing operations for all the tests. After the setting time, a monotonic growth of E_c characterizes the specimens. However, as with the CEM II cement paste, the experimental outcomes in this case also confirm a significant effect of storage on the material mechanical response after the setting point, characterized by lower E_c growth with time as a larger prehydration period is considered. The difference between E_c of the specimen with freshly opened cement, i.e., cIVd0, and the other specimens increases with time until 48–72 h, after which it remains almost constant. At 24 h, E_c values were 1.40 GPa, 1.27 GPa, 0.95 GPa, and 0.93 GPa for specimens cIVd0, cIVd30, cIVd240, and cIVd360, respectively. After 48 h from the end of the cement paste fabrication, the corresponding values were 4.14 GPa, 3.85 GPa, 3.08 GPa, and 2.84 GPa. At the end of the third day, the values evolved to 5.27 GPa, 4.85 GPa, 4.29 GPa, and 4.00 GPa. Finally, at 120 h, the final values were 6.85 GPa, 6.55 GPa, 5.83 GPa, and 5.57 GPa.

During the first 12 h, the difference among the tests in terms of E_c was almost negligible, and this aspect represents a substantial difference from the CEM II cement pastes, which exhibited different setting times and Young's modulus evolution from the beginning of the hardening phase. Afterward, a significant difference could be appreciated among specimens cIVd0, cIVd30, and cIVd240; conversely, cIVd240 and cIVd360 showed very similar hardening curves until 36 h were reached.

The modulus E_c of the specimens referring to cement stored for 30, 240, and 360 days was 96%, 85%, and 81% of the E_c modulus of specimen cIVd0 at 120 h. It is noteworthy that, unlike the Portland cement paste, after five days from the end of the mix, the slope of the curves does not tend to zero, and this represents a further significant difference attributable to the presence of pozzolan.

Definitely, prehydration caused a general deterioration of mechanical performance, which is manifested by a common reduction in the hardening rate, i.e., stiffness reached within a certain period, as the storage period increases. However, this effect was more marked for Portland cement paste, due to the higher clinker content, in line with the results available in the literature [3, 4].

The two E_c versus t curves shown in Figure 14.8 are related to the results of two tests carried out on pozzolanic cement paste under two different constant temperatures, 15°C and 20°C, respectively. The reliability of EMM-ARM for such a typology of experimental tests has been argued in Section 14.1 [29, 33, 35]. The curves are different not only in terms of hardening rate but also in their shape, i.e., the hardening speed or the time necessary to achieve a certain value of E_c. Indeed, for example, E_c equal to 0.5 GPa was reached after 17 h by the specimen cured at 20°C (assumed as the control temperature), whereas the same value was attained by the specimen cured at 15°C after 23 h. Similarly, E_c equal to 1.0 GPa was obtained after 26 h and 36 h for the specimens at 20°C and 15°C, respectively. Hence, although referring to the same material, a reduction of 5°C of the control temperature generated delays of 6 h and 10 h to achieve E_c equal to 0.5 GPa and 1.0 GPa, respectively.

After five days, the modulus E_c of the specimen cIVt15 was equal to 4.35 GPa, which is 75% of the control specimen cIVt20. This result is consistent with previous findings in the literature that recognize a faster initial progress of hydration at increased temperatures based on chemical studies [49].

14.5 CONCLUSIONS

The reliability of EMM-ARM for assessing the early-age Young's modulus evolution with time has been discussed through an extensive and updated state of the art. The experimental procedure, slightly improved in terms of the elaboration algorithm, has been implemented in this study for investigating the influence of prehydration associated with storage and curing temperature on the development of the E modulus with time during the early age. Two commercial binders were used in the mix: Portland cement (type CEM II/A LL 42.5 R) and pozzolanic cement (type CEM IV/B (P) 32.5 R). The latter has been scarcely investigated concerning prehydration and has never been considered in EMM-ARM-based research.

Six prehydration tests have been carried out at different storage times, from the opening of the cement bag up to one year later, requiring considerable time and effort. Two additional tests have been performed at two different constant temperatures, selected to be consistent with the typical temperature range for in-situ casting and curing.

From the experimental tests, the following trends can be outlined:

1. Prehydration due to the storage of binders results in a deterioration of the hardening rate, i.e., the elastic E modulus reached in a given time, and this trend increases with storage time. Furthermore, a delay in setting time is observed.
2. The detrimental effect of prehydration is much more critical for Portland cement paste compared to pozzolanic cement paste, with the Young's modulus values after five days of experimental observation equal to 62% and 81% of the reference values after one year of binder storage. No delay in reaching the setting point is identified.
3. Regardless of the prehydration level, after five days, Portland cement paste approaches its final stiffness, whereas pozzolanic cement paste is still in a phase of increasing stiffness.
4. The reduction of the curing temperature from 20°C to 15°C results in a considerable increase in the time required to achieve a certain E modulus value. After five days, the cement paste cured at the lower temperature reaches 75% of the Young's modulus of the control specimen at 20°C.

The phenomenological results reported earlier are in agreement with the results of chemical-physical experimental studies available in the literature; therefore, EMM-ARM can be exploited as a further NDT method to verify the evolution of the performance of binders over time, also in view of fitness for use or commercialization.

14.6 AKNOWLEDGMENTS

The present study has been carried out within the framework of PRIN 2022 "SUCCESS—SUstainable self-sensing Cement-based CompositES for dynamic monitoring of civil Structures" (code "2022L9PSNW"), funded by the European Union—Next Generation EU, Missione 4 "Istruzione e Ricerca", Componente C2, Investimento 1.1, "Fondo per il Programma Nazionale di Ricerca e Progetti di Rilevante Interesse Nazionale (PRIN)", CUP H53D23001330006, CUP B53D23006210006, whose financial support is gratefully acknowledged.

REFERENCES

1. Silva, D. A., Thomas, J., Kazmierczak, D., Cheung, J. 2018. Pre-hydration of cement: Global survey and laboratory results. *ZKG Int.*, 71(6): 55–60.
2. Stoian, J., Oey, T., Bullard, J. W., Huang, J., Kumar, A., Balonis, M., Sant, G. 2015. New insights into the prehydration of cement and its mitigation. *Cem. Concr. Res.*, 70: 94–103.
3. Dubina, E., Black, L., Sieber, R., Plank, J. 2010. Interaction of water vapour with anhydrous cement minerals. *Adv. Appl. Ceram.*, 109: 260–268.
4. Whittaker, M., Dubina, E., Al-Mutawa, F., Leon, L, F., Arkless, L., Plank, J., Black, L. 2013. The effect of prehydration on the engineering properties of CEM I Portland cement. *Adv. Cem. Res.*, 25: 12–20.
5. Starinieri, V., Hughes, D. C., Gosselin, C., Wilk, D., Bayer, K. 2013. Pre-hydration as a technique for the retardation of Roman cement mortars. *Cem. Concr. Res.*, 46: 1–13.
6. Ferraz, D. F., Martho, A. C. R., Burns, E. G., Romano, R. C. D. O., Pileggi, R. G. 2022. Effect of prehydration of Portland cement on the superplasticizer consumption and the impact on the rheological properties and chemical reaction. *Revista IBRACON de Estruturas e Materiais*, 16(2): e16210.
7. Abrams, D. A. 1920. *Effect of Storage of Cement*, Bulletin 6; Structural Materials Research Laboratory, Lewis Institute: Chicago, IL, USA.
8. Ozersky, A., Khomyakov, A., Peterson, K. 2021. Extended shelf life cement: Principles, microstructural analysis, and physical–mechanical properties of the cement and concrete. *Constr. Build. Mater.*, 266: 121202.
9. Dubina, E., Plank, J. 2012. Influence of moisture-and CO2-induced ageing in cement on the performance of admixtures used in construction chemistry. *ZKG Int.*, 10: 60–68.
10. Adamtsevic, A., Eremin, A., Pustovgar, A., Pashkevich, S., Nefedov, S. 2014. Research on the effect of prehydration of Portland cement stored in normal conditions. *Appl. Mech. Mater.*, 670: 376–381.
11. Maltese, C., Pistolesi, C., Bravo, A., Cella, F., Cerulli, T., Salvioni, D. 2007. A case history: Effect of moisture on the setting behaviour of a Portland cement reacting with an alkali-free accelerator. *Cem. Concr. Res.*, 37(6): 856–865.
12. Getachew, B., Adugna, T., Assefa, S. 2018. Investigation on the effect of storage time due to moisture absorption of cement on the properties of the concrete. *Int. J. Dev. Res.*, 8(1): 18558–18561.
13. Adami, C. E., Vintzileou, E. 2008. Interventions to historic masonries: Investigation of the bond mechanism between stones or bricks and grouts. *Mater. Struct.*, 41, 255–267.
14. Massazza, F., Oberti, G. Durability of pozzolanic cements and Italian experience in mass concrete. In *Proceedings of the Durability of Concrete, Second International Conference*, Montreal, QC, Canada, 4–9 August 1991.

15. Alecci, V., De Stefano, M., Focacci, F., Luciano, R., Rovero, L., Stipo, G. 2017. Strengthening masonry arches with lime-based mortar composite. *Build.*, 7: 49.
16. Vailati, M., Mercuri, M., Angiolilli, M., Gregori, A. 2021. Natural-fibrous lime-based mortar for the rapid retrofitting of heritage masonry buildings. *Fibers*, 9: 68.
17. Ramaglia, G., Lignola, G. P., Balsamo, A., Prota, A., & Manfredi, G.2017. Seismic strengthening of masonry vaults with abutments using textile-reinforced mortar. *J. Compos. Constr.*, 21: 04016079.
18. Follador, V., Carpanese, P., Donà, M., da Porto, F. 2023. Effect of retrofit interventions on seismic fragility of Italian residential masonry buildings. *Int. J. Disaster Risk Reduct.*, 91: 103668.
19. Bentz, D. P. 2008. A review of early-age properties of cement-based materials. *Cem. Concr. Res.*, 38(2): 196–204.
20. Malhotra, V. M., Carino, N. J. 2003. *Handbook on Nondestructive Testing of Concrete*, CRC Press: Boca Raton, FL, USA.
21. Serdar, M., Gabrijel, I., Schlicke, D., Staquet, S., Azenha, M. 2020. *Advanced Techniques for Testing of Cement-Based Materials*, Springer: Berlin and Heidelberg, Germany.
22. Azenha, M., Magalhães, F., Faria, R., Cunha, Á. 2010. Measurement of concrete E-modulus evolution since casting: A novel method based on ambient vibration. *Cem. Concr. Res.*, 4: 1096–1105.
23. Azenha, M., Faria, R. Magalhães, F., Ramos, L., Cunha, A. 2011. Measurement of the E-modulus of cement pastes and mortars since casting, using a vibration based technique. *Mater. Struct.*, 45: 81–92.
24. Delsaute, B., Boulay, C., Granja, J. Carette, J., Azenha, M., Dumoulin, C., Staquet, S. 2016. Testing concrete E-modulus at very early ages through several techniques: An inter-laboratory comparison. *Strain*, 52: 91–109.
25. Azenha, M., Ramos, L., Aguilar, R., Granja, J. 2012. Continuous monitoring of concrete E-modulus since casting based on modal identification: A case study for in situ application. *Cem. Concr. Compos.*, 34: 881–890.
26. Maia, L., Azenha, M., Faria, R., Figueiras, J. 2011. Influence of the cementitious paste composition on the E-modulus and heat of hydration evolutions. *Cem. Concr. Res.*, 41: 799–807.
27. Azenha, M., Ramos, L., Aguilar, R., Granja, J. 2012. Continuous monitoring of concrete E-modulus since casting based on modal identification: A case study for in situ application. *Cem. Concr. Compos.*, 34: 881–890.
28. Silva, J., Azenha, M., Correia, A., Granj, J. 2014. Continuous monitoring of sand–cement stiffness starting from layer compaction with a resonant frequency-based method: Issues on mould geometry and sampling. *Soils Found.*, 54: 56–66.
29. Granja, J., Azenha, M., de Sousa, C., Ferreira, C. 2014. Comparison between different experimental techniques for stiffness monitoring of cement pastes. *J. Adv. Concr. Technol.*, 12: 45–61.
30. Azenha, M., Silva, J., Granja, J., Gomes-Correia, A. 2016. A retrospective view of EMM-ARM: Application to quality control in soil-improvement and complementary development. *Procedia Eng.*, 143: 339–346.
31. Granja, J., Azenha, M. 2017. Towards a robust and versatile method for monitoring E-modulus of concrete since casting: Enhancements and extensions of EMM-ARM. *Strain.*, 53: e12232.
32. Velay-Lizancos, M., Azenha, M., Martínez-Lage, I., Vázquez-Burgo, P. 2017. Addition of biomass ash in concrete: Effects on E-modulus, electrical conductivity at early ages and their correlation. *Constr. Build. Mater.*, 157: 1126–1132.

33. Velay-Lizancos, M., Martinez-Lage, I., Azenha, M., Granja, J., Vazquez-Burgo, P. 2018. Concrete with fine and coarse recycled aggregates: E-modulus evolution, compressive strength and non-destructive testing at early ages. *Constr. Build. Mater.*, 193: 323–331.
34. Garijo, L., Azenha, M., Ramesh, Lourenco, P. B., Ruiz, G. 2019. Stiffness evolution of natural hydraulic lime mortars at early ages measured through EMM-ARM. *Constr. Build. Mater.*, 216: 405–415.
35. Ramesh, M., Azenha, M., Lourenço, P. B. 2019. Mechanical properties of lime–cement masonry mortars in their early ages. *Mater. Struct.*, 52: 13.
36. Dunant, C. F., Granja, J., Muller, A., Azenha, M., Scrivener, K. L. 2020. Microstructural simulation and measurement of elastic modulus evolution of hydrating cement pastes. *Cem. Conc. Res.*, 130: 106007.
37. Ramesh, M., Azenha, M., Lourenço, P. B. 2019. Study of early age stiffness development in lime–cement blended mortars. In *Structural Analysis of Historical Constructions*, Aguilar, R., Torrealva, D., Moreira, S., Pando, M. A., Ramos, L. F., Eds., RILEM Bookseries, Vol. 18, pp. 397–404, Springer: Cham, Switzerland.
38. Gupta, M., Igarashi, G., Takahashi, Y., Granja, J., Azenha, M., Ishida, T. 2024. Effect of expansive additives and external restraint on the early age mechanical properties and microstructure of cement paste. *Constr. Build. Mater.*, 422: 135655.
39. Araghi, A., Ribeiro, R. R., Russo, T, Granja, J., Azenha, M., de Melo Lameiras, R. 2024. New impulse-based test method for early-age elastic modulus measurement in cementitious materials. *Case Stud. Constr. Mater.*, 21: e03971.
40. Russo, T., Ribeiro, R. R., Araghi, A., Lameiras, R. D. M., Granja, J., Azenha, M. 2023. Continuous monitoring of elastic modulus of mortars using a single-board computer and cost-effective components. *Build.*, 13(5): 1117.
41. Rocha Ribeiro, R., Russo, T., Araghi, A., Granja, J. L., Azenha, M., Lameiras, R. 2025. Design and validation of a minimalist, open-source system for measuring elastic modulus evolution in cementitious materials. *Nondestruct. Test. Eval.*, 1–25.
42. Rocha Ribeiro, R. microEMMARM repository in GitHub: v.0.2 release. [Online] https://github.com/renr3/microEMMARM/tree/main/v0.2-microEMMARM.
43. Rocha Ribeiro, R. EMM-ARM post-process. [Online] https://github.com/renr3/postProcess-EMMARM.
44. European Committee for Standardization (CEN). *Cement—Part 1: Composition, Specifications and Conformity Criteria for Common Cements*, EN 197-1:2011.
45. Rainieri, C., Fabbrocino, G. 2014. *Operational Modal Analysis of Civil Engineering Structures: An Introduction and Guide for Applications*, Springer: New York, NY, USA.
46. S2X, S.R.L. 2021. *S2-DDA User's Manual*, S2X S.R.L.: Campobasso, Italy.
47. LabVIEW. National Instruments, Austin, TX. https://download.ni.com/support/manuals/320999b.pdf.
48. Final Report of RILEM TC 185-ATC. 2005. *Advanced Testing of Cement-Based Materials during Setting and Hardening*, Reinhardt, H. W., Grosse, C. U. (Eds.) RILEM: Paris, France. ISBN 2-912143-81-0.
49. Lothenbach, B., Winnefeld, F., Alder, C., Wieland, E., Lunk, P. 2007. Effect of temperature on the pore solution, microstructure and hydration products of Portland cement pastes. *Cem. Concr. Res.*, 37(4): 483–491.

Index

For Product Safety Concerns and Information please contact our EU representative GPSR@taylorandfrancis.com
Taylor & Francis Verlag GmbH, Kaufingerstraße 24, 80331 München, Germany

www.ingramcontent.com/pod-product-compliance
Ingram Content Group UK Ltd.
Pitfield, Milton Keynes, MK11 3LW, UK
UKHW020159200926
475702UK00004B/532
* 9 7 8 1 0 3 2 8 5 1 7 5 4 *